Antioxidants in Food, Vitamins and Supplements

Antioxidants in Food, Vitamins and Supplements

Prevention and Treatment of Disease

Amitava Dasgupta, PhD, DABCC

Professor of Pathology and Laboratory Medicine,
University of Texas Medical School at Houston

Kimberly Klein, MD

Assistant Professor of Pathology and Laboratory Medicine,
University of Texas Medical School at Houston

ELSEVIER

AMSTERDAM • BOSTON • HEIDELBERG • LONDON • NEW YORK • OXFORD
PARIS • SAN DIEGO • SAN FRANCISCO • SINGAPORE • SYDNEY • TOKYO

Elsevier
525 B Street, Suite 1900, San Diego, CA 92101-4495, USA
32 Jamestown Road, London NW1 7BY, UK
225 Wyman Street, Waltham, MA 02451, USA

British Library Cataloguing-in-Publication Data
A catalogue record for this book is available from the British Library

Library of Congress Cataloging-in-Publication Data
A catalog record for this book is available from the Library of Congress

ISBN: 978-0-12-810104-9

Printed and bound in the United States of America
14 15 16 17 18 10 9 8 7 6 5 4 3 2 1

Contents

Preface

We all know about antioxidants. Some of us take antioxidant vitamins and supplements to promote health. As such, this industry is now a multibillion dollar business. Currently, it is estimated that more than 40% men and 50% women age 60 years or older take at least one vitamin or mineral supplement daily. In popular health magazines, antioxidant supplements are advertised as magic bullets capable of preventing all chronic illnesses. However, free radicals are not villains because the human body requires modest amounts of them for normal physiological functions. Therefore, scavenging all free radicals by taking too many antioxidant supplements actually causes more harm than good. In fact, a balanced diet rich in fruits and vegetables and a healthy lifestyle is the best approach to combat oxidative stress. A healthy diet supplies more than enough antioxidants. Vitamins and antioxidant supplements may be needed only for special populations suffering from specific illnesses, and pregnant women may take folate supplements.

The purpose of this book is to present current medical knowledge regarding the role of antioxidants in health and disease in a concise manner so that health care professionals and medical students can get a balanced overview of this broad topic. We have attempted to provide an enjoyable reading experience, and we avoided detailed discussion of molecular pathways by which oxidative stress triggers individual diseases. However, we provide an overview of such molecular pathways and a long list of references in each chapter so that more advanced readers or those interested in antioxidant research will also find this book helpful.

We also deliberately avoid chemical structures and detailed discussion of the chemistry of antioxidants, but we cover all important aspects of the chemistry of free radicals and methods used by investigators in the field of antioxidant research. In order to understand what elevated malondialdehyde concentration in a disease implies, it is important to know basic chemistry of malondialdehyde and how it is measured in a biological matrix. We cover basic chemistry of free radicals in Chapter 1, and in Chapter 2 we discuss various analytical methods used in lipid peroxidation and free radical chemistry research.

Air pollution, sunlight, cigarette smoke, and even household chemicals can induce oxidative stress. How our environment induces oxidative stress is addressed in Chapters 3–5. We provide a detailed discussion of the link between oxidative stress and various diseases in Chapters 7–11. In Chapter 12, we discuss various foods that are good sources of antioxidants and present several tables for easy reference on the antioxidant potential of various fruits and vegetables. In this chapter, the glycemic index of various fruits is also given because many people throughout the world suffer from diabetes. Tea, coffee, and chocolate are full of antioxidants, and drinking tea or coffee or both is good for health. Whereas chocolate is rich in calories, tea and coffee are almost devoid of calories if milk and sugar are not added (Chapter 13). In Chapter 14, the benefits of drinking alcoholic beverages in moderation are discussed. However, consuming excessive alcohol is detrimental to health. Wine, especially red wine, is rich in antioxidants, especially resveratrol, which is an excellent antioxidant. In Chapter 15, antioxidant vitamins are discussed, and herbal supplements that are antioxidants and nonherbal supplements with antioxidant properties are discussed in Chapter 16. However, current research indicates that taking antioxidant supplements or antioxidant vitamins has no added health benefits but may cause harm. In Chapter 17, the best way to combat oxidative stress—a healthy lifestyle—is discussed, and references are provided that clearly indicate that daily exercise, pet ownership, interaction with friends and family, and even yoga or mediation can significantly reduce oxidative stress.

The purpose of this book is to provide health care professionals with an in-depth knowledge of the current state of research in oxidative stress and antioxidants. This book will provide clinicians, nurses, medical students, nursing students, and pharmacists with a clear understanding of this important aspect of health and nutrition. This is not a reference book, and it is not intended for experts in this field because there are already excellent reference books on antioxidants. However, students entering the field of antioxidant research will find this book useful.

We thank Professor Robert Hunter, our department chair, for encouraging us to write the book and for his support during preparation of the manuscript. Furthermore, we thank our loving spouses, Alice and Chad, respectively. Preparation of the manuscript required an exorbitant amount of time and energy; without their love and support, this work would not have been possible. We hope you, the readers, will find this book useful; if you do, our effort will be duly rewarded.

Amitava Dasgupta
Kimberley Klein

CHAPTER 1

Introduction to Free Radicals and the Body's Antioxidant Defense

CONTENTS

1.1 INTRODUCTION

The "oxygen paradox" is defined by the fact that aerobic organisms require oxygen for survival but oxygen is also inherently toxic to these organisms due to its association with free radical generation and oxidative stress. Various free radicals are common products of respiration and other biochemical reactions in cells that are normal physiological processes essential for survival. To survive in an unfriendly oxygen environment, living organisms generate water- and lipid-soluble antioxidants that can neutralize these highly reactive free radicals [1]. For healthy living, a delicate balance must be maintained between oxidative stress and antioxidant defense of the body. If the body's antioxidant mechanism does not operate optimally, excess free radicals can damage various biomolecules, including lipids, proteins, carbohydrates, and nucleic acids.

1.2 FREE RADICALS

A *free radical* is defined as an atom or molecule containing one or more unpaired electrons that are capable of free existence. Free radicals can be generated as products of homolytic, heterolytic, or redox reaction, and they usually consist of reactive oxygen species or reactive nitrogen species. Reactive oxygen species include oxygen-carrying free radicals as well as other reactive oxygen species such as hydrogen peroxide, which is not a free radical. Similarly, reactive nitrogen species include both nitrogen-containing free radicals and other reactive molecules in which the reactivity center is nitrogen [2]. Common free radicals and oxidants are summarized in Box 1.1.

Free radicals are generated during normal respiration and cellular functions. Under normal physiological conditions, approximately 2% of oxygen consumed by the human body during respiration is converted into superoxide

A. Dasgupta and K. Klein: Antioxidants in Food, Vitamins and Supplements. DOI: http://dx.doi.org/10.1016/B978-0-12-405872-9.00001-X

Box 1.1 COMMON FREE RADICALS AND OXIDANTS ENCOUNTERED IN HUMAN PHYSIOLOGY

Free Radicals

- Superoxide anion radical
- Hydroxyl radical
- Hydroperoxyl radical
- Peroxyl radical
- Lipid radical
- Lipid peroxyl radical
- Lipid alkoxyl radical
- Nitric oxide
- Nitrosyl cation
- Thiyl radical
- Protein radical

Oxidants

- Singlet oxygen
- Ozone
- Hydrogen peroxide
- Hypochlorite
- Nitrous acid
- Peroxynitrous acid
- Nitrous oxide

anion free radical, which is negatively charged ($O_2^{\bullet -}$) [3]. The human body also contains approximately 4.5 mg of iron, most of which is found in hemoglobin and other proteins. However, a small amount of iron that is found forming complexes with a variety of small molecules can react with hydrogen peroxide, producing the hydroxyl radical (Fenton reaction). In general, oxygen-centered free radicals (reactive oxygen species), such as superoxide, the hydroxyl radical, peroxyl radical, and alkoxy radical, and nitrogen oxide, play an important role in inducing oxidative stress [4].

Nitric oxide ($NO^{\bullet}$) is synthesized from L-arginine by many cell types through nitric oxide synthesis. Nitric oxide and superoxide are the major reactive species produced in cells. Both superoxide and nitric oxide can react with other species, producing reactive oxygen species and reactive nitrogen species, respectively. Nitric oxide can also bind to transition metals such as ferrous ions, and it plays an important role in the formation of cyclic guanosine monophosphate, a second messenger [5]. In general, superoxide anion radicals and nitric oxide are primary free radicals generated by cells during normal physiological functions.

1.2.1 Various Sources of Free Radicals

There are two sources of free radicals: endogenous sources and exogenous sources. Major endogenous sources of free radicals are summarized in Table 1.1. The most common reactive oxygen species are superoxide anion, hydrogen peroxide, hydroxyl radicals, peroxyl radicals, singlet oxygen, and ozone. The production of superoxide occurs mostly within mitochondria. The mitochondrial electron transport chain is the main source of energy that is stored in adenosine triphosphate (ATP) molecules. Movement of electrons

Table 1.1 Endogenous Sources of Free Radicals

Physiological Process	Comment
Mitochondrial respiration	Essential process of life that generates superoxide anion radical.
Autoxidation	Autoxidation of many biological molecules (hemoglobin, myoglobin, catecholamines, etc.) in the human body can produce free radicals. Superoxide is the primary free radical formed.
Enzymatic reaction	Many enzymatic reactions involving xanthine oxidase, lipoxygenase, aldehyde oxidase, etc. can generate free radicals.
Respiratory burst	This is a process in which phagocytes consume a large amount of oxygen during phagocytosis.
Metal ions	Metal ions such as copper ion and ferrous ion, which are essential for the body, can react with hydrogen peroxide to produce free radicals.
Strenuous exercise	May activate xanthine oxidase, producing free radicals.
Infection	May produce free radicals because the immune system may try to neutralize invading microorganisms with a burst of free radicals.
Ischemia/reperfusion	May activate xanthine oxidase, causing free radical generation.

from oxidizable organic molecules to molecular oxygen is responsible for ATP production by the mitochondrial electron transport system. During this process, superoxide anions are generated due to leaking of electrons to oxygen. Superoxide is also formed enzymatically in the process of reduction of molecular oxygen mediated by nicotinamide adenine dinucleotide phosphate oxidase (NADPH oxidase) and xanthine oxidase in mitochondria. Superoxide has a relatively longer half-life than that of other free radicals, but it is less reactive than the hydroxyl radical. Inflammatory cells also produce relatively large amounts of superoxide. Superoxide is membrane impermeable but can diffuse within cells [5].

Peroxisomes are another significant source for free radicals. Peroxisomes are specialized cytoplasmic organelles that carry out important physiological functions such as β-oxidation of long-chain and very-long-chain fatty acids and degradation of uric acid. Peroxisomal oxidase is capable of generating hydrogen peroxide. The reaction of xanthine and xanthine oxidase produces the superoxide anion and hydrogen peroxide through one-electron and two-electron reduction of molecular oxygen, respectively, to form uric acid. Hydrogen peroxide is a relatively stable agent that is permeable to cell membranes. Hydrogen peroxide can generate free radicals such as the hydroxyl radical but cannot directly oxidize lipid or DNA. Therefore, cytotoxicity of hydrogen peroxide is due to its ability to produce hydroxyl radicals through metal-catalyzed reactions such as the Fenton reaction. However, hydroxyl radicals are highly reactive and can damage any biomolecules close to their site of generation. These radicals are some of the most dangerous free radicals encountered in physiology [6]. Singlet oxygen is an electronically excited form of oxygen, but it is not a free radical. However, singlet oxygen is a reactive oxygen species generated during dismutation of superoxide anion in water.

Another endogenously produced oxygen free radical is the peroxyl radical ($ROO^{\bullet}$), the simplest form of which is $HOO^{\bullet}$, which is the protonated form of superoxide. Hydroperoxy radical is known to initiate lipid peroxidation. As mentioned previously, nitric oxide ($NO^{\bullet}$) is synthesized from L-arginine through nitric oxide synthases (NOS) in many cell types. Nitric oxide is a water-soluble, short-lived free radical that plays an important role as a signaling molecule in the body. Mammalian cells contain three genes encoding NOS (*NOS1*, *NOS2*, and *NOS3*), with 51–57% homology between isomers. *NOS1*, also known as nNOS (isoform first purified from neuronal tissue), and *NOS3*, also known as eNOS (isoform first found in endothelial cells), are also termed *constitutive* because they are expressed continuously in neurons and endothelial cells, respectively. These enzymes also require an increase in calcium concentration in tissues for their activity, thus producing low transient concentration of nitric oxide. In contrast, *NOS2* is an inducible calcium-independent form also known as iNOS [7]. The reaction of superoxide with nitric oxide produces peroxynitrite, which is a strong oxidizing agent capable of damaging protein and DNA.

Respiratory burst, a process by which phagocytic cells such as white blood cells consume a large amount of oxygen during phagocytosis, is another source of endogenous free radicals. Phagocytosis is an important function of immune defense that produces free radicals, mostly superoxide, and also hydrogen peroxide (respiratory burst). In addition, myeloperoxidase is also secreted. Myeloperoxidase, which is highly expressed in neutrophils, catalytically reacts with hydrogen peroxide in the presence of physiological concentration of chloride, producing hypochlorous acid (HOCl). This short-lived and diffusible oxidant plays an important role in killing invading microorganisms [8]. During ischemia, xanthine oxidase enzymes may be activated, causing oxidative stress by generating free radicals. In addition, the concentration of antioxidant enzymes may be reduced. During reperfusion, oxidative stress is also increased. Strenuous exercise also increases oxidative stress due to production of free radicals, but the body has the capability to neutralize these excess free radicals. Therefore, regular exercise has health benefits. Wagner *et al.* [9] reported that in 42 male subjects who participated in Ironman triathletes, markers of oxidative stress in blood increased moderately after the race and returned to normal levels after 5 days. They concluded that despite a temporary increase in oxidative stress, because there was appropriate antioxidant intake and protective alteration in antioxidant defense of the body, there was no DNA damage.

There are also external sources of free radicals. For example, breathing polluted air may expose people to free radicals. These external sources are summarized in Table 1.2. Epidemiological studies indicate that fine particulates found in polluted air that are less than 10 μm in diameter (PM 10) can increase the risk of asthma attack and chronic obstructive pulmonary disease. These small particulates in polluted air have free radical activities causing lung inflammation and damage to lung cells [10]. Organic compounds and metal ions, which are major components of these particulates, may originate from industrial source. Air pollution due to automobile exhaust may also contain harmful polycyclic aromatic hydrocarbons that are not only capable of generating free radicals but also act as carcinogens. Reducing exposure to air pollution by using face masks is recommended for patients who are more susceptible to the harmful effects of air pollution, such as patients with coronary heart diseases [11]. If possible, it is preferable to stay indoors as much as possible, especially when air quality is poor on hot days, when free radicals containing particulates may be at the ground level. In general, air quality is superior in rural areas compared to urban areas (see Chapter 3).

Table 1.2 External Sources of Oxidative Stress

External Source	Comment
Air pollution	Exposure to particulate matters in polluted air can produce significant oxidative stress, increasing the risk for asthma, cardiovascular diseases, chronic pulmonary obstructive disease (COPD), and lung cancer.
Inorganic particles in air	Ingestion of mineral particles from dust in individuals working in industry may cause oxidative stress, particularly if air contains fine mineral dust (quartz, silica, and asbestos).
Tobacco smoking	Oxidants present in tobacco smoke can damage lungs, causing COPD and even raising risk of lung cancer.
Some medications	Medications such as bleomycin, adriamycin, and sulfasalazine may produce oxidative stress.
Industrial solvents	Some industrial solvents, such as chloroform and carbon tetrachloride, if inhaled may cause oxidative damage.
Exposure to radiation	Exposure to excessive ultraviolet light, prolonged exposure to the sun (sunbathing), and treatment with radiation as part of cancer therapy increase oxidative stress.

Many health hazards associated with tobacco smoke are related to high concentrations of oxidizing compounds and free radicals present in the smoke. Even passive smoking can cause significant health hazards. Cigarette smoking is the leading cause of preventable death in the United States. Smoking is responsible for approximately 440,000 premature deaths in the United States annually, with an estimated $157 billion in annual health-related economic losses [12]. Faux *et al.* [13] conclude that the mechanism by which cigarette smoke causes health damage is mostly through oxidative stress because cigarette smoke contains more than 400 compounds, including oxidants and reactive oxygen species (see Chapter 4).

Certain medications may also produce oxidative stress. Anticancer drugs such as methotrexate, bleomycin, adriamycin, doxorubicin, topoisomerase inhibitors, and menadione may produce oxidative stress. Immunosuppressant drugs are used to prevent organ rejection in transplant recipients. Mycophenolic acid, cyclosporin A, and sirolimus can induce oxidative stress, whereas the immunosuppressant tacrolimus produces oxidative stress of a lesser magnitude. Antiretroviral drugs used in treating patients with HIV infection and even common drugs such as aspirin can induce oxidative stress. Common drugs that are oxidants are listed in Table 1.3. For most people, taking these drugs under medical supervision does not pose any problem; however, individuals with glucose 6-phosphate deficiency may be at much

Table 1.3 Drugs That May Induce Oxidative Stress

Type of Drug	Examples
Pain reliever	Aspirin, phenacetin
Anticancer	Methotrexate, bleomycin, adriamycin, doxorubicin, topoisomerase inhibitors, menadione
Antimalarial	Chloroquine, mefloquine, primaquine, pamaquine
Immunosuppressants	Cyclosporin A, sirolimus, tacrolimus, mycophenolic acid
Antibiotics	Chloramphenicol, ciprofloxacin, moxifloxacin, nalidixic acid, norfloxacin, ofloxacin, sulfacetamide, Bactrim (sulfamethoxazole and trimethoprim), sulfadiazine, sulfanilamide, sulfasalazine, sulfafurazole
Antituberculosis	*para*-Aminosalicylic acid, dapsone
Antiretroviral agents	Indinavir, atazanavir, azidothymidine
Diuretic	Spironolactone
Drugs of abuse	Cocaine, MDMA (Ecstasy)
Other agents	Ethyl alcohol (alcohol)

higher risk of toxicity from some of these medications. Glucose 6-phosphate dehydrogenase is an important enzyme in cellular redox balance. It is the major antioxidant enzyme present in erythrocytes, and it is important for stability and proper function of red blood cells. Glucose 6-phosphate deficiency is a hereditary disease that is more common in the black population than in white or Mediterranean populations. Glucose 6-phosphate deficiency occurs when a person does not have a sufficient number of enzymes in red blood cells. Such deficiency may cause premature destruction of red blood cells, and hemoglobin is released in the blood (hemolysis). Red blood cell destruction in these patients is triggered by ceratin drugs (which should be avoided in these patients), foods such as fava beans, infection, and severe stress [14].

Drinking alcohol beverages (ethanol) can also induce oxidative stress. Drinking alcohol in moderation has many health benefits, and minor oxidative stress induced by alcohol during metabolism can be easily counteracted by the body's antioxidant defense mechanism. In addition, red wine is full of antioxidants. However, alcohol abuse is a serious health hazard because alcohol is a known liver toxin capable of causing liver cirrhosis. Excessive alcohol consumption may induce substantial oxidative stress, especially in alcoholics [15] (see Chapter 4).

Many industrial solvents, such as carbon tetrachloride, chloroform, trichloroethylene, benzene, xylene, and toluene, can induce oxidative stress, and some of these chemicals are also carcinogenic. It has been proposed that these solvents may activate cytochrome P450 enzymes (especially CYP2E1), which are responsible for metabolism of the majority of drugs by the liver, thus inducing oxidative stress [16].

Ultraviolet light is present in sunlight and is essential for the production of vitamin D, a fat-soluble vitamin essential for proper absorption and retention of calcium and phosphorus, which are the building blocks of bone. Vitamin D deficiency may lead to increased risk of fractures and may also cause rickets. Adequate exposure to sunlight is the best way to prevent vitamin D deficiency. Human skin is capable of producing enough vitamin D from dehydrocholesterol when exposed to the ultraviolet light of the sun, which is then stored in body fat. Exposure of arms and legs for 5–30 minutes between 10 AM and 4 PM twice a week is considered sufficient to get enough vitamin D. However, too much exposure to sun can cause sun burn, skin problems, or even skin cancer. Ultraviolet light with wavelength between 320 and 400 nm (also known as ultraviolet A or UVA), present in sunlight, provokes an increase in intracellular labile iron concentration responsible for ultraviolet light-induced oxidative stress of the skin, causing damage [17]. Therefore, it is important to cover the body during the summer months to protect skin from direct exposure to sunlight (see Chapter 3). It is essential to use sunscreen while sunbathing, which can act as a filter to harmful ultraviolet radiation. Oxidative stress also plays an important role in cataract formation. When sunlight penetrates the eye, it generates reactive oxygen species such as the superoxide radical, singlet oxygen, hydrogen peroxide, and hydroxyl radicals, and it causes damage to the lens of the human eye. Drinking coffee (which contains caffeine) and ingesting vitamin C can prevent such damage [18].

X-rays and other radiation can also cause oxidative stress. Radiation exposure during airport security checks is usually considered safe, whereas radiation therapy in treating cancer induces oxidative stress. It is well established that some anticancer drugs and radiation therapy generate reactive oxygen species in patients during cancer therapy, and these reactive oxygen species are essential for apoptosis of cancer cells. It has also been speculated that mode of cell death depends on the severity of oxidative stress induced by cancer therapy [19].

1.2.2 Damage of Biomolecules by Free Radicals

As mentioned previously, due to high reactivity, free radicals can damage virtually all biomolecules, including lipids, proteins, carbohydrates, and nucleic acids. Lipids are major targets for oxidative damage induced by free radicals. Lipid peroxidation can cause cell membrane damage by alteration of membrane fluidity and permeability. Lipid peroxidation mediated by free radicals is a series of chain reactions involving initiation, propagation, and termination. Free radicals that can initiate lipid peroxidation include the hydroxyl radical (most reactive), alkoxyl radicals, peroxyl radicals, and peroxynitrite. Metal ions such as cuprous and ferrous ions can also contribute catalytically in chain initiation. The first step in lipid peroxidation is abstraction of bis-allylic hydrogen (hydrogen atom bonded to the same carbon atom that is in allylic positions with respect to two different carbon–carbon double-bonds in a polyunsaturated fatty acid) from a polyunsaturated fatty acid molecule to form a radical that rearranges into a pentadienyl radical. This radical then combines with oxygen molecules forming the peroxyl radical, which can further react with another polyunsaturated fatty acid moiety of another lipid molecule forming conjugated diene lipid hydroperoxide and a new pentadienyl radical, thus propagating the chain reaction. All polyunsaturated fatty acids can undergo lipid peroxidation, but the rate of reaction for lipid peroxidation is docosahexaenoic acid > eicosapentaenoic acid > arachidonic acid > linoleic acid. In addition, cyclooxygenase and lipoxygenase are two families of enzymes also responsible for peroxidation of polyunsaturated fatty acids. These enzymes catalyze various reactions involving polyunsaturated fatty acids, producing various lipid derivatives including prostaglandins and thromboxanes [20]. However, major end products of lipid peroxidation are aldehydes such as malondialdehyde, acrolein, and 4-hydroxy-2-nonenal. Acrolein and, to some extent, 4-hydroxy-2-nonenal are highly reactive compounds that can damage proteins, DNA, and phospholipids. Secondary oxidation products are a series of prostaglandin-like products known as isoprostanes as well as monocyclic and serial cyclic peroxides. Isoprostanes, especially F_2-isoprostane, are excellent markers of endogenous lipid peroxidation and oxidative stress [21].

Amino acids and proteins are susceptible to oxidative damage by hydroxyl radicals. Oxidation of proteins by reactive oxygen species and reactive nitrogen species involves side chains of all amino acid residues of proteins. In particular, sulfur-containing cysteine and methionine residues of proteins are very susceptible to oxidation by both reactive oxygen and reactive nitrogen species. Oxidation of cysteine residue leads to the formation of disulfides, and methionine is oxidized to methionine sulfoxide residues. However, disulfide reductase enzymes present in the human body can repair such damage, but these are the only oxidized form of protein that can be repaired. In addition, aromatic amino acid residues are susceptible to free radical damage. Tryptophan residue is readily oxidized into formylkynurenine, kynurenine, and various hydroxy derivatives. Phenylalanine and tyrosine residues are also oxidized, yielding various hydroxy derivatives. Peroxynitrite free radicals can also damage methionine, cysteine, tyrosine, and tryptophan. Peptide bond cleavage may occur due to reaction of reactive oxygen species with glutamyl, aspartyl, and prolyl side chain. As a result of protein oxidation, protein carbonyl derivatives are generated that can be used as a marker of the extent of free radical-induced protein damage. In addition, the carbonyl group may be introduced into the protein by the reaction with aldehydes (4-hydroxy-2-nonenal and malondialdehyde) or by reaction with ketoamines, ketoaldehydes, and deoxyosones generated as a result of reaction of reducing sugars or their oxidation products with lysine residue of proteins [22].

Advanced glycation end products are formed due to reaction between carbohydrate and amino group of proteins. Hemoglobin A1c, a marker of blood glucose control, is a glycation product of hemoglobin. These advanced glycation products are formed through various intermediate products, such as Amadori, Schiff base, and Maillard products. Hydroxyl radicals can damage DNA and are implicated in mutagenesis, carcinogenesis, and aging. Reactive oxygen and reactive nitrogen species can lead to DNA oxidation including direct modification of nucleotide base, formation of apurinic/apyrimidinic sites, DNA single-strand breaks, and, less frequently, DNA double-strand breaks. Of all the nucleotides, guanine is the most susceptible to oxidative damage where the hydroxyl radical can interact with C4, C5, and C8 positions in the imidazole ring of guanine, and formation of 8-hydroxyguanine (also known as 8-oxo-7,8-dihydroguanine) can be utilized as a marker of DNA damage. Peroxynitrite can also react with guanine with formation of 8-nitroguanine, which is considered a marker of nitrosative DNA damage. Single-strand break is a result of interaction of hydroxyl radicals with deoxyribose and subsequent generation of peroxy radicals responsible for nicking phosphodiester bonds that form the backbone of each helical strand of DNA [23]. Oxidative stress can also damage mitochondrial DNA, and such damage has been associated with aging and cancer development in various tissues including human skin [24]. Oxidation of RNA by reactive oxygen species has also been reported, but it has not been thoroughly studied like oxidation of DNA. Both ribosomal RNA and messenger RNA can be oxidized by reactive oxygen species.

Mitochondrial DNA is more vulnerable to oxidative damage than is nuclear DNA because mitochondrial respiration generates reactive oxygen species. In the aging process, accumulation of mitochondrial DNA mutations and impairment of oxidative phosphorylation, as well as imbalance in the expression of antioxidant enzymes result in additional production of reactive oxygen species. This mitochondrial dysfunction-induced overproduction of reactive oxygen species is the basis of the mitochondrial free radical theory of aging [25]. See Chapter 10 for an in-depth discussion of this topic, especially the link between various age-related disorders and oxidative stress.

1.2.3 Physiological Role of Free Radicals

Whereas at high concentrations free radicals and radical-derived reactive species can cause damage to all biomolecules, at moderate concentrations these free radicals, especially superoxide anion and related reactive oxygen species as well as nitric oxide, play important roles as regulatory mediators in the cell signaling process. Under normal physiological conditions, the amount of free radicals generated is tightly controlled by the body's antioxidant defense so that a steady state is maintained for free radicals. In general, under normal physiological conditions, cells prefer to be in redox state, and its oscillation determines cellular functions. Cells communicate with each other and respond to external stimuli through the biological mechanism known as cell signaling or cell transduction, which is usually stimulated by hormones, growth factors, cytokines, and neurotransmitters. Most cell types elicit a small oxidative burst generating a low concentration of reactive oxygen species when they are stimulated by cytokines, growth factors, various interleukins, tumor necrosis factor α, angiotensin II, platelet-derived growth factors, nerve growth factors, transforming growth factor β_1, granulocyte–macrophage colony-stimulating factor, and fibroblast growth factor. Therefore, initiation and function of signal transduction rely on the action of reactive oxygen species as signaling molecules that may act on different levels in the signal transduction process. Thus, reactive oxygen species play a very important physiological role as secondary messengers [26]. Reactive oxygen species also play an important role in immunity. Oxidative burst is characterized by massive production of reactive oxygen species by immune cells that kill invading

Box 1.2 IMPORTANT PHYSIOLOGICAL ROLES OF REACTIVE OXYGEN SPECIES

- Free radicals help with cell growth and cell proliferation
- Free radicals help with cell division
- Regulate redox balance of the cell
- Signal transduction
- Activate protein kinases that regulate gene functions
- Regulate immune function

pathogens. Oxidative burst is also encountered in the inflammatory environment. Reactive oxygen species can also regulate cell differentiation, and they play an important role in aging. Autophagy is a process by which cells engulf and break down intracellular proteins and organelles in the lysosome and recycle the constituents for new biosynthesis. This physiological process is important for removing and recycling damaged proteins and organelles. Reactive oxygen species are known to regulate autophagy [27]. Various physiological functions of reactive oxygen species are summarized in Box 1.2.

1.3 THE BODY'S ANTIOXIDANT DEFENSE

The body's antioxidant defense consists of both endogenous and diet-derived exogenous compounds, which can be classified into three broad categories: antioxidant enzymes, chain-breaking antioxidants, and metal binding proteins. Major antioxidant enzymes are superoxide dismutase (SOD), catalase, and peroxidases, which are endogenous in origin. Antioxidants that interfere with chain reactions initiated by free radicals are known as chain-breaking antioxidants. These antioxidants are small molecules that can be either water soluble or lipid soluble. Some of these antioxidants derive from diet, such as carotenoids, flavonoids, and antioxidant vitamins. Endogenous proteins such as ferritin, transferrin, and ceruloplasmin are also important antioxidant proteins because these are capable of binding metal ions such as copper and iron so that free radicals are not generated through the Fenton reaction. Usually, antioxidant enzymes provide the strongest antioxidant defense, although all antioxidants are important for proper neutralization of oxidative stress (see Figure 1.1).

1.3.1 Enzymes as Antioxidants

Probably the most important antioxidant enzyme is superoxide dismutase, which neutralizes the harmful superoxide anion radical by converting it to hydrogen peroxide. There are three forms of superoxide dismutase. SOD1 requires both copper and zinc for its function and is found in mitochondria as well as cytosol. Manganese superoxide dismutase (SOD2) is found mostly in

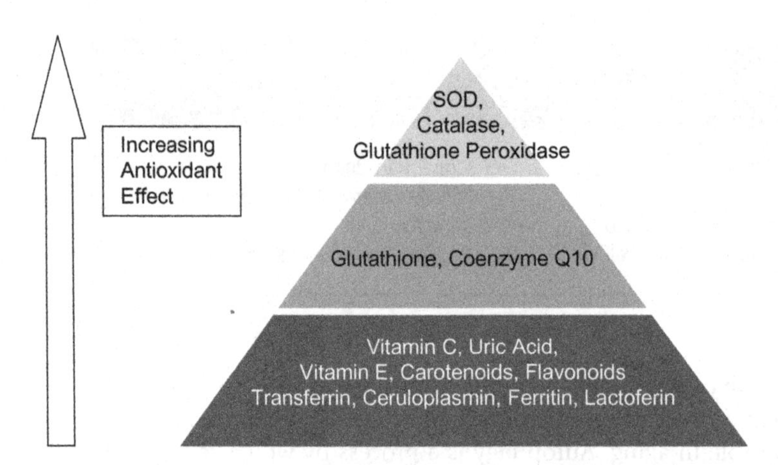

FIGURE 1.1 Effectiveness of various antioxidants in the human body in neutralizing free radicals.
SOD, superoxide dismutase.

mitochondria. The third form of superoxide dismutase (SOD3) is found outside cells. SOD3 also requires copper–zinc as a cofactor. Concentrations of these enzymes are regulated by appropriate genes [28]. SOD1 is a dimer, whereas SOD2 and -3 are tetramers with a higher molecular weight than SOD1.

Catalase, another important tetrameric antioxidant enzyme, converts hydrogen peroxide into water and oxygen, and thus completely neutralizes the superoxide free radical. Catalase is a homotetramer with a molecular weight of 240 kDa that is widely distributed within cells. Catalase requires iron as a cofactor, which is attached to the active site [29].

Glutathione peroxidase (GPx) belongs to a family of phylogenetically related enzymes comprising eight glutathione peroxidases (GPx1–GPx8). Collectively, selenium-containing GPx (GPx1–4 and -6) as well as nonselenium congeners (GPx5, -7, and -8) play important roles far beyond detoxifying hydroperoxides. In humans, GPx isoforms 1–4 and 6 contain selenocysteine residues that are important for their catalytic activities, and all members of this group use reduced glutathione to catalyze reduction of hydrogen peroxide into water [30]. GPx1 was the first selenoprotein identified. It is a homotetramer and one of the most abundant and ubiquitously expressed selenoproteins, which play an important role in overall recovery of cells after oxidative stress. GPx1 can react with hydrogen peroxide and soluble low-molecular-weight hydroperoxides such as *tert*-butyl hydroperoxide, hydroperoxy fatty acids, and lysophosphatides but not with more complex lipids. However, GPx can react with these more complex lipid hydroperoxides. In the process of reacting with hydroperoxides, glutathione is converted into glutathione disulfide. Glutathione reductase then converts it back to reduced glutathione for further reaction [31]. GPx1 plays a role in protecting individuals from cancer and neurodegenerative disease and is expressed in the gastrointestinal tract as well as the liver.

GPx2 is a homotetramer and closely related to GPx1. GPx2 is expressed mostly in the gastrointestinal system and the liver. GPx3 is a tetramer and is found in plasma. It may play a role in regulating the bioavailability of nitric oxide produced from platelets and vascular cells. GPx4 is a monomer but can react with complex lipid hydroperoxides such as phospholipids, cholesterol, and cholesterol hydroperoxides, even when they are inserted in membranes or lipoproteins. GPx exists in three different isoforms: a cytosolic isoform, a mitochondrial isoform, and a sperm nuclear isoform. However, all three isoforms are derived from the same gene. GPx4 seems to play a protective role in cardiovascular diseases along with GPx1 and GPx3. GPx6 is believed to be restricted in expression in developing embryo and olfactory epithelium of adults [32]. GPx5 is an epididymis-specific cysteine glutathione in mice. It is the closest homolog of GPx3 and seems to play a role in fertility. GPx7 was first described as a novel glutathione peroxidase with a cysteine instead of selenium in the catalytic center. GPx8 is the most recently discovered mammalian glutathione peroxidase. GPx8 is a membrane protein of the endoplasmic reticulum [33].

In addition to the previously mentioned antioxidant enzymes, peroxiredoxin and thioredoxin are involved in detoxifying intracellular hydrogen peroxides [34]. Peroxiredoxins are a family of small (22–27 kDa) nonselenothiol peroxidases that degrade hydroperoxides to water. Catalase, glutathione peroxidase, and peroxiredoxins are major enzymes responsible for protecting cells from hydroperoxide-induced damage. However, peroxiredoxins can be easily inactivated by hydrogen peroxide, thus limiting their ability as antioxidant enzymes. Peroxiredoxins are sensitive to oxidation and may act as redox sensors. The accumulation of oxidized peroxiredoxins may indicate disruption of cellular redox homeostasis [35]. Thioredoxin is a ubiquitous redox protein, and human thioredoxin is a 105-amino acid protein with two redox reactive cysteine residues in the active center. This protein plays important roles in the cellular response, including cell growth, cell cycle, gene expression, and apoptosis, to maintain redox balance. In addition, redox regulation by thioredoxin is involved in the pathogenesis of various oxidative stress-associated disorders [36].

1.3.2 Chain-Breaking Antioxidants

Chain-breaking antioxidants are small molecules capable of neutralizing free radicals by breaking the chain reaction initiated by free radicals. Chain-breaking antioxidants can be either endogenous or exogenous in origin. The classic example of a chain reaction initiated by free radicals is lipid peroxidation. Chain-breaking antioxidants act either by donating an electron or by receiving an electron from a free radical species, thus converting it into a stable species. Chain-breaking antioxidants can be further classified into two broad categories: water soluble and fat soluble. The most important water-soluble chain-breaking antioxidant is vitamin C, which is also known as ascorbic acid. Vitamin C is capable of scavenging superoxide, hydroxyl, and aqueous peroxyl radicals, as well as other oxidants, including hydrogen peroxide, hypochlorous acid, and singlet oxygen. In this process, vitamin C is converted into dehydroascorbic acid, which eventually breaks down into oxalic acid. Uric acid is present in human blood and acts as an antioxidant. Uric acid also complexes with iron, thus preventing the harmful Fenton reaction. Glutathione molecules contain three amino acids. This molecule may exist in the reduced state or the oxidized state. In the reduced state, glutathione is an excellent antioxidant capable of scavenging a variety of free radicals, and the reaction is catalyzed by glutathione peroxidase, as mentioned previously. Inside the cell, most glutathione is found in the reduced state because it is also useful for proper functioning of glutathione peroxidase. Bilirubin, the final product of heme catabolism, is an excellent antioxidant that can protect cells from toxic levels of hydrogen peroxide. Bilirubin can also neutralize peroxy radicals [37].

The most important fat-soluble chain-breaking antioxidant is vitamin E (tocopherols and tocotrienols), which exists in eight different states. However, α-tocopherol, the common form of vitamin E, is very efficient at breaking the chain reaction of lipid peroxidation. Carotenoids are also important lipid-soluble antioxidants, and the most common form is β-carotene. These antioxidants can also neutralize peroxy radicals as well as singlet oxygen. In addition, β-carotene is a precursor of vitamin A, which also has antioxidant activity (see Chapter 14). Flavonoids are antioxidants found in plants and food such as fruits and vegetables. In addition, drinking tea or coffee can provide sufficient flavonoids for adequate antioxidant defense of the human body (see Chapters 12 and 13). The reduced form of coenzyme Q10 (also known as ubiquinol-10) is an effective lipid-soluble chain-breaking antioxidant that is capable of scavenging lipid peroxy radical. It is an effective antioxidant that prevents harmful oxidation of low-density lipoprotein cholesterol [38].

1.3.3 Exercise and Antioxidant Status of Blood

Regular exercise is critical for good health, but exercise can also produce oxidative stress. The elevation of metabolism during exercise results in greater production of superoxide radicals. It has been postulated that during exercise, generation of superoxide radicals occurs mostly in heart and skeletal muscle and such radicals may also be released in the extracellular space. During exercise, reactive oxygen species are generated by electron leakage at the mitochondrial electron transport system. In addition, ischemia–reperfusion and activation of xanthine oxidase, the inflammatory response, and auto-oxidation of catecholamines may induce oxidative stress during exercise [39]. However, superoxide dismutase converts superoxide into hydrogen peroxide, and hydrogen peroxide can be detoxified by catalase and glutathione peroxidase, water and oxygen being formed. Interestingly, chronic exercise provides protection against exercise-induced oxidative stress by upregulation of the endogenous antioxidant defense of the body.

Using 34 untrained male volunteers, Berzosa *et al.* [40] observed that when volunteers performed three cycloergometer-based exercises—continuous progressive, strenuous, and submaximal—total antioxidant status of the blood was increased compared to basal level after all three exercises. The strenuous protocol resulted in the highest increase in catalase activity, although other exercise protocols also resulted in increased catalase activity compared to the resting value. Other antioxidant enzyme activities (superoxide dismutase and glutathione peroxidase) as well as glutathione levels were also increased following all exercise protocols. The authors concluded that even acute exercise may play a beneficial role because of its ability to increase the antioxidant defense mechanism through a redox-sensitive pathway [40]. Regular exercise also increases antioxidant defense of the human body. El Abed *et al.* [41] observed that competitive judo athletes have higher endogenous antioxidant protection compared to sedentary subjects. However, ultraendurance exercise such as a 143-mile running event causes oxidative stress that persists for up to 1 month [42]. Many investigators have studied the potential benefit of antioxidant supplementation during endurance exercise, but current recommendations are that adequate intake of vitamins and minerals through a varied and balance diet remains the best approach to maintain optimal antioxidant status in exercising individuals [43].

1.3.4 Markers for Oxidative Stress in Human Blood

To estimate oxidative stress using animal models or human subjects, many markers of oxidative stress are used. One of the oldest but still widely used assays for the determination of oxidative stress in serum is the TBARS (thiobarbituric acid reactive substances) assay, which measures the concentration of malondialdehyde produced due to degradation of unstable lipid peroxides. This assay can be used not only in human blood but also in urine and tissue samples. Other markers of oxidative stress in blood include isoprostane (formed due to oxidation of arachidonic acid), oxysterol (oxidation product of cholesterol), and other markers of protein and DNA damage. Moreover, designer probe molecules (several endogenous subunits of molecules attached covalently to form molecules that do not exist endogenously in humans) can be used as exogenous markers for assessing oxidative stress [44]. In addition, antioxidant compounds in blood, such as vitamin C, vitamin E, β-carotene, and glutathione, can be directly measured to evaluate the antioxidant capacity of blood. The activity of antioxidant enzymes can also be determined. Moreover, some of the assays used for measuring antioxidant capacity of various foods can also be used for measuring antioxidant capacity of human blood. Total antioxidant capacity of blood can be used to estimate antioxidant status of an individual [45]. Yucel *et al.* [46] used plasma total antioxidant capacity assay for evaluation of oxidative status in patients with slow coronary flow. Various assays used for measuring lipid peroxidation, other markers of oxidative stress, and plasma antioxidant capacity are discussed in Chapter 2.

1.4 CONCLUSION

The delicate balance between oxidative stress and antioxidant defense is maintained by the human body for optimal health. Reactive oxygen species and reactive nitrogen species generated during the normal physiological process are also needed in low to moderate concentration for signal transduction and other important biological functions. If excess reactive oxygen species and reactive nitrogen species cannot be neutralized properly by the body's antioxidant defense, oxidative stress is generated that is linked to many diseases, including cardiovascular diseases, various types of cancer, diabetes, and neurodegenerative diseases. Although exercise may induce oxidative stress, a healthy body can easily counteract such added stress by upregulation of antioxidant defense of the body. Regular exercise (at least 30 minutes a day) is highly recommended for good health. Current research indicates that antioxidant supplements are not needed; rather, a healthy balanced diet with generous servings of fruits and vegetables is essential for obtaining adequate antioxidant defense of the human body.

REFERENCES

[1] Davies KJ. Oxidative stress: the paradox of aerobic life. Biochem Soc Symp 1995;61:1–31.

[2] Turpaev KT. Reactive oxygen species and regulation of gene expression. Biochemistry (Moscow) 2002;67:281–92.

[3] Kunwar A, Priyadarsini KI. Free radicals, oxidative stress and importance of antioxidants in human health. J Med Allied Sci 2011;1:53–60.

[4] Gulcin I. Antioxidant activity of food constituents: an overview. Arch Toxicol 2012;86:345–91.

[5] Powere SK, Jackson MJ. Exercise induced oxidative stress: cellular mechanism and impact on muscle force production. Physiol Rev 2008;88:1243–76.

[6] Halliwell B. How to characterize an antioxidant: an update. Biochem Soc Symp 1995;61:73–101.

[7] Korde Choudhari S, Chaudhary M, Bagde S, Gadbail AR, et al. Nitric oxide and cancer: a review. World J Surg Oncol 2013;11:118.

[8] Catmona-Gutierrez D, Alvain-Ghavanini A, Habernig L, Bauer MA, et al. The cell cycle death protease kex 1p is essential for hypochlorite-induced apoptosis in yeast. Cell Cycle 2013;12:1704–12.

[9] Wagner KH, Reichhold S, Holzl C, Knasmuller S, et al. Well trained healthy triathletes experience no adverse health risks regarding oxidative stress and FNA damage by participating in an ultra-endurance event. Toxicology 2010;278:211–16.

[10] Li XY, Gilmour PS, Donaldson K, MacNee W. Free radical activity and pro-inflammatory effects of particulate air pollution (PM10) *in vivo* and *in vitro*. Thorax 1996;51:1216–22.

[11] Langrish JP, Li X, Wang S, Lee MM, et al. Reducing personal exposure to particulate air pollution improves cardiovascular health in patients with coronary heart disease. Environ Health Perspect 2012;120:367–72.

[12] Centers for Disease Control and Prevention. Annual smoking attributable mortality, years of potential life lost and economic costs—United States, 1995–1999. MMWR 2002;51 (14):300–3.

[13] Faux SP, Tai T, Thorne D, Xu Y, et al. The role of oxidative stress in the biological responses of lung epithelial cells to cigarette smoke. Biomarkers 2009;15(Suppl. 1):90–6.

[14] Ho HY, Cheng ML, Chiu DT. Glucose 6-phosphate dehydrogenase: from oxidative stress to cellular functions and degenerative diseases. Redox Rep J 2007;12:109–18.

[15] Das SK, Vasudevan DM. Alcohol induced oxidative stress. Life Sci 2007;81:177–87.

[16] Al-Ghamdi SS, Rafter MJ, Yaqoob MM. Acute solvent exposure induced activation of cytochrome P4502E1 causes proximal tubular cell necrosis by oxidative stress. Toxicol In Vitro 2003;17:335–41.

[17] Aroun A, Zhong JL, Tyrrell RM, O'Pourzand C. Iron, oxidative stress and the example of solar ultraviolet A radiation. Photochem Photobiol Sci 2012;11:118–34.

[18] Varma SD, Kovtun S, Hegde KR. Role of ultraviolet irradiation and oxidative stress in cataract formation: medical prevention by nutritional antioxidants and metabolic agonists. Eye Contact Lens 2011;37:233–45.

[19] Ozben T. Oxidative stress and apoptosis: impact on cancer therapy. J Pharm Sci 2007;96:2181–96.

[20] Xu Y, Gu Y, Qian SY. An advanced electron spin resonance (ESR) spin trapping and LC (ESR) MS technique for the study of lipid peroxidation. Int J Mol Sci 2012;13:14648–66.

[21] Milne GL, Yin H, Hardy KF, Davies SS, et al. Isoprostane generation and function. Chem Rev 2011;111:5973–96.

[22] Berlett BS, Stadtman E. Protein oxidation and aging, disease and oxidative stress. J Biol Chem 1997;33:20313–16.

[23] Smith JA, Park S, Krause JS, Banik NL. Oxidative stress, DNA damage and the telomeric complex as therapeutic targets in acute neurodegeneration. Neurochem Int 2013;62:764–75.

[24] Birch-Machin MA, Russell EV, Latimer JA. Mitochondrial DNA damage as a biomarker for ultraviolet radiation exposure and oxidative stress. Br J Dermatol 2013;169(Suppl. 2):9–14.

[25] Wang CH, Wu SB, Wu YT, Wei YH. Oxidative stress response elicited by mitochondrial dysfunction: implications in the pathophysiology of aging. Exp Biol Med 2013;238: 450–60.

[26] Valko M, Leibfritz D, Moncol J, Cronin M, et al. Free radicals and antioxidants in normal physiological functions and human diseases. Int J Biochem Cell Biol 2007;39:44–84.

[27] Sena LA, Chandel NS. Physiological roles of mitochondrial reactive oxygen species. Mol Cell 2012;48:158–67.

[28] Miao L, St. Clair DK. Regulation of superoxide dismutase genes: implications in disease. Free Rad Biol Med 2009;47:344–56.

[29] Kirkman HN, Gaetani GF. Mammalian catalase: a venerable enzyme with new mysteries. Trends Biochem Sci 2007;32:44–50.

[30] Tappel AL. Glutathione peroxidase and hydroperoxides. Methods Enzymol 1978;52:506–13.

[31] Lubos E, Loscalzo J, Handy DE. Glutathione peroxidase-1 in health and disease: from molecular mechanism to therapeutic opportunities. Antioxid Redox Signal 2011;15:1957–97.

[32] Reeves MA, Hoffmann PR. The human selenoproteome: recent insights into functions and regulation. Cell Mol Life Sci 2009;66:2457–78.

[33] Brigelius-Flohe R, Mariorino M. Glutathione peroxidases. Biochim Biophys Acta 2013;830:3289–303.

[34] Kalyanaraman B. Teaching the basics of redox biology to medical students and graduate students: oxidants, antioxidants and disease mechanisms. Redox Biol 2013;8:244–57.

[35] Poynton RA, Hampton MB. Peroxiredoxins as biomarker of oxidative stress. Biochim Biophys Acta 2013 [E-pub ahead of print].

[36] Kondo N, Nakamura H, Masutani H, Yodoi J. Redox regulation of human thioredoxin network. Antioxid Redox Signal 2006;8:1881–90.

[37] Stocker R, Glazer AN, Ames BN. Antioxidant activity of albumin bound bilirubin. Proc Natl Acad Sci USA 1987;84:5918–22.

[38] Young IS, Woodside JV. Antioxidants in health and disease. J Clin Pathol 2001;54:176–86.

[39] Gomes EC, Silva AN, de Oliveira MR. Oxidants, antioxidants and the beneficial roles of exercise induced production of reactive species. Oxid Med Cell Longev 2012;2012:756132.

[40] Berzosa C, Cebrian I, Fuentes-Broto I, Gomez-Trullen E, et al. Acute exercise increases plasma total antioxidant status and antioxidant enzyme activities in untrained men. J Biomed Biotech 2011;540458.

[41] El Abed K, Rebai H, Bloomer RJ, Trabelsi K, et al. Antioxidant status and oxidative stress at rest and in response to acute exercise in judokas and sedentary men. J Strength Cond Res 2011;25:2400–9.

[42] Turner JE, Hodges NJ, Bosch JA, Aldred S. Prolonged depletion of antioxidant capacity after ultraendurance exercise. Med Sci Sports Exerc 2011;43:1770–6.

[43] Peternelj TT, Coombes JS. Antioxidant supplementation during exercise training: beneficial or detrimental? Sports Med 2011;41:1043–69.

[44] Vaya J. Exogenous markers for the characterization of human disease associated with oxidative stress. Biochimie 2013;95:578–84.

[45] Yeum KJ, Russell RM, Krinsky NI, Aldini G. Biomarkers of antioxidant capacity in the hydrophilic and lipophilic compartments of human plasma. Arch Biochem Biophys 2004;430:97–103.

[46] Yucel H, Ozaydin M, Dogan A, Erdogan D, et al. Evaluation of plasma oxidative status in patients with slow coronary flow. Kardiol Pol 2013;71:588–94.

CHAPTER 2

Methods for Measuring Oxidative Stress in the Laboratory

CONTENTS

2.1 INTRODUCTION

In general, oxygen-centered free radicals (reactive oxygen species), such as superoxide anion, hydroxyl radical, peroxyl radical, and alkoxy radical, and reactive nitrogen species such as nitric oxide and related free radicals, play an important role in inducing oxidative stress and damaging most biomolecules, including lipids, proteins, nucleic acids, and carbohydrates [1]. In the human body, various antioxidant enzymes as well as antioxidant molecules are responsible for neutralizing these free radicals. As mentioned in Chapter 1, small amounts of reactive oxygen and nitrogen species are also required for normal physiological functions because they play an important role in various biological processes including signal transduction. In the laboratory, various parameters are measured in order to evaluate oxidative stress in human or animal models. These parameters can be broadly classified into two categories: measurement of total antioxidant capacity of a food or total antioxidant capacity of blood; and measurement of individual parameters of lipid peroxidation products or marker compounds representing DNA or protein damage. This chapter provides a brief overview of the various methods used to assess antioxidant capacity or measure markers of oxidative damage.

2.2 MEASURING TOTAL ANTIOXIDANT CAPACITY

Fruits, vegetables, chocolate, tea, coffee, and various other foods are excellent sources of antioxidants. There are various ways of measuring antioxidant capacity of foods and supplements. A common approach is the oxygen radical absorbance capacity (ORAC) value, or "ORAC score," which is a method of measuring the antioxidant capacity of different foods and supplements that was initially developed by scientists at the National Institutes of Health. Later, scientists at the US Department of Agriculture (USDA) perfected the assay. Although the exact relationship between the ORAC value of a food

A. Dasgupta and K. Klein: Antioxidants in Food, Vitamins and Supplements. DOI: http://dx.doi.org/10.1016/B978-0-12-405872-9.00002-1

and its health benefit has not been established, ORAC values are widely used in many papers published in the field of antioxidant research. In the past, the USDA published ORAC values of many foods, but such values were deleted from the available database in 2012 because ORAC values determined *in vitro* cannot be accurately applied to determine *in vivo* antioxidant capacity after consumption of such foods. Moreover, ORAC values have also been misused in advertisements of food and supplements.

Various methods are used for measuring total antioxidant capacity of a compound or a mixture of compounds such as food, but many of these methods can also be used for determining antioxidant capacity of human serum or plasma. In general, hydrogen atom transfer (HAT)-based or single electron transfer (SET)-based assays are widely used for measuring total antioxidant capacity of a compound or mixture, and some of these methods are also widely used for measuring total antioxidant capacity of human serum or plasma. Interestingly, the end results of HAT- and SET-based methods are the same, but the kinetics and potential side reactions may vary between these two methods. In HAT-based methods, the ability of antioxidants to quench free radicals by donating hydrogen atoms is measured. These reactions are pH and solvent independent and are usually fast reactions because they are completed in seconds to minutes. The relative reactivity is determined by the bond dissociation energy of the hydrogen donating group of the antioxidant. The majority of HAT reaction assays apply a competitive reaction scheme in which antioxidant and substrate used in the assay compete for thermally generated peroxyl radicals through the decomposition of an azo compound. If the antioxidant can trap peroxyl radicals, signal is not decreased. Therefore, the magnitude of reduction of signal (mostly fluorescence) inversely correlates with antioxidant capacity of the specimen. In SET-based methods, the ability of an antioxidant is measured by its capacity to transfer one electron to reduce any compounds, including free radicals, carbonyls, or metal ions. Relative reactivity of an antioxidant in SET-based assays depends on the ionization potential of the reactive functional group in the molecule. Therefore, these reactions are pH dependent, and in general ionization potential decreases with increasing pH, indicating increased electron donation capacity with deprotonation [1]. In electron transfer reaction-based assays, when the oxidant is reduced, it changes color, and the degree of color change correlates with antioxidant capacity [2]. HAT- and electron transfer reaction-based assays employed to measure antioxidant capacity of a compound or mixture of compounds or human plasma or serum are summarized in Table 2.1. Moreover, other assays are also used to measure total antioxidant capacity of a compound. Such assays are listed in Box 2.1. In addition, methods such as electron spin resonance (ESR) and related methods capable of directly measuring free radicals are used in the field of antioxidant research [3].

Table 2.1 Common Hydrogen Atom Transfer- and Electron Transfer Reaction-Based Assays Used for Assessing Antioxidant Capacity of a Compound or Mixture of Compounds or Body Fluids

Chemical Principle of the Assay	Individual Assays
Hydrogen atom transfer reaction	Oxygen radical absorption capacity (ORAC)
	TOSC (total oxyradical scavenging capacity)
	Total radical trapping antioxidant parameter (TRAP)
	Inhibition of low-density lipoprotein oxidation
	Inhibition of linoleic acid oxidation
	Crocin bleaching acid
	Inhibited oxygen uptake
Electron transfer reaction	2,2-Azino-bis 3-ethylbenzthiazoline-6-sulfonic acid radical ($ABTS^{\bullet+}$) scavenging assay
	Trolox equivalent antioxidant capacity
	N,*N*-dimethyl-*p*-phenylenediamine radical ($DMPD^{\bullet+}$) scavenging assay
	Ferric ion reducing antioxidant power
	Cupric ion reducing antioxidant power
	DPPH[a] reduction capacity
	Total phenols assay using Folin–Ciocalteu reagent

[a]*DPPH, diphenyl-1-picrylhydrazyl.*

Box 2.1 OTHER ASSAYS USED FOR ASSESSING ANTIOXIDANT CAPACITY

- Superoxide anion radical scavenging assay
- Hydroxyl radical scavenging assay
- Hydrogen peroxide scavenging assay
- Nitric oxide radical scavenging capacity
- Peroxynitrite scavenging assay
- Singlet oxygen scavenging capacity assay
- Metal chelating assay

2.2.1 ORAC Assay and TOSC Assay

Two commonly used HAT-based assays are the ORAC assay and the total oxyradical scavenging capacity (TOSC) assay. The ORAC assay was originally developed by Cao *et al.* [4]. Later, various investigators modified this assay, but the basic principle is the same. The ORAC assay is very useful in investigating the effectiveness of chain-breaking antioxidants, the major antioxidants present in various foods. In the ORAC assay, the peroxyl radicals generated by thermal decomposition of 2,2′-azobis(2-amidinopropane) dihydrochloride (ABAP) decrease the fluorescence of a compound such as fluorescein. The antioxidant capacity of a specimen is measured by the ability of the specimen to prevent loss of fluorescence signal by neutralizing peroxyl radicals. Decrease of fluorescence signal should be minimal if the specimen is rich in antioxidant compounds. In contrast, the signal may decay fast if the antioxidant component of the specimen tested is poor. The antioxidant capacity is usually expressed as Trolox (6-hydroxy-2,5,7,8-tetramethylchroman-2-carboxylic acid) equivalent. Trolox is a water-soluble derivative of vitamin E, which is an excellent antioxidant. Due to difficulty in measuring the antioxidant capacity of individual components of fruits, vegetables, or supplements, Trolox equivalent is used for total antioxidant capacity of such complex mixtures of antioxidants present in food. Usually, Trolox equivalent is expressed as micromoles of Trolox/100 g of specimen. ORAC values are used to describe antioxidant capacity of various foods, beverages, and supplements in many research papers published. Some researchers also use a slightly different unit (millimoles of Trolox/100 g) to express ORAC values.

In the TOSC assay, the thermal hydrolysis of 2,2′-azobis(2-methyl-propionamidine) dihydrochloride generates peroxyl radicals that then oxidize α-ketomethiol butyric acid to produce ethylene gas, which can be measured by gas chromatography. When antioxidant is present in the reaction mixture, it should quench peroxyl radicals, thus inhibiting ethylene gas production, and the degree of inhibition is related to the antioxidant capacity of the compound present in the reaction mixture. The antioxidant capacity of a compound in the TOSC assay can also be expressed as Trolox equivalent similar to the ORAC assay. The ORAC assay can estimate both inhibition time and degree of inhibition, thus accurately measuring total antioxidant capacity of a compound or a mixture of compounds for their ability to quench peroxyl radicals. The TOSC assay can distinguish between fast- and slow-acting antioxidants, and it is capable of measuring antioxidant capacity of both water- and fat-soluble antioxidants. Tomer *et al.* [5] measured the antioxidant activity of 11 phytochemicals using both ORAC and TOSC assays and observed that quercetin, pycnogenols, grape skin extract, and green tea polyphenols had the highest ORAC and TOSC values, whereas lemon fruit and citrus bioflavonoids had low ORAC and TOSC values. Although there was some correlation ($r = 0.77$) between TOSC and ORAC values, rutin and α-lipoic acid showed low ORAC values but high TOSC values [5]. Garrett *et al.* [6] reviewed both ORAC and TOSC methods in detail.

2.2.2 TRAP Assay

The total radical trapping antioxidant parameter (TRAP) assay was developed by Wayner *et al.* [7], who measured TRAP values of human plasma obtained from seven subjects. In this assay, water-soluble azo compound such as ABAP is added to plasma and the antioxidant capacity of plasma is measured by the amount of time oxygen uptake is inhibited by plasma antioxidant due to quenching of peroxyl radicals generated by the azo compound at 37°C. The induction period was measured by an oxygen electrode. TRAP is defined as the number of moles of peroxyl radicals trapped per liter of fluid, and values can be expressed as Trolox equivalent. The authors observed that the total plasma antioxidant capacities of the seven human subjects studied were similar, but values were 10–20 times larger than the effect attributable to plasma vitamin E content. The authors speculated that 57–73% of the total antioxidant capacity of plasma cannot be explained by urate and vitamin C concentration, indicating an unrecognized large reserve antioxidant capacity attributable to plasma [7]. Instead of measuring oxygen uptake, fluorescence of R-phycoerythrin or absorbance of 2,2′-azino-bis(3-ethylbenzothiazoline-6-sulfonate) as the reaction probe can be used in the TRAP assay. The oxidation of dichlorofluorescein diacetate by peroxyl radicals initiated by thermal decomposition of ABAP results in the production of highly fluorescent dichlorofluorescein product, and this compound can also be used as a probe compound in the TRAP assay. In this case, antioxidants present in the reaction mixture inhibit the increase in fluorescence signal. In addition, dichlorofluorescein has strong absorption at 504 nm, and measurement can also be done spectrophotometrically instead of fluorometrically [8]. The TRAP assay involves the initiation of lipid peroxidation by peroxyl radicals, and this assay is sensitive to chain-breaking antioxidants and capable of estimating antioxidant capacity of antioxidants toward peroxyl radicals, hydroxyl radicals, and peroxynitrite. However, this assay is also relatively complex and time-consuming. This method is most commonly used for measuring the antioxidant capacity of plasma [1].

2.2.3 Inhibition of Low-density Lipoprotein Oxidation, Linoleic Acid Oxidation, and Other Assays Based on Hydrogen Atom Transfer Reaction

Inhibition of low-density lipoprotein (LDL) oxidation assay is also used to estimate total antioxidant capacity of extracts of natural products. Many antioxidants, including phenolic antioxidants present in red wine, inhibit LDL oxidation *in vitro*. Chu and Liu [9] measured free radical scavenging capacity of antioxidants and extracts of natural products by their ability to inhibit oxidation of human LDL by peroxyl radicals generated by thermal decomposition of water-soluble ABAP at 37°C. The authors extracted human LDL from human plasma and measured LDL peroxidation by measuring hexanal using headspace gas chromatography. Hexanal is produced due to oxidation of polyunsaturated fatty acids in LDL and correlates with LDL oxidation. The authors observed that antioxidant capacity of 100 g of fresh apple extract measured with this assay was equivalent to 1470 mg of vitamin E or 402 mg of vitamin C. The authors commented that oxidation of LDL can also be initiated by copper ions. Instead of using LDL oxidation *in vitro*, oxidation of methyl linoleate *in vitro* in the presence of peroxyl radicals generated by decomposition of an azo compound can be used to measure antioxidant capacity of food or natural product extract. This assay is called the inhibition of linoleic acid *in vitro* assay [10].

In the crocin bleaching assay, inhibition capacity of an antioxidant in bleaching of crocin, a naturally occurring carotenoid derivative, is measured in the presence of peroxyl radicals. Usually, peroxyl radicals are generated by thermal decomposition of ABAP [11]. This method has limited application in measuring antioxidant capacity of food constituents but is very useful in measuring total antioxidant capacity of human plasma.

In the inhibited oxygen uptake method, the rate of oxygen uptake or conjugated diene peroxide formation is applied to determine inhibition of peroxyl radicals by an antioxidant. Using styrene as a substrate and azoisobutyronitrile as a radical initiator, Ingold and coworkers measured the oxygen consumption rate in the presence or absence of tocopherols in chlorobenzene using a pressure transducer system under a 1-atmosphere pressure of oxygen [12]. However, this method is not widely used due to the difficulty of accurately measuring the oxygen uptake rate and other issues.

2.2.4 Electron Transfer-Based Assays for Measuring Total Antioxidant Capacity

Various assays for measuring total antioxidant capacity of antioxidants in food constituents or biological samples are also based on electron transfer reaction. The basis of this assay is the following reaction:

$$\text{Probe} + \text{electron from antioxidant} \longrightarrow \text{reduced probe} + \text{oxidized antioxidant}$$

In the $ABTS^{\bullet+}$ radical scavenging assay (an electron transfer-based assay), the 2,2′-azino-bis(3-ethylbenzothiazoline-6-sulfonate) radical cation ($ABTS^{\bullet+}$), which has a dark blue color, is reduced by an antioxidant into colorless ABTS, which can be measured spectrophotometrically. In the original assay, metmyoglobin reacted with hydrogen peroxide to generate ferrylmyoglobin, which reacted with ABTS to form the $ABTS^{\bullet+}$ radical cation. However, other antioxidants can also be used for oxidation of ABTS, including manganese dioxide potassium persulfate and horseradish peroxidase. In the original Trolox equivalent antioxidant capacity (TEAC) assay, which is based on reduction of the total antioxidant capacity of body fluids, drug solution measurements being based on reduction of $ABTS^{\bullet+}$ radical by an antioxidant, the reduction of absorbance of $ABTS^{\bullet+}$ radical cation by an antioxidant is measured and the reactivity relative to 1.0 mmol/L Trolox standard is calculated. This assay can be easily automated using a centrifugal analyzer and was originally described by Miller *et al.* [13], who also used this method to measure total antioxidant capacity of plasma collected from premature neonates and compared the values obtained with those in adult human plasma. Later, Erel [14] described an improved version of the assay in which ABTS was dissolved in water to achieve a concentration of 7 mmol/L and then radical cation $ABTS^{\cdot+}$ was produced by treating ABTS solution with potassium persulfate and allowing the mixture to stand in the dark at room temperature for 12–16 h to produce a dark blue solution. This solution was diluted with ethanol or phosphate buffer (pH 7.4) until the absorbance was 0.7 at 734 nm. Then, 1 mL of solution was incubated with 10 μL of the specimen, and absorbance was read at 30°C 5 min after mixing. The TEAC value was calculated by comparison of change of absorbance of 1 mmol of Trolox. In the presence of antioxidant in the specimen, the blue green color of the $ABTS^{\cdot+}$ radical cation changed to colorlessness (decrease in absorption) because antioxidant can reduce this radical cation into ABTS and decolorization of the sample was linear with increasing antioxidant capacity of the sample studied. TEAC is an end point assay, and, due to its simplicity, it is widely used in various laboratories for measuring total antioxidant capacity of various compounds and food constituents.

An assay similar to $ABTS^{\cdot+}$ radical cation assay is $DMPD^{\cdot+}$ radical assay, in which the $ABTS^{\cdot+}$ radical is changed for a more stable $DMPD^{\cdot+}$ radical cation derived from *N,N*-dimethyl-*p*-phenylenediamine (DMPD). Fogliano *et al.* [15] described this assay, in which DMPD, a colorless compound, was oxidized by ferric chloride to colored $DMPD^{\cdot+}$ radical cation. In the presence of an antioxidant, $DMPD^{\cdot+}$ radical cation was reduced back to colorless DMPD, and the decrease in absorbance was monitored at 505 nm to measure antioxidant capacity of a compound. The authors measured total antioxidant capacity of various wines using this assay, and antioxidant capacity was expressed as TEAC. The values obtained using this assay were very similar to those obtained using $ABTS^{\cdot+}$ radical cation. However, DMPD is only soluble in water, and this assay cannot be used for measuring antioxidant capacity of hydrophobic compounds. Later, this assay was used for measuring total antioxidant capacity of human plasma, which was expressed as hydrogen peroxide equivalent. The assay was automated and was operated in a kinetic mode [16].

The ferric ion reducing antioxidant power assay (FRAP) is based on reducing power of an antioxidant to reduce a ferric salt (Fe^{3+}) to ferrous salt (Fe^{2+}) by electron transfer reaction. The ferric salt usually used is ferric (TPTZ; 2,4,6-tripydidyl-*s*-triazine) chloride. When this ferric salt is reduced to ferrous salt, a blue color is formed, the intensity of which can be measured spectrophotometrically. The redox potential of ferric salt is comparable to that of $ABTS^{\cdot+}$ radical cation. Thus, FPAP is similar to TEAC assay, except TEAC assay is conducted at neutral pH, whereas FPAP assay is usually performed in acidic pH (3.6) [1]. However, FRAP assay can also be carried out using ferricyanide

$[Fe^{3+}(CN^-)_6]$, in which an antioxidant donates an electron and converts ferricyanide to ferrocyanide $[Fe^{2+}(CN^-)_6]$, which in the presence of excess ferric ion (ferric chloride) produces an intense blue color (Perl's Prussian blue). The intensity of the color can be measured at 700 nm [17].

Cupric ion reducing power of an antioxidant can also be used to measure total antioxidant capacity of a compound. In this assay, cupric ion is reduced to cuprous ion by an antioxidant. Apak *et al.* [18] developed cupric reducing antioxidant capacity (CUPRAC) assay using the copper (II) neocuproine reagent (neocuproine is 2,9-dimethyl-1,10-phenanthroline, a heterocyclic organic compound). When an antioxidant reduces copper (II) neocuproine to copper (I) neocuproine, a blue color develops with absorption maxima at 450 nm. The intensity of the blue color is related to the antioxidant capacity of the compound tested. This method can be applied for determination of antioxidant capacity of food constituents. However, in addition to determining antioxidant capacity of polyphenols and flavonoids, this method can be used for measuring antioxidant capacity of human serum. Using ammonium acetate buffer medium for the reaction, total antioxidant capacity of serum due to ascorbic acid, α-tocopherol, and glutathione can be directly estimated by recording absorbance at 450 nm. For serum antioxidants such as uric acid, bilirubin, and bilirubin, an incubation time of 20 min at 50°C is required before recording absorbance. This method also takes into account the hydroxyl radical quenching ability of an antioxidant.

DPPH (2,2-diphenyl-1-picrylhydrazyl) radical scavenging assay is based on reduction of this free radical by an antioxidant whereby the DPPH radical with an absorption maxima of 515 nm is converted into a colorless compound after reduction. This reaction can be monitored spectrophotometrically. DPPH is a stable radical, and solution of DPPH can be made in methanol. After mixing with a sample solution containing the antioxidant of interest, absorption is monitored for 30 min or until the absorption is stable. The concentration of antioxidant needed to reduce DPPH concentration by 50% is termed EC_{50} (moles of antioxidant per mole of DPPH). Jimenez-Escrig *et al.* [19] used 580 nm instead of 515 nm for DPPH assay to determine the antioxidant capacity of dietary carotenoids, and they obtained a mean EC_{50} value of 0.16 for lycopene, a good antioxidant.

The Folin–Ciocalteu reducing capacity (FCR) assay, which employs the Folin–Ciocalteu method, was originally developed for analysis of protein taking advantage of reactivity of this reagent toward protein tyrosine (which has a phenolic group) and was expanded by Singleton *et al.* [20] to the analysis of total phenols in wine samples. The FCR assay is based on the transfer of electron from a phenolic compound to Folin–Ciocalteu reagent in alkaline medium (carbonate), and during this process phenolic antioxidant is oxidized. The Folin–Ciocalteu reagent is a mixture of phosphomolybdate and phosphotungstate that contains molybdotungstophosphate heteropolyanion. The molybdenum center in this complex is reduced from Mo(VI) to Mo(V) by an electron donated by an antioxidant, and during this process the light yellow color of the complex changes to an intense blue color with absorption maxima at 765 nm. Gallic acid is often used as the reference standard compound for this assay, and the antioxidant value is expressed as gallic acid equivalent. However, this reagent is not soluble in organic solvent. Berker *et al.* [21] modified this assay to measure lipophilic antioxidant by diluting aqueous Folin–Ciocalteu reagent with isobutanol (1:2 by volume) and measuring absorption at 665 nm.

2.2.5 Other Assays for Measuring Antioxidant Capacity

Superoxide anion radical can be generated by xanthine/xanthine oxidase at pH 7.4. This superoxide anion radical reduces nitroblue tetrazolium into blue formazan, which can be monitored spectrophotometrically at 560 nm [22]. Quick *et al.* [23] adopted a microplate format and reduction of cytochrome *c* by superoxide anion radical as the basis of measuring superoxide scavenging ability of extracts from two different neuronal cell cultures. The reaction was measured spectrophotometrically at 550 nm. The authors generated superoxide anion radical by xanthine oxidase metabolism of hypoxanthine. Other investigators have used nonenzymatic reaction in which phenazine methosulfate in the presence of nicotinamide adenine dinucleotide (NADH) and nitroblue tetrazolium was used as a probe. The riboflavin–methionine–illuminate system can also be used to produce superoxide anion radical, and nitroblue tetrazolium can be used as a probe [1].

Hydroxyl radicals are formed when hydrogen peroxide reacts with ferrous ions (Fenton reaction), but the ferrous ion hydrogen peroxide system cannot be used in hydroxyl radical scavenging assay because many antioxidants are metal chelators, thus interfering with this assay. However, Zhu *et al.* [24] described an organic Fenton reaction in which a mixture of tetrachlorohydroquinone (a major metabolite of the widely used biocide pentachlorophenol) and hydrogen peroxide hydroxylates salicylic acid yielding 2,3-dihydroxybenzoic and 2,5-dihydroxybenzoic acid, which can be measured using high-performance liquid chromatography with electrochemical detection. This hydroxylation reaction is markedly inhibited by hydroxyl radical scavenging antioxidants, and a metal chelator does not affect this inhibition. Ou *et al.* [25] developed a novel fluorometric method to evaluate hydroxyl radical scavenging capacity using fluorescein as a probe. Hydroxyl radicals were generated by cobalt (II)-mediated Fenton-like reaction, and hydroxyl radical formation was indirectly confirmed by the hydroxylation of *p*-hydroxybenzoic acid. The fluorescence decay curve of fluorescein was monitored in the absence and presence of an antioxidant, and hydroxyl radical scavenging capacity was calculated by subtracting the area under the curve (AUC) of fluorescence observed with blank sample from the AUC of the sample containing antioxidant. Gallic acid was used as a reference standard, and values were expressed as gallic acid equivalent. In the hydrogen peroxide scavenging assay, intrinsic absorption of hydrogen peroxide in the UV region at 230 nm was used, and because an antioxidant successfully scavenges hydrogen peroxide, absorption at 230 nm was reduced [26].

The nitric oxide radical scavenging capacity of antioxidants such as sulfur-containing compounds can be measured in aqueous solution using an amperometric nitric oxide radical sensor. Nitric oxide radical scavenging capacity can also be determined by ESR spectroscopy. The Griess reaction is a two-step diazotization reaction in which nitrosating agent derived from the acid-catalyzed formation of nitrous acid from nitrite reacts with sulphanilic acid to produce a diazonium ion. Then, this ion is coupled with *N*-(1-naphthyl) ethylenediamine to form a chromophoric azo compound that absorbs strongly at 548 nm. This reagent is frequently used to determine the *in vitro* nitric oxide radical scavenging capacity of antioxidants [1]. Superoxide and nitric oxide react to form the peroxynitrite radical, and scavenging capacity of an antioxidant toward this free radical is also used for assessing its antioxidant capacity. The scavenging capacity of an antioxidant toward peroxynitrite radical can be measured by inhibition of tyrosine nitration by peroxynitrite radical or inhibition of oxidation of dihydrorhodamine 123 [27]. Singlet oxygen is not a free radical but is an excited state of oxygen generated in the presence of light, and it is believed to be responsible for UV light-induced skin damage. Singlet oxygen has characteristics of phosphorescence at 1270 nm, and decay of this signal by an antioxidant due to quenching of singlet oxygen can be used as the basis of singlet oxygen quenching capacity of an antioxidant. A more sensitive method is based on monitoring scavenging of singlet oxygen delayed fluorescence of *tert*-butyl-phthalocyanine [1]. Costa *et al.* [28] also developed a singlet oxygen quenching capacity assay in which singlet oxygen was selectively generated by thermal decomposition of the endoperoxide 3,3′-(1,4-naphthalene) bispropionate. Singlet oxygen then oxidized dihydrorhodamine 123 into fluorescent form (rhodamine 123).

The most common metal chelating assay is based on ferrous ion chelating capacity. Minimizing ferrous ion (Fe^{2+}) may prevent production of reactive oxygen species because ferrous ion is known to react with hydrogen peroxide, producing highly reactive hydroxyl radicals (Fenton reaction). Ferrozine forms a complex with free ferrous ion but not ferrous ion already chelated by a compound. The ferrozine–ferrous ion complex forms a red color that can be monitored spectrophotometrically at 562 nm [1].

2.3 MARKERS OF OXIDATIVE STRESS IN HUMAN BLOOD AND OTHER PARAMETERS

Another approach in antioxidant research is to measure various markers of oxidative damage to biomolecules. Because lipid molecules are significantly damaged by reactive oxygen species, various degradation products of lipid peroxidation are used for estimating oxidative damage and the degree of protection provided by an antioxidant. In addition, markers of oxidative damage to DNA are also commonly measured. There are also markers of oxidative damage to proteins. Moreover, antioxidant compounds in blood, such as vitamin C, vitamin E, β-carotene, and glutathione, can be directly measured to evaluate the antioxidant capacity of blood. The activity of antioxidant enzymes can also be determined. Various markers of oxidative stress and antioxidant enzymes are listed in Table 2.2.

Lipid peroxidation is a complex process involving uptake of oxygen, formation and propagation of lipid radicals, and rearrangement of double bonds in unsaturated and polyunsaturated fatty acids, eventually producing a mixture of complex breakdown products including alcohols, aldehydes, ketones, and ethers. For example, peroxidation of linoleic acid alone produces at least 20 breakdown products. The first step in lipid peroxidation involves abstraction of a hydrogen atom from an unsaturated fatty acid, yielding a lipid

Table 2.2 Various Markers of Oxidative Stress

Assay Type	Individual Assays
Biomarkers of oxidative stress	
Markers of lipid damage	Malondialdehyde (thiobarbituric acid reactive substances)
	4-Hydroxy 2-nonenal
	Conjugated dienes
	Lipid hydroperoxide
	Isoprostane
	Oxidized low-density lipoprotein
	Exhaled hydrocarbon in breath
Markers of protein damage	Various protein carbonylation products
Marker of DNA damage	8-Oxo-7,8-dihydro-2′-deoxyguanosine(8-hydroxy-2′-deoxyguanosine)
Marker of RNA damage	8-Oxo-7,8-dihydroguanosine
Antioxidant enzymes	Superoxide dismutase
	Catalase
	Glutathione peroxidase
	Glutathione reductase
	Thioredoxin reductase
Other markers	Glutathione
	Antioxidant vitamins

radical. The rearrangement of double bonds in polyunsaturated fatty acids produces conjugated dienes, a marker of lipid peroxidation. Peroxyl radicals are formed due to attack by molecular oxygen, which can abstract a hydrogen atom from a double bond in fatty acid side chains, producing a lipid hydroperoxide (another marker of lipid peroxidation) or lipid endoperoxide. The formation of endoperoxide in an unsaturated fatty acid containing at least three methylene interrupted double bonds (e.g., linolenic acid and arachidonic acid) can yield malondialdehyde, another marker compound of lipid peroxidation [29]. One of the oldest but still widely used assays for determination of oxidative stress in serum is the TBARS (thiobarbituric acid reactive substances) assay, which measures the concentration of malondialdehyde that is produced due to degradation of unstable lipid peroxides. This assay can be used not only using human serum but also using urine and tissue samples. Other markers of oxidative stress in blood include isoprostane (formed due to oxidation of arachidonic acid), oxysterol (oxidation product of cholesterol), and other markers of protein and DNA damage [30]. In addition, some of the assays used for measuring antioxidant capacity of various foods can also be used for measuring antioxidant capacity of human blood. Total antioxidant capacity of blood is a popular assay that physicians can order to evaluate antioxidant capacity of blood in patients. In one commercially available assay that can be adopted easily in clinical laboratories, antioxidant potential of a specimen (serum, saliva, and urine) is determined by its capacity to inhibit oxidation of a probe ABTS and the result can be expressed as Trolox equivalent [31].

2.4 MARKERS OF LIPID PEROXIDATION

TBARS is probably the oldest and one of the most widely used assays for measuring lipid peroxidation end product malondialdehyde, a reactive aldehyde produced by lipid peroxidation of polyunsaturated fatty acids. Malondialdehyde forms adduct with two thiobarbituric acid molecules to produce a pink color species that absorbs at 532–535 nm. In order to form TBARS, a biological specimen must be heated with the reagent (containing thiobarbituric acid, trichloroacetic acid, and hydrochloric acid) for 20 min at 50°C. After cooling, the specimen can be centrifuged if necessary and absorbance can be measured spectrophotometrically, with absorbance being proportional to the concentration of malondialdehyde. However, this method is subject to interferences. In human plasma, sialic acid present in glycoprotein can interfere, but such interference can be minimized by carrying out the reaction in phosphoric acid medium [32]. However, a variety of other compounds, such as oxidized lipids, saturated and unsaturated aldehydes, sucrose, and urea, interfere with the assay, causing overestimation of malondialdehyde concentration. Chromatographic separation of malondialdehyde–thiobarbituric acid adduct can produce more precise results than using spectrophotometric estimation of TBARS without any separation. Acid precipitation of lipoproteins may improve performance of the assay. However, the best estimation of malondialdehyde concentration is achieved by direct estimation of this compound using a specific chromatographic method. Moselhy *et al.* [33] described two chromatographic methods—high-performance liquid chromatography (HPLC)–diode array detection combined with fluorometry and HPLC–diode array–mass spectrometric detection—for detection of plasma malondialdehyde levels and compared values with TBARS values. Up to a threefold difference was noted between TBARS values measured in human plasma and malondialdehyde values obtained by specific chromatographic methods. In addition to malondialdehyde, other reactive aldehydes are formed during peroxidation of unsaturated fatty acid, including 4-hydroxy-2-nonenal and 4-oxo-2-nonenal. In general, 4-hydroxy-2-nonenal is usually measured as a marker of lipid peroxidation. However, these aldehydes are very reactive and unstable in serum. Nevertheless, 4-hydroxy-2-nonenal can be analyzed by thin-layer chromatography combined with densitometry (both fluorescence and reflectance mode) using dinitrophenylhydrazine or cyclohexanedione for derivatization [34]. However, gas chromatography combined with mass spectrometry and HPLC-based methods are superior for accurate determination of 4-hydroxy-2-nonenal and related aldehydes [35]. In addition, a urinary stable metabolite of 4-hydroxy-2-nonenal, such as 1,4-dihydroxynonane-mercapturic acid, can be used as a biomarker of oxidative stress. 4-Oxo-2-nonenal has been less studied than 4-hydroxy-2-nonenal as a biomarker of lipid peroxidation because 4-oxo-2-nonenal is more reactive and unstable in human serum. Acrolein is the most reactive of all aldehydes from lipid peroxidation, and attempts have been made to use this reactive aldehyde as a potential biomarker of oxidative stress [36].

For conjugated diene, absorbance at 233 nm can be measured, and the intensity of the signal is proportional to the concentration of conjugated diene. Membrane lipid or human serum can be extracted with chloroform:methanol (2:1 by volume), and then the upper layer containing proteins should be discarded. The lower layer (organic layer) can be dried under an inert gas, and residue remaining after reconstructing with cyclohexane can be used for absorbance measurement at 233 nm against cyclohexane blank [32]. Although the organic layer is usually the top layer and aqueous phase is the bottom layer during organic solvent extraction, chloroform and chlorinated hydrocarbons have higher specific gravity than water and, for extraction using chloroform or related solvents, the aqueous layer is the top layer and the organic layer is the bottom layer. For estimation of lipid hydroperoxide, the specimen of interest can be extracted with chloroform:methanol (2:1 by volume) similarly to analysis of conjugated diene, but organic extract after evaporation should be incubated with potassium iodide in the dark for 6–8 h. After incubation, acetic acid:chloroform mixture (3:2 by volume) should be added, followed by 0.25% cadmium acetate. After centrifugation, the upper layer can be transferred to a tube followed by absorbance measurement at 373 nm. However, lipid hydroperoxide can also be conveniently analyzed by using Fox reagent, which consists of 250 μmol ammonium sulfate, 100 μmol xylenol orange, 25 mmol sulfuric acid, and 4 mmol butylated hydroxytoluene [32]. Vassalle *et al.* [37] reported adaptation of a colorimetric assay for lipid hydroperoxide on a Synchron CX9 Pro automated analyzer (Beckman Coulter, Brea, CA). The colorimetric assay (d-ROM) was obtained from Diacron (Grosseto, Italy). This assay allows indirect estimation of lipid hydroperoxide in human serum, and it is based on the capacity of transition metals to catalyze peroxides present in the specimen into alkoxy and peroxy radicals, which then react with an amine (*N,N*-diethyl-*para*-phenylenediamine), forming a colored complex that can be measured spectrophotometrically at 520 nm. The lipid hydroperoxide concentration can be expressed as an arbitrary unit (AU) [37]. Vassalle *et al.* [38] later applied this automated assay to demonstrate that serum hydroperoxide concentrations were significantly elevated in patients with multivessel coronary artery disease (412 ± 16 AU) compared to controls (341 ± 12 AU). Patients with diabetes, dyslipidemia, and elevated C-reactive protein also showed an elevated level of serum hydroperoxide, indicating that these patients were subjected to elevated *in vivo* oxidative stress compared to controls [38]. However, liquid chromatography–mass spectrometry is a more specific analytical method for analysis of lipid hydroperoxide in human serum [39].

F_2-Isoprostane (8-iso-$PGF_{2\alpha}$, where PGF is prostaglandin) is produced by nonenzymatic oxidation of arachidonic acid by reactive oxygen species including free radicals. Depending on the position where the oxygen molecule is added to the arachidonic acid (arachidonic acid has four double bonds), four region isomers are formed, resulting in four isoprostane series, and each series has 16 stereoisomers, making isoprostane a complex class of molecules (64 isomers of isoprostane from oxidation of arachidonic acid). F_2-Isoprostane can be measured in human serum and urine, and it is the most commonly used isoprostane marker to evaluate *in vivo* oxidative stress. F_2-Isoprostane and its metabolites are found in urine and are stable and can be used as markers of oxidative stress. Three main assays are used for measuring F_2-isoprostane in serum and urine: enzyme-linked immunosorbent assay (ELISA), gas chromatography combined with mass spectrometry, and liquid chromatography combined with mass spectrometry or tandem mass spectrometry. In general, ELISA methods are less sensitive and less specific compared to chromatography combined with mass spectrometry-based assays [36]. In addition, oxidation of γ-linoleic acid produces F_1 series of isoprostanes and oxidation of eicosapentaenoic acid produces F_3 series of isoprostanes, but these isoprostanes are not usually utilized as markers of lipid peroxidation.

Oxidized low-density lipoprotein (ox-LDL) is an important biomarker of oxidative stress, and studies have linked ox-LDL with risk of cardiovascular diseases. Immunoassays are commercially available for analysis of ox-LDL and are widely used to study oxidative stress in a variety of diseases. Pawlak *et al.* [40] used a commercially available sandwich ELISA (Mercodia, Uppsala, Sweden) for analysis of ox-LDL in dialysis patients. This assay is based on murine monoclonal antibody directed against a conformational epitope in the apoB-100 moiety generated due to reaction of lysine residues with aldehydes, and the assay is specific for ox-LDL. Lipid peroxidation also generates hydrocarbons such as ethane and pentane, which are excreted in breath and can be analyzed in breath condensate (commercially available instruments are available for breath collection) using gas chromatography or gas chromatography combined with mass spectrometry. Such methods can detect these markers in picomolar concentrations. In general, alkanes along with monomethylated alkanes are formed due to lipid peroxidation that can be used as markers of oxidative stress [41]. Leukotrienes and F_2-isoprostane can also be used as markers of oxidative stress in breath condensate. Caballero *et al.* [42] measured both leukotrienes and isoprostane in breath condensate using specific enzyme immunoassays capable of determining concentration at the picogram level. In addition, 4-hydroxy-2-nonenal and other reactive aldehydes, such as malondialdehyde, are also excreted in breath and can be analyzed as markers of oxidative stress preferably using gas chromatography or liquid chromatography combined with tandem mass spectrometry. Peroni *et al.* [43] analyzed 4-hydroxy-2-nonenal and malondialdehyde in breath using HPLC combined with mass spectrometry after derivativation with 2,4-dinitrophenylhydrazine.

2.5 MARKERS OF PROTEIN AND DNA DAMAGE

Reactive oxygen species can also damage proteins and DNA. Markers of proteins and DNA damage are also used for investigating *in vivo* oxidative stress in various diseases and in animal models. Reactive oxygen species can cause oxidation of amino acid side chains, and, in general, cysteine and methionine are especially sensitive to such oxidation. Aromatic amino acids such as tyrosine, tryptophan, and phenylalanine are also sensitive to oxidation by reactive oxygen species. Moreover, methionine, tryptophan, cysteine, and tyrosine are targets of reactive nitrogen species, especially peroxynitrite.

Oxidation of the carbon skeleton of protein also occurs due to oxidative stress. In addition, formation of carbonyl groups on proteins either by direct oxidation of amino acid side chains (lysine, arginine, proline, and threonine) or due to fragmentation of carbon chains by oxidation of a glutamyl residue or α-amidation, may also occur. Introduction of carbonyl groups to proteins might also occur due to reaction of reactive aldehydic end products of lipid peroxidation. Dicarbonyl compounds are also formed due to glycation and glycoxidation. Various methods are available for detection of protein carbonyls. Because these products do not have UV or visible spectrophotometric absorption or fluorescence properties, detection of these compounds requires specific chemical probes, such as 2,4-dinitrophenylhydrazine, tritiated sodium borohydride, biotin-containing probes, and *N*′-aminooxymethylcarbonylhydrazine. With the exception of tritiated sodium borohydride probe, the probes have a hydrazine-like moiety that can react specifically with carbonyl groups of oxidized proteins, producing stable hydrazones that can be analyzed spectrophotometrically (hydrazone formed due to reaction of carbonyl groups with 2,4-dinitrophenylhydrazine has a characteristic absorption at 365–375 nm), or, by taking advantage of commercially available antibodies against 2,4-dinitrophenyl, analysis can also be performed using ELISA assay or Western blot. Proteomics approaches such as fluorescence-based two-dimensional gel electrophoresis and mass spectrometric methods have recently been used for analysis of these complex carbonylated proteins [44]. Polac *et al.* [45] commented that carbonyl groups in protein coupled to 2,4-dinitrophenylhydrazine are thought to be a good marker of oxidative stress. The authors used ELISA for quantitation of carbonyl groups in oxidized proteins by first derivatizing these proteins using 2,4-dinitrophenylhydrazine followed by analysis using 2,4-dinitrophenyl antibody commercially available from Sigma–Aldrich (St. Louis, MO). The authors showed that hormonal therapy reduces the level of carbonyl proteins in postmenopausal women. Hypochlorous acid generated by phagocyte enzyme myeloperoxidase during inflammation can cause oxidative damage to residues such as lysine as well as chlorination of aromatic residues such as tyrosine, yielding 3-chlorotyrosine, which can be used as a marker of oxidative stress. Free 3-chlorotyrosine can be measured using gas chromatography–mass spectrometry. Similarly, peroxynitrite free radical can also damage amino acids and proteins, yielding nitrotyrosine. Various mass spectrometric methods including soft ionization techniques such as matrix-assisted laser desorption/ionization (MALDI) combined with time-of-flight (TOF) mass spectrometry can be used for analysis of complex products generated by oxidative damage to proteins [46]. Ahsan [47] commented that 3-nitrotyrosine is a biomarker of endogenous peroxynitrite radical activity, and peroxynitrite-derived oxidation and nitration of biomolecules may lead to autoimmune disease such as systemic lupus.

Oxidative stress is also known to damage DNA. The most studied marker of DNA damage due to oxidative stress is 8-oxo-7,8-dihydro-2′-deoxyguanosine (8 oxodG), a stable end product of DNA oxidation. This compound is also known as 8-hydroxy-2′-deoxyguanosine, and this biomarker for oxidative stress can be measured in both serum and urine. Immunoassays such as ELISA or chromatography-based methods can be used for analysis of 8-oxodG in serum or urine. Topic *et al.* [48] described a liquid chromatography combined with tandem mass spectrometric assay for analysis of 8-oxodG in urine specimens to establish a gender-specific reference range of this marker of DNA oxidation in the Serbian population. Oxidation of RNA produces 8-oxo-7,8-dihydroguanosine (8-oxoGuo), and the urine level of this marker is often used for investigating *in vivo* oxidative stress. This marker, along with 8-oxodG, can be measured by ultra performance liquid chromatography combined with tandem mass spectrometry [49].

2.6 MEASUREMENT OF ANTIOXIDANT ENZYMES AND OTHER COMPOUNDS

For analysis of common antioxidant enzyme such as superoxide dismutase, glutathione peroxidase, glutathione reductase, and catalase, several methods have been described in the literature. Kits for analysis of these enzymes are commercially available. Kaplan *et al.* [50] used commercially available test kits from Cayman Chemical Company (Ann Arbor, MI) for analysis of erythrocyte glutathione peroxidase, superoxide dismutase, and catalase in their study to investigate the effect of antioxidant enzyme levels and exercise on patients with syndrome X (metabolic syndrome) and slow coronary flow in comparison to controls. Cayman's superoxide dismutase kit utilizes a tetrazolium salt for detection of superoxide anion radicals generated by xanthine oxidase and hypoxanthine. One unit of superoxide dismutase activity is defined as the amount of enzyme needed to exhibit 50% dismutation of superoxide radical. Assays for analysis of glutathione peroxidase, glutathione reductase, and superoxide dismutase are also available from Randox (Kearneysville, WV). Kits are also commercially available for assay of thioredoxin reductase. For example, the thioredoxin assay kit available from Sigma–Aldrich is based on reduction of 5,5′-dithio-bis(2-nitrobenzoic acid) (DTNB) and NADPH to 5-thio-2-nitrobenzoic acid (TNB), which produces a yellow color that is measured spectrophotometrically at 412 nm. An assay kit available from Cayman is based on the same principle.

Glutathione and various antioxidant vitamins, such as vitamin C, vitamin E, and various carotenoids, are also measured to evaluate the effect of oxidative stress in both human and animal studies. Assay kits are commercially available for analysis of glutathione. For example, in the assay kit from

Sigma–Aldrich, the sample is first deproteinized with the 5% 5-sulfosalicylic acid solution and the glutathione content of the sample is then assayed using a kinetic assay in which catalytic amounts of glutathione cause a continuous reduction of DTNB to TNB. The oxidized glutathione formed is recycled by glutathione reductase and NADPH, and TNB is analyzed spectrophotometrically at 412 nm.

Assay kits are commercially available for analysis of vitamin C. However, for accurate analysis of antioxidant vitamins, chromatography-based methods provide better precision and accuracy. Many reference clinical laboratories provide analysis of vitamins in human serum or plasma. Faure *et al.* [51] used HPLC for analysis of α-tocopherol (vitamin E), retinol, and β-carotene and a spectrofluorometric method for analysis of vitamin C in human serum. Andreoli *et al.* [52] described a liquid chromatography combined with tandem mass spectrometric method for simultaneous analysis of retinol, α-tocopherol, and β-carotene in human serum. The authors used atmospheric pressure chemical ionization for mass spectrometric analysis.

2.7 ELECTRON SPIN RESONANCE AND RELATED METHODS TO STUDY FREE RADICALS

The ESR (also called electron paramagnetic resonance) technique allows unambiguous detection of free radicals, but the major limitation of this method in the study of free radicals in cells and tissues is related to its poor sensitivity to detect very low (1 nmol to 1 pmol) levels of free radicals generated under normal physiological conditions or stress. To overcome this limitation, the spin trapping technique is used, in which free radicals are added to a chemical agent to form a more stable radical that can be studied by ESR due to its higher concentration. Many spin trap agents that are nitroso and nitrone compounds have been used by many investigators, and one popular spin trap agent is 5,5-dimethyl-1-pyrroline-*N*-oxide (DMPO). The lifetime of a spin trapped radical dictates the success of ESR spin trapping experiments: The greater the stability of the radical adduct, the higher the concentration of the radical adduct for a given rate of free radical generation. Both polyclonal and monoclonal antibodies are available to detect DMPO (anti-DMPO antibodies). Therefore, DMPO adducts can be detected by heterogeneous immunoassays, mass spectrometry, and molecular magnetic resonance imaging [53]. When ESR is used to study spin trapped radicals, various radical adducts exhibit a distinctive spectrum that can be used to identify such radicals. Using α-phenyl-*tert*-butylnitrone as a spin trapping agent, Davison *et al.* [54] demonstrated elevated levels of free radicals in venous blood of young male patients with type 1 diabetes compared to controls. The hyperfine coupling constants recorded from ESR spectra suggested the presence of oxygen-

and carbon-centered radicals in patients with diabetes. Higher lipid hydroperoxide levels were also observed in diabetic patients, which was consistent with the observation of higher concentrations of free radicals in these patients as measured by ESR.

2.8 CONCLUSION

Many methods are available to measure the total antioxidant capacity of food components, and cell and tissue extract, as well as serum of plasma of humans or animals for investigating the role of free radicals in biology and medicine. However, measurement of individual markers of lipid peroxidation, protein oxidation, and DNA or RNA damage is most commonly performed in studies of serum, plasma, or urine. Many methods are also available for analysis of various antioxidant enzymes and antioxidant vitamins. Serum or erythrocyte levels of antioxidant enzymes are commonly measured to evaluate the effect of oxidative stress by a stressor or a disease state. ESR using spin trapping is also used to study free radicals in various disease conditions. The choice of technique depends on the scope of investigation as well as on the investigators.

REFERENCES

[1] Gulcin I. Antioxidant activity of food constituents: an overview. Arch Toxicol 2012;86:345−91.

[2] Huang D, Ou B, Prior RL. The chemistry behind antioxidant capacity assays. J Agric Food Chem 2005;53:1841−56.

[3] Lastawska K. Evaluation of selected food supplements containing antioxidants. Rocz Panstw Zalk Hyg 2010;61:151−4.

[4] Cao G, Alessio HM, Cutler RG. Oxygen radical absorption capacity assay for antioxidants. Free Radic Biol Med 1993;14:303−11.

[5] Tomer DP, McLeman LD, Ohmine SZ, Scherer PM, et al. Comparison of the total oxyradical scavenging capacity and oxygen radical absorption capacity antioxidant assays. J Med Food 2007;10:337−44.

[6] Garrett AR, Murray BK, Robinson RA, O'Neil KL. Measuring antioxidant capacity using ORAC and TOSC assays. Methods Mol Biol 2010;594:251−62.

[7] Wayner DDM, Burton KU, Locke S. Quantitative measurement of total, peroxyl radical trapping antioxidant capability of human blood plasma by controlled peroxidation. FEBS Lett 1985;187:33−7.

[8] Valkonen M, Kuusi T. Spectrophotometric assay for total peroxyl radical trapping antioxidant potential in human serum. J Lipid Res 1997;38:823−33.

[9] Chu YF, Liu RH. Novel low density lipoprotein (LDL) oxidation model: antioxidant capacity for the inhibition of LDL oxidation. J Agric Food Chem 2004;52:6818−23.

[10] Niki E, Sato T, Kawakami A, Kamiya Y. Inhibition of oxidation of methyl linoleate in solution by vitamin E and vitamin C. J Biol Chem 1984;259:4177−82.

[11] Bors W, Michel C, Saran M. Inhibition of bleaching of carotenoid crocin, a rapid test for quantifying antioxidant activity. Biochim Biophys Acta 1984;796:312–19.

[12] Burton GW, Ingold KU. Autooxidation of biological molecules I: the autooxidation of vitamin E and related chain breaking antioxidant *in vitro*. J Am Chem Soc 1981;103:6472–7.

[13] Miller NJ, Rice-Evans C, Davies MJ, Gopinathan V, et al. A novel method for measuring antioxidant capacity and its application to monitoring the antioxidant status in premature neonates. Clin Sci (Lond) 1993;84:407–12.

[14] Erel O. A novel automated direct measurement method for total antioxidant capacity using a new generation more stable ABTS radical cation. Clin Biochem 2004;37:277–84.

[15] Fogliano V, Verde V, Randazzo G, Ritieni A. Method for measuring antioxidant activity and its application to monitoring the antioxidant capacity of wine. J Agric Food Chem 1999;47:1035–40.

[16] Verde V, Gogliano V, Ritieni A, Maiani G, et al. Use of *N,N-p*-phenylenediamine to evaluate the oxidative status of human plasma. Free Radic Res 2002;36:869–73.

[17] Gulcin I. Antioxidant activity of L-adrenaline: a structure–activity insight. Chem Biol Interact 2009;179:71–80.

[18] Apak R, Guclu K, Ozyurek M, Bektasoglu B, et al. Cupric ion reducing antioxidant capacity assay for antioxidants in human serum and for hydroxyl radical scavenger. Methods Mol Biol 2010;594:215–39.

[19] Jimenez-Escrig A, Jimenez-Jimenez I, Sanchez-Moreno C, Saura-Calixto F. Evaluation of free radical scavenging of dietary carotenoids by stable radical 2,2-diphenyl-1-picrylhydrazyl. J Sci Food Agric 2000;80:1686–90.

[20] Singleton VL, Orthofer R, Lamuela-Raventos RM. Analysis of total phenolics with phosphomolybdic–phosphotungstic acid reagent. Methods Enzymol 1999;299:152–78.

[21] Berker KL, Ozdemir Olgun FA, Ozyurt D, Demirata B, et al. Modified Folin–Ciocalteu antioxidant capacity assay for measuring lipophilic antioxidant. J Agric Food Chem 2013;61:4783–91.

[22] Parejo I, Viladomat F, Bastida J, Rosas-Romero A, et al. Comparison between the radical scavenging activity and antioxidant activity in six distilled and non-distilled Mediterranean herbs and aromatic plants. J Agric Food Chem 2002;50:6882–90.

[23] Quick KL, Hardt JI, Dugan LL. Rapid microplate assay for superoxide scavenging efficiency. J Neurosci Methods 2000;97:139–44.

[24] Zhu BZ, Kitrossky N, Chevion M. Evidence for production of hydroxyl radicals by pentachlorophenol metabolites and hydrogen peroxide: a metal independent organic Fenton reaction. Biochem Biophys Res Commun 2000;270:942–6.

[25] Ou B, Hampsch-Woodill M, Flanagan J, Deemer EK, et al. Novel fluorometric assay for hydroxyl radical prevention capacity using fluorescein as the probe. J Agric Food Chem 2002;50:2772–7.

[26] Berges A, Van Nassauw L, Timmermans JP, Vrints C. Time dependent expression pattern of nitric oxide and superoxide after myocardial infarction in rats. Pharmacol Res 2007;55:72–9.

[27] Hunag D, Ou B, Prior RL. The chemistry behind antioxidant capacity assays. J Agric Food Chem 2005;53:1841–56.

[28] Costa D, Fernandes E, Santos JLM, Pinto DCG. Noncellular fluorescence microplate screening assay for scavenging activity against singlet oxygen. Anal Bioanal Chem 2007;387:2071–81.

[29] Buege JA, Aust SD. Microsomal lipid peroxidation. Methods Enzymol 1987;52:302–10.

[30] Vaya J. Exogenous markers for the characterization of human disease associated with oxidative stress. Biochimie 2013;95:578–84.

[31] Yeum KJ, Russell RM, Krinsky NI, Aldini G. Biomarkers of antioxidant capacity in the hydrophilic and lipophilic compartments of human plasma. Arch Biochem Biophys 2004;430:97–103.

[32] Devasagayam TPA, Bollor KK, Ramasarma T. Methods for estimating lipid peroxide: an analysis of merits and demerits. Indian J Biochem Biophys 2003;40:300–8.

[33] Moselhy HF, Reid RG, Yousef S, Boyle SP. A specific accurate and sensitive measure of total plasma malondialdehyde by HPLC. J Lipid Res 2013;54:852–8.

[34] Beckman JK, Morley SA, Greene HL. Analysis of aldehydic lipid peroxidation products by TLC/densitometry. Lipids 1991;26:155–61.

[35] Esterbauer H, Zollner H. Methods for determination of aldehydic lipid peroxidation products. Free Radic Biol Med 1989;7:197–203.

[36] Ii'yasova D, Scarbrough P, Spasojevic I. Urinary biomarkers of oxidative stress. Clin Chim Acta 2012;413:1446–53.

[37] Vassalle C, Boni C, Cecco PD, Ndreu R, et al. Automation and validation of a fast method for the assessment of *in vivo* oxidative stress level. Clin Chem Lab Med 2006;44:1372–5.

[38] Vassalle C, Landi P, Boni C, Zucchelli G. Oxidative stress evaluated using an automated method for hydroperoxide estimation in patients with coronary artery disease. Clin Chem Lab Med 2007;45:367–71.

[39] Ferreiro-Vera C, Ribeiro JP, Mata-Granados JM, Priego-Capote F, et al. Standard operation protocol for analysis of lipid hydroperoxide in human serum using a fully automated method based on extraction and liquid chromatography–mass spectrometry in selected ion monitoring. J Chromatogr A 2011;1218:6720–6.

[40] Pawlak K, Mysliwiec M, Pawlak D. Oxidized low-density lipoprotein (oxLDL) plasma levels and oxLDL to LDL ration: are they real oxidative stress markers in dialyzed patients? Life Sci 2013;92:253–8.

[41] Phillips M, Cataneo RN, Cheema T, Greenberg J. Increased breath biomarkers of oxidative stress in diabetes mellitus. Clin Chim Acta 2013;344:189–94.

[42] Caballero S, Martorell A, Escribano A, Belda J. Markers of airway inflammation in the exhaled breath condensate of preschool wheezers. J Investig Allergol Clin Immunol 2013;23(1):7–13.

[43] Peroni DG, Bodini A, Corradi M, Coghi A, et al. Markers of oxidative stress is increased in exhaled breath condensate of children with atopic dermatitis. Br J Dermatol 2012;166:839–43.

[44] Baraibar MA, Ladouc R, Friguet B. Proteomic quantification and identification of carbonylated proteins upon oxidative stress and during cellular aging. J Proteom 2013;92:63–70.

[45] Polac I, Borowiecka M, Wilamowska A, Nowak P. Oxidative stress measured by carbonyl groups level in postmenopausal women after oral and transdermal hormone therapy. J Obstet Gynaecol 2012;38:1177–81.

[46] Mouls L, Silajdzic E, Haroune N, Spickett C, et al. Development of novel mass spectrometric methods for identifying HOCL induced modification of proteins. Proteomics 2009;9:1617–31.

[47] Ahsan H. 3-Nitrotyrosine: a biomarker of nitrogen free radical species modified proteins in systematic autoimmunogenic condition. Hum Immunol 2013;74:1392–9.

[48] Topic A, Francuski D, Markovic B, Stankovic M, et al. Gender related reference intervals of urinary 8-oxo-7,8-dihydro-2′-deoxyguanosine determined by liquid chromatography–tandem mass spectrometry in Serbian population. Clin Biochem 2013;46:321–6.

[49] Jorgensen A, Broedback K, Fink-Jensen A, Knorr U, et al. Increased systematic oxidatively generated DNA and RNA damage in schizophrenia. Psychiatr Res 2013;209:417–24.

[50] Kaplan O, Meric M, Acar Z, Kale A, et al. The effect of antioxidant enzyme levels and exercise on syndrome X and coronary slow flow phenomenon: an observational study. Anadolu Kardiyol Derg 2013; [Epub ahead of print].

[51] Faure H, Preziosi P, Roussel AM, Bertrais S, et al. Factors influencing blood concentration of retinol, alpha-tocopherol, vitamin C and beta-carotene in French participants of the SU. VI. MAX trial. Eur J Clin Nutr 2006;60:706–17.

[52] Andreoli R, Manini P, Poli D, Bergamaschi E, et al. Development of a simplified method for simultaneous determination of retinol, alpha-tocopherol, and beta-carotene in serum by liquid chromatography–tandem mass spectrometry with atmospheric pressure chemical ionization. Anal Bioanal Chem 2004;378:987–94.

[53] Gomez-Mejiba S, Zhai Z, Della-Vedova M, Munoz M, et al. Immuno-spin trapping from biochemistry to medicine: advances, challenges and pitfalls—focus on protein centered radicals. Biochim Biophys Acta 2013; [Epub ahead of print].

[54] Davison GW, George L, Jackson SK, Young IS, et al. Exercise, free radicals and lipid peroxidation in type 1 diabetes mellitus. Free Radic Biol Med 2002;33:1543–51.

CHAPTER 3

Oxidative Stress Induced by Air Pollution and Exposure to Sunlight

CONTENTS

3.1 INTRODUCTION

Breathing polluted air may increase oxidative stress because polluted air contains small particles capable of generating free radicals. In addition, various free radicals may already be present in polluted air. In general, the air quality of urban areas is poorer than that of rural areas. Industrial smoke also contains free radicals and small particles. Sunlight contains ultraviolet (UV) radiation, and prolonged exposure to sunlight or sunbathing without application of sunscreen may cause sunburn, skin damage, and even skin cancer including melanomas, which are linked to oxidative stress. During the summer months when the UV index is high, exposure to sunlight should be reduced or protective clothing should be worn. Moreover, use of sunglasses is recommended because UV light can induce oxidative stress in eyes, causing cataract formation and other eye diseases such as macular degeneration.

3.2 COMPOSITION OF POLLUTED AIR

Air pollution can be defined as the presence of gaseous and particulate matter in the air that is not a natural constituent of the air. Smog, which is a combination of fog and smoke, may also cause significant air pollution, especially during the winter months in large cities. Sources of air pollution include natural phenomena, such as wildfires, volcano eruptions, and land dust, in addition to human activities. Air pollution attributable to human activities is caused by motor vehicles, power plants, and industrial sources, as well as household combustion devices. In addition to primary pollutants, which are directly emitted from a source, secondary pollutants are formed by chemical reactions of primary pollutants with the atmosphere. For example, in the troposphere, ozone is produced by the reaction of sunlight with volatile hydrocarbons [1]. Ozone, nitrogen oxide, carbon monoxide, sulfur

A. Dasgupta and K. Klein: Antioxidants in Food, Vitamins and Supplements. DOI: http://dx.doi.org/10.1016/B978-0-12-405872-9.00003-3

Table 3.1 Common Air Pollutants

Air Pollutant	Comment
Ozone	Ozone is found mostly in the upper atmosphere (stratosphere), but if ozone is present in ground air, it acts as an air pollutant.
Nitrogen oxide	Nitrous oxide and nitric oxide are formed from fossil fuel combustion and industrial processes. Nitrogen oxides are also produced by thunderstorms.
Sulfur dioxide	Sulfur dioxide is produced during factory emissions (burning of coal or natural gas) and naturally during volcano eruptions.
Carbon monoxide	Carbon monoxide is emitted from motor vehicle exhaust due to incomplete combustion and also from industrial sources.
Volatile organics	Volatile organics such as methane, benzene, toluene, and xylene may be found in urban air due to industrial processes. Chlorofluorocarbon emitted by products such as old refrigerators is currently banned in the United States but may still be a problem in developing countries.
Ammonia	Ammonia may pollute indoor air from some household chemicals. It is also emitted during agricultural processes.
Particulates	Particulates or particulate matter (PM) and fine particles are produced naturally or through human activities such as burning fuel. These particles may be solid or liquid, with a diameter between 0.1 and 10 μm.
Heavy metals	Heavy metals such as lead, cadmium, copper, antimony, mercury, chromium, cobalt, copper, selenium, and strontium may be associated with PM.

dioxide, and particulate matter are major air pollutants. Common air pollutants are summarized in Table 3.1.

The most dangerous component of air pollution is small solid or liquid droplets that are suspended in the ground air. These droplets are called particulate matter (PM) and are usually defined as PM_{10}, $PM_{2.5}$, and $PM_{0.1}$, which correspond to diameters equal to or less than 10, 2.5 (fine), or 0.1 μm (ultrafine particle). The larger particles, such as PM_{10}, found in urban areas may come from several sources, including motor vehicles and industrial as well as construction sites, and may also be fragmentation products of larger particles such as pollen and spores. However, urban aerosols also contain finer particles ($PM_{2.5}$ and $PM_{0.1}$) that mostly consist of a core elemental carbon coming from burning of fossil fuel. Unfortunately, these finer particles can absorb inorganic (ammonium chloride, nitrates, and metal ions) as well as organic components (alkane, aliphatic acid, quinone, polycyclic aromatic hydrocarbons, etc.). The major source of PM in urban air pollution is exhaust from motor vehicles, especially diesel-operated motor vehicles, contributing approximately 40% of the pollution [2]. Therefore, PM in ambient air is probably the best indicator of the degree of air pollution. After breathing polluted air, relatively large particles such as PM_{10} are deposited in the oral cavity and the upper respiratory tract, whereas fine and ultrafine particles ($PM_{2.5}$ and $PM_{0.1}$) may penetrate farther into the respiratory tract and may be deposited in tracheobronchial and alveolar regions or even in the bloodstream [3]. Therefore, fine and ultrafine particles in polluted air cause more harm than larger particles.

3.3 OUTDOOR AIR POLLUTION IS LINKED TO MANY DISEASES

When mucociliary functions cannot neutralize PM particles, inflammation, asthma, and chronic obstructive pulmonary disease (COPD) may result. Long-term exposure to polluted air may even cause lung cancer [3,4]. An epidemiological study indicated a link between occupational exposure to exhaust from diesel engines by truck drivers and lung cancer [5]. In addition, based on long-term follow-up of 2037 urban bus drivers in the three largest cities in Denmark, Petersen *et al.* [6] found little evidence of a causal association between employment as an urban bus driver and cancer. COPD is by far the greatest risk factor for developing lung cancer [7]. Although COPD is related to tobacco smoking, a least one-fourth of patients with COPD are nonsmokers. Genetic makeup and asthma are risk factors for nonsmokers for developing COPD, but outdoor air pollution is also associated with COPD in nonsmokers [8]. Canova *et al.* [9] observed an increase in asthma- and COPD-related hospital admission due to an increase in PM_{10} concentration of polluted air. Interestingly, higher levels of vitamin C, an antioxidant in blood, modified the effect of these particulate matters, but another antioxidant vitamin, vitamin A, showed no effect. Chronic bronchitis in adults is also associated with outdoor air pollution [10]. Vieira *et al.* [11] reported that exposure to higher levels of nitric oxide and ozone was associated with increased risk of asthma and pneumonia in children aged 6–10 years.

Exposure to finer PM for a long period of time may increase the risk of cardiovascular disease and stroke [12]. Data from 50 million people living in the 20 largest cities and metropolitan areas in the United States (the National Morbidity, Mortality, and Air Pollution Study, 1987–1994) indicated that total mortality rates were associated with PM_{10} concentration in polluted air but other pollutants, such as carbon monoxide, sulfur dioxide, and nitrogen dioxide, were not significantly related to the mortality rate. The estimated increase in the relative rate of death from all causes was 0.51% for each increase in the PM_{10} concentration of 10 $\mu g/m^3$, whereas the estimated increase in the relative rate of death from cardiovascular respiratory cause was 0.68% for each increase in the PM_{10} concentration of 10 $\mu g/m^3$. There was also a weaker correlation between increased ozone levels near ground air during the summer months and increased relative rate of death, but no such association was observed during the winter months [13]. The Air Pollution and Health: A European Approach study, based on 43 million people living in 29 major European cities, demonstrated a stronger association between air pollution and mortality from cardiovascular disease, with a 0.76% increase in mortality for each 10-$\mu g/m^3$ increase in the PM_{10} level [14].

Wellenius *et al.* [15] studied the association between daily level of respirable PM_{10} particles and the rate of hospitalization for congestive heart failure in seven US cities and observed that an overall, 10-$\mu g/m^3$ increase in PM_{10} level was associated with a 0.72% increase in the rate of hospital admission due to congestive heart failure on the same day. Based on a study of elderly residents in 21 US cities, Zanobetti and Schwartz [16] reported that increased risk for hospitalization due to acute myocardial infarction was 0.65% for each 10-$\mu g/m^3$ increase in the PM_{10} level. The authors concluded that the increased concentrations of ambient PM_{10} levels were associated with increased risk of myocardial infarction among the elderly. In addition, there is a significant association between exposure to PM_{10} and adverse post-myocardial infarction outcome in patients who survive myocardial infarction [17].

A higher level of fine particles, such as of $PM_{2.5}$, is associated with higher risk of mortality from cardiovascular disease, myocardial infarction, stroke, and respiratory disease, as studied by Zanobetti and Schwartz [18] using data collected from 112 US cities. The authors observed a 0.98% increase in total mortality, a 0.85% increase in mortality due to cardiovascular disease, a 1.18% increase in myocardial infarction, a 1.78% increase in stroke, and a 1.68% increase in respiratory death for a 10-$\mu g/m^3$ increase in 2-days' averaged $PM_{2.5}$ level. The authors concluded that there was increased mortality for all and specific causes associated with raised $PM_{2.5}$ levels, and the risk was higher than that observed with PM_{10} levels. Air pollution, especially the PM_{10} level, is associated with hypercoagulability even after short-term exposure, although whether such change contributes to trigger cardiovascular events remains to be established [19].

3.4 INDOOR AIR QUALITY IS ALSO LINKED TO DISEASES

Although indoor air quality may not be a major concern in developed countries, poor indoor air quality in many developing countries due to burning of biomass fuel is a major public health concern. It is estimated that approximately 3 billion people worldwide burn solid fuels indoors. Exposure to indoor fuel smoke is strongly associated with COPD, acute respiratory tract infection, and lung cancer, whereas indoor air pollution is weakly associated with asthma, tuberculosis, and interstitial lung disease [20]. Diette *et al.* [21] commented that up to half of the world's population burns biomass fuel (wood, crop residues, animal dung, and coal) for indoor use such as cooking, lighting, and heating, which leads to indoor air pollution, and there is a strong association between biomass smoke exposure and COPD in adult women. Mortimer *et al.* [22] noted that indoor air pollution due to burning of biomass including coal and kerosene is linked to at least 2 million deaths per year worldwide, many of which occur in children younger than 5 years of age with pneumonia and women with COPD, cancer, and cardiovascular disease. Barone-Adesi *et al.* [23] observed that lung cancer mortality was substantially higher among users of smoky coal than users of smokeless coal. The absolute risk of lung cancer death before 70 years of age for men and women using smoky coal was 18 and 20%, respectively, compared to less than 0.5% among smokeless coal users of both sexes [23].

3.5 AIR POLLUTION AND DISEASES: ROLE OF OXIDATIVE STRESS

Air pollution is a health hazard, and many studies in the past 30 years have demonstrated a strong relationship between levels of exposure to PM in polluted air (both short term and long term) and health-related issues. Surface area of PM is related to the potential of PM to induce oxidative stress, inflammation, and other biological processes. The major composition of PM is metals, ions (sulfate and nitrite), organic compounds, stable radicals of carbonaceous materials, minerals, reactive gases, and other molecules of biological origin. Research has established that after entering the body, PM—especially fine and ultrafine PM, which can penetrate deep into the respiratory tract—can cause cytotoxicity by inducing oxidative stress, which may lead to oxidative damage of DNA, mutagenesis, and stimulation of proinflammatory factors [24]. Kim and Hong [25] commented that PM and ozone in ground air, the two most important components of polluted air, can produce reactive oxygen species after breathing of polluted air, which can overwhelm antioxidant defense of the body. As a result, these free radicals are capable of damaging cell walls, lipids, proteins, and DNA, leading to airway inflammation and hyperreactivity. Pollutants can also have harmful effects via epigenetic mechanisms that control the expression of genes without damaging DNA directly [25]. Risom *et al.* [26] noted that PM-induced adverse health effects are believed to involve inflammation and oxidative stress. Moreover, surface area is the most important criterion for inducing oxidative stress by PM with fine and ultrafine particles capable of generating more oxidative stress, whereas for larger particles (>100 nm in diameter), the chemical composition of these particles is more important. For example, diesel exhaust can cause significant DNA damage.

Kelly *et al.* [27] observed that PM_{10} particles collected from roadside locations in London showed higher oxidative potential than PM_{10} particles in urban areas. This may be related to higher amounts of copper, barium, and bioavailable iron found in PM_{10} particles collected from roadsides compared to particles collected from urban areas. As expected, studies have shown increased morbidity and mortality related to living near major highways because oxidative potential of PM near major urban roads is highly elevated compared with urban and suburban ground locations. The oxidative potential of PM_{10} particles in specimens collected from major streets was 3.6 times higher than that of PM_{10} particles collected from urban background locations, and oxidative potential was highly correlated with soot, barium, chromium, copper, iron, and manganese compositions of these PM particles. Interestingly, the oxidative potential of PM_{10} particles was higher than that of $PM_{2.5}$ particles [28]. Organic extract of $PM_{2.5}$ particles collected from urban areas has genotoxic effects, and reactive oxygen species play a vital role in such toxicity. These organic extracts are also capable of oxidative DNA damage in cultured human lung bronchial epithelial cells [29]. In addition to oxidative stress (due to oxygen free radicals), air pollution may also produce nitrosative stress (stress caused by reactive nitrogen species such as nitrous oxide).

Several studies have also documented increased concentrations of markers of oxidative stress following the breathing of polluted air. Vehicle emissions are a known source of air pollution. An occupational risk for bus drivers is breathing such polluted air, and bus drivers may experience higher oxidative stress. In one study, the authors assessed levels of various oxidative markers in 50 bus drivers from Prague, Czech Republic, and 50 matching controls and observed that 8-oxodeoxyguanosine (also known as 8-hydroxy-2′-deoxyguanosine, a maker of oxidative DNA damage) levels were elevated in bus drivers (mean, 7.79 nmol/mmol of creatinine) compared to controls (mean, 6.12 nmol/mmol of creatinine). Other markers of oxidative stress and nitrosative stress were also elevated in bus drivers, and such markers positively correlated with exposure to $PM_{2.5}$ and PM_{10} particles [30]. In another study, the authors observed that markers of oxidative damage to lipids (isoprostane) and proteins (protein carbonyl levels) in bus drivers were increased in winter months compared to summer months [31].

Rossner *et al.* [32] observed a positive relationship between air pollution and lipid peroxidation while protein oxidation was correlated with levels of $PM_{2.5}$ particles. In a study of 120 schoolchildren, Bae *et al.* [33] observed statistically significant increases in urinary malondialdehyde (a marker of lipid peroxidation) with ambient level of PM (PM_{10} and $PM_{2.5}$ particles). In addition, urinary levels of 1-hydroxypyrene, a marker for exposure to polycyclic aromatic hydrocarbons, also showed a positive association with urinary levels of malondialdehyde. Some metal bound to PM_{10} particles (aluminum, iron, strontium, manganese, silicon, arsenic, barium, zinc, copper, and cadmium) and $PM_{2.5}$ particles (magnesium, iron, strontium, arsenic, cadmium, zinc, aluminum, mercury, barium, and copper) also had significant association with urinary malondialdehyde levels. Urban air pollution is also associated with oxidative stress, inflammation, blood coagulation, and autonomic dysfunction in healthy young adults, as revealed by association between levels of PM_{10} and $PM_{2.5}$ particles, sulfate, nitrite, and ozone and increased concentration of 8-hydroxy-2′-deoxyguanosine, high-sensitivity C-reactive protein (marker for inflammation), and plasminogen activator fibrinogen inhibitor 1 (marker for blood coagulation) in 76 healthy young students [34]. Miller *et al.* [35] concluded that increased oxidative stress due to breathing of polluted air is responsible for adverse cardiovascular effect of air pollution. Moulton and Yang [36] speculated that exposure to polluted air leads to chronic oxidative stress in the body, which is involved in the pathogenesis of Alzheimer's disease.

3.6 AIR QUALITY: SEASONAL VARIATION AND URBAN VERSUS RURAL AREA

Air quality may vary from winter to summer months. In one study, higher levels of PM_{10} particles in winter months compared to summer months were observed in air samples collected near the Taj Mahal, Agra, India [37]. Kan *et al.* [38] studied the association of ambient air pollution in different seasons with mortality in Shanghai, China, and observed significant association between levels of PM_{10} particles, sulfur dioxide, nitric oxide, and ozone and all-cause daily mortality from natural causes and cardiopulmonary disease. In addition, women, older people, and people with a low level of education were more affected by air pollution than men, younger people, and people with a high level of education. Interestingly, the association between air pollution and daily mortality appeared to be more pronounced in cooler seasons than in warmer seasons. Qui *et al.* [39] observed an increase in the detrimental effects of air pollution in cooler seasons and on low-humidity days compared to warm and humid seasons. Air-conditioning could decrease the penetration of air pollutants from outdoors to indoors. In addition, certain air-conditioning systems may remove a fraction of PM entering from outdoor to indoor air. Therefore, in areas and in developing countries where air-conditioning is not common, PM levels in indoor air may be higher compared to those of indoor air where use of air-conditioning is common. Bell *et al.* [40] observed that communities with higher air-conditioning prevalence had lower PM effects, and each additional 20% of households with central air-conditioning was associated with a 43% decrease in $PM_{2.5}$ effects on hospitalization due to cardiovascular diseases.

In general, rural areas have better air quality than urban areas. Callen *et al.* [41] observed that concentrations of polycyclic aromatic hydrocarbons bound to PM_{10} particles were 10 times greater in urban areas than in rural areas. The concentrations of polycyclic aromatic hydrocarbons associated with PM_{10} particles were also higher during cold periods. Hulin *et al.* [42] reported that urban dwellings were associated with exposure to more polluted air than were rural dwellings. Trees can significantly reduce air pollution. Norwak *et al.* [43] commented that urban trees and shrubs offer the ability to remove significant amounts of air pollutants, thus improving the quality of air and human health in urban areas where trees are abundant. Trees are also capable of removing PM from polluted air. Although the percentage of air quality improvement by trees is modest, the combined total effects of trees on removal of air pollutants (PM, ozone, nitric oxide, sulfur dioxide, and carbon monoxide) are significant enough that urban tree management may provide a viable means to improve air quality in the United States.

3.7 AIR QUALITY INDEX: CLEANEST AND DIRTIEST US CITIES

Air quality of cities is reported daily in local newspapers, and this information is also available from local weather stations. Air quality is expressed by the air quality index (AQI). In the United States, the US Environmental Protection Agency (EPA) calculates the AQI by measuring five major air pollutants (ground-level ozone, PM, carbon monoxide, sulfur dioxide, and nitrous oxide) regulated by the Clean Air Act and then applying a formula. The AQI scale ranges from 0 to 500, where values below 50 represent good air quality and a value of 100 usually corresponds to the desirable national air quality, which is the level the EPA has set to protect public health. Higher AQI values indicate pollution (Table 3.2). When air quality is poor, it is better to avoid outside air as much as possible by staying indoors. The American Lung Association publishes an extensive report of the dirtiest and cleanest US cities based on air pollution data (American Lung Association, State of the Air, 2013). The dirtiest cities are ranked according to worst ozone layer, year-round particulate pollution, and short-term particulate pollution. Ozone is found in the higher atmosphere (stratosphere), but the presence of ozone at ground level (troposphere) is a health hazard because it can induce free radical production. However, PM in polluted air is a major concern

Table 3.2 Air Quality Index

Air Quality Index	Color Code	Comment
0–50	Green	Good air quality; air pollution poses no risk.
51–100	Yellow	Air quality is acceptable, but a small number of people may be sensitive to certain air pollutants.
101–150	Orange	Air quality may be unhealthy for some people, but healthy people may also experience some effects.
151–200	Red	Unhealthy air quality where everyone may be affected. Sensitive people may feel adverse effects more seriously.
201–300	Purple	Very unhealthy air quality affecting entire population.
301–500	Maroon	Health hazard for everyone.

because it comprises the primary pollutants. In Table 3.3 are listed the most polluted cities in the United States, and in Table 3.4, the cleanest cities.

3.8 EXPOSURE TO SUNLIGHT AND THE ULTRAVIOLET INDEX

Ultraviolet radiation of sunlight is composed of ultraviolet A (UVA; wavelength, 320–400 nm), ultraviolet B (UVB; 280–320 nm), and ultraviolet C (UVC; 100–280 nm). Only UVA and UVB reach the earth's atmosphere because the earth's higher atmosphere absorbs most of the harmful UVC, which has the shortest wavelength. UVB is more damaging than UVA, and exposure to sunlight may induce oxidative stress in the body and is linked with skin cancer. The amount of UV radiation in sunlight is expressed by the UV index, an international standard to measure the strength of the ultraviolet radiation of the sun at a particular place or on a particular day. The US National Weather Service calculates the UV index of various US cities by using a computer model that relates the ground-level strength of UV radiation to forecasted stratospheric ozone concentration, forecasted cloud amount, and elevation of the ground. The scale of the UV index ranges from 0 to 11+; the higher the index, the greater the danger of exposure to sunlight (Table 3.5). Usually, the UV index is higher in countries located near the Equator and decreases with increasing latitude. Therefore, the number of days per year with high and extremely high UV index in Boston, Massachusetts, is significantly lower than that in Houston, Texas. Cities in the southern United States typically have a greater number of days per year when UV indices are very high or extremely high. Anchorage, Alaska,

Table 3.3 Most Polluted US Cities

Polluted Based on Year-Round Particle Pollution (Rank)	Polluted Based on Ozone Layer (Rank)
Bakersfield–Delano, CA (1)	Los Angeles–Long Beach–Riverside, CA (1)
Merced, CA (1)	Visalia–Porterville, CA (2)
Fresno–Madera, CA (3)	Bakersfield–Delano, CA (3)
Hanford–Corcoran, CA (4)	Fresno–Madera, CA (4)
Los Angeles–Long Beach–Riverside, CA (4)	Hanford–Corcoran, CA (5)
Modesto, CA (6)	Sacramento–Arden–Arcade–Yuba City, CA, NV (6)
Visalia–Porterville, CA (7)	Houston–Baytown–Huntsville, TX (7)
Pittsburgh–New Castel, PA (8)	Dallas–Fort Worth, TX (8)
El Centro, CA (9)	Washington, DC–Baltimore–northern Virginia, DC, MD, VA, WV (9)
Cincinnati–Middletown–Wilmington, OH, KY, IN (10)	El Centro, CA (10)
Philadelphia–Camden–Vineland, PA, NJ, DE, MD (11)	San Diego–Carlsbad–San Marcos, CA (11)
Louisville–Jefferson County–Elizabeth town–Scottsburg, KY, IN (12)	Merced, CA (11)
St. Louis–St. Charles–Farmington, MO, IL (12)	Modesto, CA (13)
Canton–Massillon, OH (14)	Cincinnati–Middletown–Wilmington, OH, KY, IN (14)
Macon–Warner Robins–Fort Valley, GA (14)	Birmingham–Hoover–Cullman, AL (14)
Allentown–Bethlehem–Easton, PA, NJ (14)	Las Vegas–Paradise–Pahrump, NV (16)
	Louisville–Jefferson County–Elizabeth town–Scottsburg, KY, IN (17)

Source: *American Lung Association. State of the Air 2013. Available at http://www.stateoftheair.org/2013/city-rankings/most-polluted-cities.html. Accessed November 2013.*

has the lowest number of days per year with a UV index scored "high." US cities with the highest UV index for most days in a year are listed in Table 3.6, and cities with the fewest days with very high UV index are listed in Table 3.7. UV index also increases with altitude because the atmosphere becomes thinner at higher altitude. Moreover, snow and sand can reflect sunlight. Therefore, protection from sunlight is essential and should include wearing light-colored clothes (preferably white), covering the body, and wearing sunglasses that can protect the eyes from the harmful UV radiation of sunlight. Moreover, for sunbathing, use of sunscreen is essential. Despite public education, Buller *et al.* [44] reported that approximately only 3 of 10 adults in the United States routinely practice sun protection behavior; women and elderly persons are more likely to use such protection. Sunburn is still an epidemic in the United States, with 34.4% of adults experiencing sunburn annually.

Table 3.4 Cleanest US Cities

Cleanest Based on Year-Round Particle Pollution	Rank
Cheyenne, WY	1
St. George, UT	2
Santa Fe–Espanola, NM	2
Prescott, AZ	4
Farmington, NM	5
Pocatello, ID	6
Redding, CA	7
Tucson, AZ	8
Flagstaff, AZ	9
Rapid City, SD	9
Colorado Springs, CO	9
Albuquerque, NM	9
Salinas, CA	13
Anchorage, AK	14
Fort Collins–Loveland, CO	15
Palm Bay–Melbourne–Titusville, FL	16
Duluth, MN	16
Reno–Sparks–Fernley, NV	18
Sarasota–Bradenton–Punta Gorda, FL	19
Bismarck, ND	20
Cape Coral–Fort Myers, FL	20

Source: *American Lung Association. State of the Air 2013. Available at http://www.stateoftheair.org/2013/city-rankings/cleanest-cities.html. Accessed November 2013.*

3.9 EXPOSURE TO SUNLIGHT, OXIDATIVE STRESS, AND SKIN CANCER

The main types of human skin cancer are nonmelanoma (basal and squamous cell cancer) and melanoma skin cancers. The origin of nonmelanoma skin cancer is keratinocytes undergoing malignant transformation, whereas melanoma is due to transformation of melanocytes in the skin. Although skin cancer is the most common type of cancer, more dangerous melanoma accounts for less than 10% of skin cancers. The available epidemiological studies clearly establish a relationship between exposure to solar UV radiation and skin cancer. The causative relationship between nonmelanoma skin cancer and exposure to UVB (290–320 nm) has been well documented. However, the link between melanoma and exposure to the sunlight is less clear and is

Table 3.5 UV Index and Protection Needed to Avoid Sunburn and Other Complications of Exposure to Sunlight

UV Index	Exposure Category	Sun Protection Needed
0–2	Low	Low level of danger from sun exposure, but wearing sunglasses if it snows is advisable.
3–5	Moderate	Using sunglasses and wearing a hat to reduce exposure is recommended. Use of sunscreen is also advisable.
6–7	High	Protection against sun damage is needed, including sunglasses, protective clothing, and sunscreen for exposed areas.
8–10	High	All necessary protection from sunlight is required. Avoid direct exposure to the sunlight from 10 a.m. to 4 p.m. if possible. For beachgoers, white sand reflects UV radiation and may double exposure to UV radiation. Use of sunscreen is highly advisable.
11 +	Extreme	Try to avoid sunlight as much as possible. Taking necessary protection and trying to minimize exposure between 10 a.m. and 4 p.m. is strongly recommended. Avoiding sun bathing is also strongly advised.

Source: *US Environmental Protection Agency.*

Table 3.6 US Cities with the Highest Number of Days with the UV Index in the Extreme or Very High Range

Rank	City	Days per Year with Extreme/High UV Index
1	Sun Juan, PR	312
2	Honolulu, HI	278
3	Miami, FL	226
4	Tampa Bay, FL	200
5	New Orleans, LA	185
6	Houston, TX	185
7	Jacksonville, FL	182
8	Albuquerque, NM	173
9	Los Angeles, CA	169
10	Atlanta, GA	167
11	Dallas, TX	166
12	Phoenix, AZ	165

Source: *National Weather Service 2010 annual data on UV index.*

Table 3.7 Top Five US Cities with the Fewest of Days with the UV Index in the Extreme or Very High Range

Rank	City	Days per Year with Extreme/High UV Index
1	Anchorage, AK	0
2	Seattle, WA	24
3	Portland, OR	48
4	Burlington, VT	55
5	Portland, ME	58

Source: *National Weather Service 2010 annual data on UV index.*

controversial. Free radicals, especially reactive oxygen species, can be generated by UV radiation that can damage DNA, producing various types of DNA photoproducts. It has been speculated that the formation of *cis-syn*-cyclobutane pyrimidine dimers, mostly by UVB, is the most prominent DNA lesion that can initiate nonmelanoma and possibly also melanoma skin cancer [45]. Air pollution can also intensify damage caused by UV radiation on the skin. When ozone exposure in ground air precedes exposure to sunlight, there is an enhancement of UV-induced depletion of antioxidant vitamin E from the skin's stratum corneum. In addition, several environmental pollutants are photosensitizers that can produce reactive oxygen species when the skin is exposed to UV radiation of sunlight. Therefore, air pollution may act in synergy with UV radiation of sunlight in causing skin damage through the production of oxidative stress [46]. Certain medication can also cause photosensitivity, and diseases such as porphyria can make the skin very sensitive to sunlight.

Both nonmelanoma and melanoma skin cancers are the most common type of cancer in Caucasians, although skin cancer may also occur in nonwhite populations. Intense exposure to sunlight in childhood and adolescence may increase the risk of squamous cell carcinoma, although exposure to sunlight also increases the risk of other types of skin cancer [47]. Han *et al.* [48] concluded that higher sun exposure while wearing a bathing suit is an independent risk factor for basal cell carcinoma, squamous cell carcinoma, and melanoma. In addition, sunlamp usage or tanning salon attendance are risk factors for melanoma. Wehner *et al.* [49] concluded that indoor tanning is associated with both basal and squamous cell carcinoma. The risk is higher with use in early life (<25 years of age). Actinic keratoses, also known as solar keratoses, appears as rough scaly patches with discoloration of the skin (pink or flesh color to dark brown) when the skin is exposed to the sun. This phenomenon is more common in fair-skinned people. If a skin rash or skin problem due to sunburn does not respond to home therapy and persists more than 1 week or if the rash is painful, it is advisable to seek medical attention immediately.

3.10 EXPOSURE TO SUNLIGHT, OXIDATIVE STRESS, AND OPHTHALMOLOGICAL DISORDERS

After exposure to UV radiation, the cornea is susceptible to damage in the epithelial, stromal, and endothelial cellular layers. In addition, UV exposure can lead to DNA damage, apoptosis, and altered protease expression within the cornea. Moreover, increased oxidative stress due to production of reactive oxygen species can damage the lens. Cataract formation, a significant cause of visual disability, is related to oxidative stress caused by intraocular penetration of ultralight and consequent photochemical generation of reactive oxygen species such as superoxide and singlet oxygen and their derivatization to other oxidants such as hydrogen peroxide and the hydroxyl radical (see Chapter 10). Nutritional antioxidants such as ascorbic acid, vitamin E, pyruvate, and caffeine may counteract such oxidative stress [50]. Age-related macular degeneration is a leading cause of blindness in the Western World. Although cornea and lens can block a major portion of UV radiation from reaching the retina, short-wavelength radiation and blue light induce significant oxidative stress to the retinal pigment epithelium. Epidemiological evidence indicates a trend toward a link between the severity of light exposure and age-related macular degeneration [51]. Therefore, preventing harmful UV radiation from reaching eyes is a good way to prevent eye diseases, especially cataract formation and age-related macular degeneration. One way to easily achieve this is to wear sunglasses that block wavelengths below 400 nm (marked 400 on the glasses). Similar protection can be achieved by wearing contact lenses that absorb significant amounts of UV radiation. For people older than 50 years, it is advisable to wear specially designed sunglasses or contact lenses to protect eyes from harmful UV radiation [52]. Natural tears have antioxidant capacity as well as UV absorption capacity. Dry eye syndrome is a common age-related syndrome in which decreased antioxidant and ultraviolet protection of human tears may be the cause of corneal stress. Unfortunately, artificial tears do not have antioxidant- or UV-blocking capacity [53].

3.11 CONCLUSION

Free radicals are produced not only by the body's normal metabolism causing oxidative stress but also by other sources such as pollution and exposure to sunlight, especially UV radiation present in sunlight. Although it is not possible to completely avoid polluted air, staying indoors as much as possible during days when the air quality index is high may be a practical way to minimize the effects of air pollution. It may also be advisable to avoid sunbathing or going to the beach on days when the UV index is extremely high. A healthy lifestyle and eating balanced meals rich in fruits and vegetables is the best way to counteract oxidative stress induced by various external factors.

REFERENCES

[1] Franchini M, Guida A, Tufano A, Coppola A. Air pollution, vascular disease and thrombosis: Linking clinical data and pathogenic mechanism. J Thromb Haemost 2012;10:2438–51.

[2] Sydbom A, Blomberg A, Parina S, Stenfors N, et al. Health effect of diesel exhaust emission. Eur Respir J 2001;17:733–46.

[3] Andreau K, Leroux M, Bouharrour A. Health and cellular impacts of air pollutants: from cytoprotection to cytotoxicity. Biochem Res Int 2012 [Article ID 493894].

[4] Yang W, Omaye ST. Air pollutants, oxidative stress and human health. Mutat Res 2009;674:45–54.

[5] Boffetta B, Stellman SD, Garfinkel L. Diesel exhaust exposure and mortality among males in American Cancer Society prospective study. Am J Ind Med 1988;14:403–15.

[6] Petersen A, Hansen J, Olsen JH, Netterstrom B. Cancer morbidity among Danish male urban bus drivers: a historical cohort study. Am J Ind Med 2010;53:757–61.

[7] Wang ZL. Association between chronic obstructive pulmonary disease and lung cancer: the missing link. Chin Med J (Engl) 2013;126:154–65.

[8] Zeng G, Sun B, Zhong N. Non-smoking related chronic obstructive pulmonary disease: a neglected entity? Respirology 2012;17:908–12.

[9] Canova C, Dunster C, Kelly FJ, Minelli C, et al. OM10 induced hospital admission for asthma and chronic obstructive pulmonary disease: the modifying effect of individual characteristics. Epidemiology 2012;23:607–15.

[10] Salameh O, Salame J, Khayat G, Akhdar A, et al. Exposure to outdoor pollution and chronic bronchitis in adults: a case controlled study. Int J Occup Environ Med 2012;3:165–77.

[11] Vieira SE, Stein RT, Ferraro AA, Pastro LD, et al. Urban air pollutants are significant risk factors for asthma and pneumonia in children: the influence of location of the measurement of pollutant. Arch Bronconeumol 2012;48:389–95.

[12] Wellenius GA, Burger MR, Schwartz J, Suh HH, et al. Ambient air pollution and the risk of acute ischemic stroke. Arch Intern Med 2012;172:229–34.

[13] Samet JM, Dominici F, Curriero FC, Coursac I, et al. Fine particulate in air pollution and mortality in 20 cities, 1987–1994. N Engl J Med 2000;343:1742–9.

[14] Analitis A, Katsouyanni K, Dimikopoulo K, Samoli E, et al. Short term effects of ambient particles on cardiovascular and respiratory mortality. Epidemiology 2006;17:230–3.

[15] Wellenius GA, Schwartz J, Mittleman MA. Particulate air pollution hospital admission for congestive heart failure in seven United States cities. Am J Cardiol 2006;97:404–8.

[16] Zanobetti A, Schwartz J. The effect of particulate air pollution on emergency admission for myocardial infarction: a multicity case-crossover analysis. Environ Health Perspect 2005;113:978–82.

[17] Zanobetti A, Schwartz J. Particulate air pollution, progression and survival after myocardial infarction. Environ Health Perspect 2007;115:769–75.

[18] Zanobetti A, Schwartz J. The effect of fine and coarse particulate air pollution on mortality: a national analysis. Environ Health Perspect 2009;117:898–903.

[19] Baccarelli A, Zanobetti A, Martinelli I, Grillo P, et al. Effects of exposure to air pollution on blood coagulation. J Thromb Haemost 2007;5:252–60.

[20] Sood A. Indoor fuel exposure and ling in both developing and developed countries. Clin Chest Med 2012;33:649–65.

[21] Diette GB, Accinelli RA, Balmes JR, Buist AS, et al. Obstructive lung disease and exposure to burning biomass fuel in the indoor environment. Globe Heart 2012;7:265–70.

[22] Mortimer K, Gordon SN, Jindal SK, Accinelli RA, et al. Household air pollution is a major avoidable risk factor for cardiorespiratory disease. Chest 2012;142:1308–15.

[23] Barone-Adesi F, Chapman RS, Silverman DT, He X, et al. Risk of lung cancer associated with domestic use of coal in Xuanwei China: retrospective cohort study. BMJ 2012; 345:e5414.

[24] Valavanidis A, Fiotakis K, Vlachogianni T. Airborne particulate matter and human health: toxicological assessment and importance of size and composition of particles for oxidative damage and carcinogenic mechanism. J Environ Health C Environ Carcinog Ecotoxicol Rev 2008;26:339–62.

[25] Kim BJ, Hong SJ. Ambient air pollution and allergic diseases in children. Korean J Pediatr 2012;55:185–92.

[26] Risom L, Moller P, Loft S. Oxidative stress induced DNA damage by particulate air pollution. Mutat Res 2005;592:119–37.

[27] Kelly F, Anderson HR, Armstrong B, Atkinson R, et al. The impact of the congestion charging scheme on air quality in London: part 2. Analysis of the oxidative potential of particulate matter. Res Rep Health Eff Inst 2011;155:73–144.

[28] Boogaard H, Janssen N, Fischer PH, Kos G, et al. Contrasts in oxidative potential and other particulate matter characteristics collected near major streets and background locations. Environ Health Perspect 2012;120:185–91.

[29] Oh SM, Kim HR, Park YJ, Lee SY, et al. Organic extracts of urban air pollution particulate matter (PM2.5) induced genotoxicity and oxidative stress in human lung bronchial epithelial cells (BEAS-2B cells). Mutat Res 2011;723:142–51.

[30] Rossner P, Svecova V, Milicova A, Lnenickova Z, et al. Oxidative and nitrosative stress markers in bus drivers. Mutat Res 2007;617:23–32.

[31] Rossner P, Svecova V, Milicova A, Lnenickova Z, et al. Seasonal variability of oxidative stress markers in city bus drivers: part II. Oxidative damage to lipids and proteins. Mutat Res 2008;642:21–7.

[32] Rossner O, Rossnerova A, Spatove M, Beskid O, et al. Analysis of biomarkers in a Czech population exposed to heavy air pollution: part II. Chromosomal aberrations and oxidative stress. Mutagenesis 2013;28:97–106.

[33] Bae S, Pan XC, Kim SY, Park K, et al. Exposure to particulate matter and polycyclic aromatic hydrocarbons and oxidative stress in schoolchildren. Environ Health Perspect 2010;118:579–83.

[34] Chuang KJ, Chan CC, Su TC, Lee CT, et al. The effect of urban air pollution on inflammation, oxidative stress, coagulation and automatic dysfunction in young adults. Am J Respir Crit Care Med 2007;176:370–6.

[35] Miller MR, Shaw CA, Langrish JP. From particles to patients: oxidative stress and the cardiovascular effects of air pollution. Future Cardiol 2012;8:577–602.

[36] Moulton PV, Yang W. Air pollution, oxidative stress and Alzheimer's disease. J Environ Public Health 2012;2012:472751.

[37] Singh R, Sharma BS. Composition, seasonal variation, and source of PM10 from world heritage site Taj Mahal. Environ Monit Assess 2012;184:5945–56.

[38] Kan H, Chen B, Zhao N, London SJ, et al. Part 1: a time series study of ambient air pollution and daily mortality in Shanghai, China. Res Rep Health Eff Inst 2010;154:17–78.

[39] Qui H, Yu IT, Wang X, Tian L, et al. Cool and dry weather enhances the effects of air pollution on emergency IHD hospital admission. Int J Cardiol 2013;168:500–5.

[40] Bell ML, Ebisu K, Peng RD, Dominici F. Adverse health effects of particulate air pollution: modification by air conditioning. Epidemiology 2009;20:682–6.

[41] Callen MS, Lopez JM, Mastral AM. Characteristics of PM-10 bound polycyclic aromatic hydrocarbons in the ambient air of Spanish urban and rural areas. J Environ Monit 2011;13:319–27.

[42] Hulin M, Caillaud D, Annesi-Maesano I. Indoor air pollution and childhood asthma: variations between urban and rural areas. Indoor Air 2010;20:502–14.

[43] Norwak DJ, Crane DE, Stevens JC. Air pollution removal by urban trees and shrubs in the United States. Urban Forestry Urban Greenings 2006;4:115–23.

[44] Buller DB, Cokkinides V, Hall HI, Hartman AM, et al. Prevalence of sunburn, sun protection and indoor tanning behaviors among Americans: review from national surveys and case studies of 3 states. J Am Acad Dermatol 2011;65(5 Suppl):S114–23.

[45] Pfeifer G, Besarantinia A. UV wavelength dependent DNA damage and human nonmelanoma and melanoma skin cancer. Photochem Photobiol Sci 2012;11:90–7.

[46] Burke KE, Wei H. Synergistic damage by UVA radiation and pollutants. Toxicol Ind Health 2009;25:219–24.

[47] Iannacone MR, Wang W, Stockwell HG, O'Rourke K, et al. Patterns and timing of sunlight exposure and risk of basal cell and squamous cell carcinomas of the skin: a case controlled study. BMC Cancer 2012;12(1):417.

[48] Han J, Colditz GA, Hunter DJ. Risk factors for skin cancers: a nested case–controlled study within the Nurses' Health Study. Int J Epidemiol 2006;35:1514–21.

[49] Wehner MR, Shive ML, Chren MM, Han J, et al. Indoor tanning and non-melanoma skin cancer: systematic review and meta-analysis. BMJ 2012;345:e5909.

[50] Varma SD, Kovtun S, Hegde KR. Role of ultraviolet irradiation and oxidative stress in cataract formation: medical prevention by nutritional antioxidants and metabolic agonists. Eye Contact Lens 2011;37:233–45.

[51] Chalam KV, Khetpal V, Rusovici R, Balaiya S. A review of ultraviolet radiation in age-related macular degeneration. Eye Contact Lens 2011;37:225–32.

[52] Roberts KE. Ultraviolet radiation as a risk factor for cataract and macular degeneration. Eye Contact Lens 2011;27:246–9.

[53] Choy CK, Cho P, Benzie IF. Antioxidant content and ultraviolet absorption characteristics of human tears. Optom Vis Sci 2011;88:507–11.

CHAPTER 4

Oxidative Stress Caused by Cigarette Smoking, Alcohol Abuse, and Drug Abuse

CONTENTS

4.1 INTRODUCTION

Many habits can significantly increase oxidative stress in the human body, including smoking, alcohol and drug abuse. It has been well publicized that cigarette smoking is a health hazard. Although consuming alcohol in moderation is beneficial for health, excess alcohol consumption can significantly increase the oxidative stress experienced and all beneficial effects of drinking alcohol in moderation quickly disappear. In reality, alcohol abuse increases the risk of cardiovascular disease, stroke, and a variety of other diseases, whereas drinking in moderation can reduce the risk of developing these diseases. Excess alcohol consumption can cause fatty liver, and chronic alcohol abuse can result in liver cirrhosis, a potentially fatal disease. Drug abuse is a serious problem not only in the United States but also worldwide. Moreover, abusing alcohol and drugs at the same time increases the risk of adverse outcomes because an abused drug can be toxic at a lower blood level in the presence of alcohol.

4.2 CIGARETTE SMOKING IN THE UNITED STATES AND RELATED HEALTH HAZARDS

In the United States, an estimated 45.8 million adults are current smokers and more than 400,000 individuals die each year from smoking-related illness. An estimated 190,000 deaths due to cardiovascular disease are related to cigarette smoking, and an alarming 37,000–40,000 deaths are related to second-hand smoke inhalation (sidestream smoke) [1]. Cigarette smoking is responsible for more deaths than those from alcohol and illegal drug abuse, motor vehicle accidents, murders, suicides, and HIV infection combined [2]. Cigarette smoking is associated with many diseases. Also, as expected, smoking is negatively associated with health-related quality of life, and this

A. Dasgupta and K. Klein: Antioxidants in Food, Vitamins and Supplements. DOI: http://dx.doi.org/10.1016/B978-0-12-405872-9.00004-5

association is determined by the number of cigarette smoked each day [3]. A British study indicated that men who smoked cigarettes died on average 10 years younger than nonsmokers. However, quitting by the age of 30 years could avoid that risk [4]. A Japanese study indicated that smoking reduced life expectancy in male smokers by an average of 8 years, whereas in female smokers, the average life expectancy was reduced by 10 years. Those who stopped smoking by age 35 years avoided most of the risks associated with smoking [5]. Hayashida *et al.* [6] observed that smokers had a higher mortality rate, shorter life expectancy, and generally higher medical expenditure than nonsmokers.

The major health hazards of smoking are various pulmonary diseases including lung cancer and noncancerous lung diseases, and atherosclerotic diseases of the heart and blood vessels, as well as various cancers and toxicity to the human reproductive system (Table 4.1). It has been well documented that the large number of free radicals and oxidants present in inhaled cigarette smoke induces adverse effects on tissues and organs through oxidative damage of key biological molecules. In addition, cigarette smoke can induce activation of inflammatory cells, thus further contributing to additional oxidative stress. Products of lipid peroxidation, protein and thiol oxidation, and oxidized DNA are increased in smokers compare to nonsmokers, further indicating increased oxidative stress in smokers. Oxidative stress is responsible for the association between various diseases and smoking. In addition, smoking may also induce other conditions, such as dyslipidemia, glucose intolerance, and nutritional abnormalities [7].

Table 4.1 Various Diseases Associated with Smoking

Disease Type	Specific Diseases/Location/Effect
Pulmonary disease	Bronchitis, pneumonia, emphysema, chronic obstructive pulmonary disease
Cardiovascular disease	Ischemic heart disease, atherosclerosis, aortic aneurysm
Cerebrovascular disease	Various strokes
Cancer	Lips, oral cavity, pharynx, esophagus, stomach, lung, kidney, cervix, uterus, prostate, colon, acute myeloid leukemia
Reproductive system	Impaired female fertility, impotence in males, poor pregnancy outcome, earlier menopause
Eye disease	Cataract, macular degeneration
Skeletal muscle dysfunction	Muscle loss and dysfunction, reduced adipose tissue
Other conditions	Dyslipidemia, glucose intolerance, anorexia, weight loss

4.2.1 Composition of Cigarette Smoke

More than 4800 compounds have been identified in cigarette smoke. Approximately 400–500 compounds are present in the gaseous phase, and approximately 300 of these compounds are classified as semivolatile compounds [8]. The composition of cigarette smoke is complex, comprising gaseous components that contain carbon monoxide, nitrogen oxide, and hydrogen cyanide as well as many volatile organic compounds and particulate matter that consists of submicrometer-sized particles. Various compounds may be associated with particulate matter, such as nicotine, tobacco-specific nitrosamines, and polyaromatic hydrocarbons. Particulate matter may also be loosely defined as cigarette tar material. Although the major danger of cigarette smoking is associated with smokers who inhale mostly mainstream smoke, nonsmokers who are in close proximity to smokers are also exposed to sidestream smoke that is emitted in the surrounding air between puffs from the ends of smoldering cigarettes. Sidestream smoke is diluted compared to mainstream smoke and has a composition different from it, but inhaling sidestream smoke is also unhealthy.

The chemical composition of both mainstream smoke and sidestream smoke generated from a burning cigarette depends on many factors, including the type of tobacco used, the type of filter used, and pyrolysis that takes place during burning of the cigarette. A burning cigarette is a good example of an incomplete combustion system. Tobacco is a plant material containing more than 3800 constituents ranging from small organic molecules to small inorganic molecules and biopolymers consisting of cellulose, hemicellulose, pectin, lignin, starch, proteins, peptides, and nucleic acids. During burning of a cigarette, all these compounds present in tobacco are subjected to very high temperatures up to 950°C in the presence of varying concentrations of oxygen in the burning zone of the cigarette [9]. Approximately 2800 compounds are found in cigarette smoke that are not present in tobacco, indicating the importance of pyrolysis and combustion in the formation of various compounds present in cigarette smoke. The burning zone of a cigarette can be divided into two subzones—the exothermic combustion zone and the endothermic pyrolysis/distillation zone. The interior of the burning zone is oxygen deficient but hydrogen rich; however, as air is drawn inside a cigarette during a puff, extra oxygen is consumed by combustion of carbonized tobacco, resulting in simple combustion products such as carbon monoxide, carbon dioxide, and water, which are present in cigarette smoke. During this process, the temperature in this zone can be very high (700–950°C), and the pyrolysis/distillation process is initiated in this region. However, pyrolysis products are also present immediately downstream of the combustion region, which has a lower temperature (200–600°C). The majority of compounds present in cigarette smoke are generated in this region, and a dense aerosol containing many compounds is formed. In general, 10^9 to 10^{10} particles are present per milliliter of such mainstream cigarette smoke emerging from the mouth end of a cigarette during burning. These particles are in fact particulate matter with a diameter between 0.1 and 10 μm [10].

The major components of the gaseous phase of cigarette smoke are carbon monoxide and carbon dioxide, and nitric and nitrous oxide and aldehydes are also present. The major component of the particulate phase in cigarette smoke is nicotine. Many toxic organic and inorganic components are present in tobacco smoke (Box 4.1). These compounds are sometimes collectively called Hoffmann analytes [11,12]. In general, small hydrocarbon molecules are produced (one to three carbon chains) in the cigarette burning region in which the temperature range is 300–550°C. Longer chain alkanes and alkenes are formed in the burning region with temperatures between 400 and 700°C. Benzenes and alkylbenzenes are formed at temperatures greater than 500°C, and naphthalenes are formed at temperatures greater than 700°C. More than 75 monocyclic aromatic hydrocarbons, such as benzene and toluene, are formed during pyrolysis of amino acids, fatty acids, sugars, and paraffins as well as compounds with an aromatic or cyclohexane ring. Polynuclear aromatic hydrocarbons are formed at 800°C or more from pyrolysis of long-chain hydrocarbons, terpenes, and phytosterol such as stigmasterol, paraffins, sugar, cellulose, and amino acids. In addition, tobacco-specific and some tobacco-nonspecific relatively nonvolatile nitrosamines are present in tobacco smoke [13].

Box 4.1 TOXIC CHEMICALS PRESENT IN CIGARETTE SMOKE

- Major components
  - Tar, nicotine, carbon monoxide
- Inorganic gas
  - Ammonia, hydrogen cyanide, nitrous and nitric oxide
- Metals
  - Lead, chromium, mercury, arsenic, cadmium, selenium
- Tobacco-specific nitrosamines
  - *N*′-nitrosonornicotine, *N*-nitrosoanabasine, *N*-nitrosoanatabine, 4-(methylnitroso)-1-(-3-pyridyl)-1-butanone
- Phenols
  - Phenol, *o*-cresol, *m*-cresol, *p*-cresol, catechol, resorcinol, hydroquinone
- Carbonyls
  - Formaldehyde, acetaldehyde, acetone, propionaldehyde, methyl ethyl ketone, acrolein, butyraldehyde, crotonaldehyde
- Aromatic amines
  - 1-Aminonaphthalene, 2-aminonaphthalene, 3-aminobiphenyl, 4-aminobiphenyl
- Polynuclear aromatic hydrocarbon
  - Benzopyrene
- Other organic compounds
  - Benzene, toluene, 1,3-butadiene, isoprene, acrylonitrile, pyridine, styrene, quinoline

Filter cigarettes were first introduced in the early 1950s in response to demand for cigarettes with lower smoke yield and less hazardous to health. Today, the majority of cigarettes sold in the United States and developed countries are filter cigarettes. Cigarette filters are usually made from monofilament cellulose acetate and are capable of reducing the amount of tar and nicotine in smoke by up to 50% compared to nonfilter cigarettes. Dual-filter cigarettes have been introduced that contain dispersed charcoal within their second segment, and these charcoal-containing filters are capable of further reducing the level of toxic components in cigarette smoke. Although aromatic amines, nitrosamines, and carbonyl compounds are present in cigarette smoke, these toxic compounds can be significantly reduced by using filters. The mutagenicity of cigarette smoke can also be reduced by approximately 35% in cigarettes with a carbon filter compared to cigarettes with no filter [14]. However, when a person smokes a cigarette with a filter, both the gas phase smoke (mainstream smoke) and the particulate matter that penetrates through the filter enter the human body through inhalation, containing substantial amounts of hazardous chemicals. Free radicals are present in both the gaseous phase and tar.

4.2.2 Nicotine: The Main Cause of Addiction

Approximately 60–80% of adults who smoke would like to quit because they are aware of the health hazards associated with smoking. However, among smokers who attempt to quit on their own (self-quitters), 80% relapse within 1 month and 87% within 6 months [15], demonstrating the highly addictive nature of smoking. The main cause of this addiction is nicotine. For example, cigarettes from which nicotine has been removed or cigarettes with very low nicotine content are unpopular among smokers. Nicotine fulfills all criteria for an abused drug that produces drug dependence due to both its psychoactive effect and reinforcement of use. Physiological dependence on nicotine is due to its action on nicotinic acetylcholine receptors in the human brain. When nicotine activates such receptors, neurotransmitters such as dopamine (major component), norepinephrine, acetylcholine, and β-endorphin are released, which produces pleasurable experiences and a reduction of anxiety and stress. With repeated smoking, these receptors are increased in number and may also become desensitized. Although nicotine is mainly responsible for addiction, components such as acetaldehyde may enhance the physiological properties of nicotine [16]. Nicotine dependence is also heritable because there is a genetic component to nicotine addiction. Smoking cessation measures are also heritable, and measured genetic variation is associated with efficacy of nicotine dependence treatment [17]. Although nicotine is not a carcinogen, it may act as a tumor promoter. With its specific binding to nicotinic acetylcholine receptors, nicotine deregulates essential biological processes such as regulation of cell proliferation, apoptosis, migration, invasion, angiogenesis, inflammation, and cell-mediated immunity in a wide variety of cells including cancer cells [18]. Nicotine has a pro-fibrotic effect on myocardial tissue that may consequently increase its susceptibility to catecholamine. Thus, nicotine may play a role in increasing the risk of various heart diseases associated with smoking [19].

4.2.3 Cigarette Smoking and Oxidative Stress

Cigarette smoke contains many free radicals and other oxidizing compounds that may induce free radical production upon entering the lungs and circulation. The vapor phase of cigarette smoke may contain up to 10^{15} free radicals per puff of a cigarette, and these reactive substances can enter the bloodstream, inducing direct oxidative stress in the circulation [20]. However, both vapor phase and particulate matter of cigarette smoke contain free radicals, including reactive oxygen species, reactive nitrogen species, superoxide, hydroxyl radicals, nitric oxide, and hydrogen peroxide. Although these free radicals and oxidants present in the vapor phase of cigarette smoke have a short half-life, they can enter the bloodstream, causing oxidative damage to macromolecules. The vapor phase of cigarette smoke also contains saturated and unsaturated aldehydes, which are more stable than free radicals and hydrogen peroxide. These compounds can also enter the bloodstream, generating reactive oxygen species, especially through interacting with enzymes such as nicotinamide adenine dinucleotide phosphate oxidase (NADPH). As a result, tissues distant from the lungs can also experience increased oxidative stress [21]. In sharp contrast to the vapor phase of cigarette smoke, radicals associated with particulate matter (tar) have a much longer half-life. Semiquinone radicals associated with tar can reduce oxygen to produce superoxide as well as hydrogen peroxide and hydroxyl radicals. These radicals can penetrate cell membranes and can induce oxidative stress. Moreover, cigarette tar contains metal ions that can generate hydroxyl radicals from hydrogen peroxide [22]. The presence of pro-oxidants semiquinone, hydroquinone, quinone, α- and β-unsaturated aldehydes such as acrolein and crotonaldehyde, ketones, and saturated aldehydes in cigarette tar has been well documented by chemical analysis of aqueous tar extract. α_1-proteinase has been implicated in the pathogenesis of emphysema. Evans *et al.* [23] demonstrated that aqueous cigarette tar extract is capable of severely compromising elastase inhibitory capacity of human α_1-proteinase, and such damage is related to oxidative stress. The authors speculated that such damage in smokers' lungs may contribute to emphysema observed in many smokers. Cigarette smoke contains approximately 500 ppm nitric oxide, and nitrogen oxide reacts quickly with superoxide radicals to form peroxynitrite and other organic peroxyl radicals. These free radicals can cause significant damage to DNA [24].

Extensive research has documented the association between significant oxidative stress and smoking. Despite overnight abstinence, smokers demonstrate higher plasma levels of biomarkers of oxidative stress, including F_2-isoprostane, hydroxyeicosatetraenoic acid products, F_4-neuroprostane, and 7-ketocholesterol, as well as 24- and 27-hydroxycholesterol. Moreover, after smoking cigarettes, additional biomarkers of oxidative stress such as plasma allantoin and certain cholesterol oxidation products are also elevated. In general, smokers tend to have a lower plasma high-density lipoprotein cholesterol level, which increases the risk of cardiovascular diseases. Cessation of smoking for 1 week can reduce urinary F_2-isoprostane levels. Cigarette smoking can also damage DNA, and it generates DNA base oxidation products such as 8-hydroxy-2′-deoxyguanosine [25]. 3-Nitrotyrosine is considered to be a promising biomarker of oxidative stress to living organisms. Szumska *et al.* [26] reported that the mean concentration of 3-nitrotyrosine was significantly higher in smokers than in nonsmokers, thus further demonstrating increased oxidative stress in smokers. Padmavathi *et al.* [27] observed elevated nitrite and nitrate levels in plasma and red cell lysate in smokers compared to nonsmokers. In addition, smokers showed elevated erythrocyte membrane lipid peroxidation, protein carbonyls, increased cholesterol-to-phospholipid ratio, and decreased Na,K-ATPase activity. The authors concluded that smoking induced the generation of reactive oxygen species and reactive nitrogen species, which possibly altered membrane physicochemical properties.

It has been speculated that elevated levels of biomarkers of oxidative stress in smokers are related to exposure of smokers to free radicals and oxidants in both the gaseous phase and the tar phase of cigarette smoke and also related to lower antioxidant capacity of blood. However, whether the lower antioxidant capacity of blood in smokers compared to nonsmokers is due to lower intake of foods rich in antioxidants, such as fruits and vegetables, or due to depletion of antioxidant nutrients as a result of overwhelming exposure to free radicals during smoking is a subject of debate. Compared to nonsmokers, smokers have more than 25% lower circulating concentrations of important antioxidant compounds, including vitamin C, β-carotene, and cryptoxanthin. The concentrations of these antioxidants in former smokers are between those of current smokers and individuals who have never smoked cigarettes. In addition, smoking can also decrease levels of other antioxidants in the circulation, including vitamin E, lycopene, lutein, and zeaxanthin [28]. Bloomer [29] studied 15 young novice smokers (mean age, 24 years) and 13 nonsmokers (mean age, 24 years) who showed similar dietary intake based on caloric intake, macronutrients, and micronutrients. Smokers showed significantly lower plasma antioxidant capacity, serum Trolox equivalent antioxidant capacity, and whole blood glutathione concentration compared to nonsmokers, indicating that the plasma antioxidant capacity of smokers was significantly compromised compared to that of nonsmokers. In addition, markers of lipid peroxidation such as plasma malondialdehyde and plasma oxidized low-density lipoprotein concentrations were significantly increased in smokers compared to nonsmokers. Bloomer concluded that young novice cigarette smokers showed lower blood antioxidant capacity and higher lipid peroxidation levels compared to nonsmokers, despite the two groups having similar dietary intake.

4.2.4 Cigarette Smoking is Related to Many Diseases: Role of Oxidative Stress

Many studies have shown an association between cigarette smoking and an increased risk of cardiovascular diseases. Exposure to cigarette smoke induces multiple pathological effects in the endothelium, and many of these effects are related to increased oxidative stress induced by reactive oxygen species, reactive nitrogen species, and other oxidants present in cigarette smoke. In addition, reactive oxygen species can upregulate inflammatory cytokines and may cause endothelium dysfunction by reducing the bioavailability of nitric oxide. Increased plaque formation and the development of vulnerable plaques also result from exposure to cigarette smoke due to the inflammatory process and activation of matrix metalloproteins. Moreover, cigarette smoke can cause platelet activation and stimulation of the coagulation cascade, thus increasing the risk of atherothrombotic disease [30].

Chronic obstructive pulmonary disease (COPD) is strongly associated with cigarette smoking. COPD has considerable impact on exercise tolerance and quality of life as well as increased mortality in these individuals. Oxidative stress induced by cigarette smoke is considered a major factor contributing to muscle dysfunction and muscle wasting in COPD. The oxidative damage to muscle proteins usually precedes the characteristic respiratory changes in COPD patients [7]. Cigarette smoke also causes direct oxidative stress and proinflammatory responses of cells in the lungs [31]. Sangani and Ghio [32] proposed that lung injury after cigarette smoking is related to particulate matter. Smoking one cigarette exposes the human respiratory tract to 15,000–40,000 μg particulate matter, and exposure to these particles produces oxidative stress, the initial step in smoking-induced lung injury.

Cigarette smoking increases the risk of cancer in various organs. Cigarette smoke contains many carcinogens, including benzene(*a*)pyrene, dibenzoacridine, various nitrosamines, aromatic amines, and aldehydes including formaldehyde as well as benzene and other aromatic hydrocarbons. In addition, free radicals associated with cigarette smoke can be involved in the chemical carcinogenesis. Cigarette tar radicals can also cause significant DNA damage [33]. Cigarette smoking may cause erectile dysfunction with impaired arterial flow to the penis or acute vasospasm of the penile arteries. Smoking probably affects penile erection through the impairment of endothelium-dependent smooth muscle relaxation by affecting nitric oxide production via increased generation of reactive oxygen species [21]. Cigarette smoking has an adverse dose-dependent effect on ovarian function. The increased miscarriage rate among mothers who smoke may be related to direct adverse effects of nicotine, cadmium, and polyaromatic hydrocarbons on trophoblast invasion and proliferation [34]. Dittrich *et al.* [35] reported that the concentration of malondialdehyde was higher in amniotic fluid of mothers who smoked compared to that of nonsmoker mothers.

Age-related macular degeneration is a leading cause of blindness among the elderly, and cigarette smoking is the strongest epidemiological risk factor for developing macular degeneration. Smoking-related oxidative damage during the early phase of macular degeneration has been documented. Moreover, oxidative stress to the retinal pigmented epithelium may also contribute to the pathogenesis of macular degeneration [36].

4.2.5 Are Electronic Cigarettes Safe?

The electronic cigarette, also known as an e-cigarette, is a battery-operated device that may look like a regular cigarette. However, e-cigarettes, like regular cigarettes, also contain nicotine. An e-cigarette has three essential components—a plastic cartridge that has a dual function of a mouthpiece and reservoir of liquid, an atomizer that vaporizes the liquid, and a battery. The atomizer contains a small heating coil and heats the liquid containing nicotine, turning it into a vapor cloud that resembles cigarette smoke and is inhaled. The liquid that produces vapor in e-cigarettes, known as e-juice, is a solution of propylene glycol and/or glycerin or a related compound. Manufacturers of e-cigarettes claim that they are a safe alternative to smoking because unlike conventional cigarettes, in which tobacco is burned, e-cigarettes vaporize nicotine along with other components present in the cartridge in the form of aerosol but do not produce thousands of chemicals, including some harmful chemicals that are produced during burning of conventional cigarettes. Moreover, the smell is eliminated. However, the US Food and Drug Administration is concerned about the safety of e-cigarettes because they contain nicotine, the major addictive substance present in tobacco. The World Health Organization is also concerned because the safety of e-cigarettes has not been documented by rigorous peer-reviewed research. Wagener *et al.* [37] commented that e-cigarettes may motivate smokers to eventually quit smoking due to the low nicotine delivery by e-cigarettes. Pellegrino *et al.* [38] reported that e-cigarettes manufactured in Italy had propylene glycol and glycerine as the main components of the e-liquid, and these e-cigarettes produced significantly less particulate matter than conventional cigarettes. Electronic cigarettes are also more environmentally friendly because they practically eliminate passive smoking [38]. However, the safety of e-cigarettes has not yet been established.

4.3 ALCOHOL AND DRUG ABUSE IN THE UNITED STATES

Alcohol and drug abuse is a significant public health issue both in the United States and in many countries worldwide. It has been estimated that 18 million Americans may suffer from alcohol abuse, and an estimated 22.6 million Americans (8.9% of the US population) aged 12 years or older are illicit drug users. Illicit drugs abused by Americans include marijuana/hashish, cocaine (including crack), heroin, hallucinogens, inhalants, and prescription-type psychoactive drugs obtained illegally without a valid prescription. Marijuana is the most commonly used illicit drug among the general population [39]. Commonly abused drugs in the United States are listed in Box 4.2. White *et al.* [40] reported an increase in the rate of hospitalization due to drug abuse in the United States. Excessive consumption of alcohol along with drug abuse is prevalent in young adults aged 18–24 years. Hospitalization for drug overdose alone was estimated to be 113,907 cases in 2008, totaling $737 million in hospitalization costs. The authors commented that major efforts are needed to educate the public regarding the risk of drug overdose, especially when drugs are consumed with alcohol.

Recently, designer drugs such as bath salts and spices have been entering the illegal drug market in the United States. Bath salts (synthetic cathinones)

Box 4.2 COMMONLY ABUSED DRUGS IN THE UNITED STATES

- Stimulants
  - Amphetamine, methamphetamine, cocaine
- Narcotic analgesics
  - Heroin, codeine, morphine, oxycodone, hydrocodone, hydromorphone, buprenorphine, meperidine, methadone, fentanyl and other synthetic or designer drugs
- Sedative–hypnotics
  - Barbiturates (e.g., secobarbital, pentobarbital, and amobarbital), benzodiazepines (e.g., diazepam, alprazolam, clonazepam, lorazepam, temazepam, zolpidem, and flunitrazepam), γ-hydroxybutyric acid
- Hallucinogens
  - LSD (lysergic acid diethylamide), marijuana
- Anesthetics
  - Ketamine, phencyclidine
- New designer drugs
  - Bath salts, spices

have been categorized as Schedule 1 drugs due to high abuse potential. These synthetic cathinones include methylenedioxypyrovalerone (MDPV), 4-methylmethcathinone, and other related synthetic drugs with intense amphetamine-like effects [41]. However, MDPV and mephedrone are major compounds sold as bath salts. Bath salts are very toxic and may even be fatal. Spiller *et al.* [42] studied 236 patients (184 males) with suspected exposure to bath salts. Clinical effects include agitation, combative behavior, tachycardia, hallucination, paranoia, confusion, chest pain, hypertension, and blurred vision. One person died, 8 suffered severe toxicity, and 130 patients experienced moderate toxicity. Since 2008, synthetic marijuana compounds sold as Spice, K2, or herbal high have been gaining popularity among drug abusers. Although less toxic than bath salts, these compounds are very addictive and may cause severe overdose, including death. The first synthetic compound in this category, JWH-018, was synthesized by Dr. John W. Huffman at Clemson University to study its effect on cannabinoid receptors. It has been speculated that someone copied his method to produce JWH-018 illegally for abuse. Synthetic cannabinoids are called herbal high or legal high, and usually plant materials are sprayed with these compounds so that these active compounds are present at the surface of the product for maximum effect. However, pure compounds that are white powders are also available for purchase from the black market. Currently, more than 100 compounds are available that belong to this class of abuse drugs, the most common of which are JWH-018, JWH-073, JWH-250, JWH-015, JWH-081, HU-210, HU-211 (synthesized at Hebrew University), and CP-47,497 (synthesized at Pfizer). When a synthetic cannabinoid compound is smoked, it produces an effect similar to or more intense than that of smoking marijuana. By 2010, there were 1057 cases of synthetic marijuana-related toxicity in 18 states and the District of Columbia. Synthetic marijuana compounds are abused due to their agonist activity at cannabinoid receptors 1 and 2 (CB1 and CB2). Therefore, the mechanism that causes euphoria and hallucination is similar to that in natural marijuana. However, users also report adverse effects, such as rapid heartbeat, irritability, and paranoia. In addition, tremor, seizure, slurred speech, dilated pupils, hypokalemia, tachycardia, hypertension, and chest pain may also be present in patients overdosed with Spice. Long-term adverse effects are not known [43]. Common designer drugs abused in the United States are listed in Table 4.2.

Unlike conventional drugs, which can be detected quickly in hospital laboratories using appropriate immunoassays, overdose with these relatively new designer drugs is difficult to identify due to lack of availability of commercial immunoassays. Currently, relatively few reference laboratories have the capability to analyze these drugs in body fluids.

Table 4.2 Common Designer Drugs Abused in the United States

Designer Drug	Analog of Drug
3,4-Methylenedioxymethamphetamine (MDMA)	Methamphetamine
3,4-Methylenedioxyamphetamine	Amphetamine
para-Methoxyamphetamine (PMMA)	Amphetamine
Mephedrone (bath salt)	β-Keto-amphetamine
3,4-Methylenedioxypyrovalerone (bath salt)	β-Keto-amphetamine
γ-Hydroxy butyrolactone (GBL)	γ-Hydroxybutyric acid (GHB)
1,4-Butanediol	GHB
JWH-018, JWH-073, JWH-250 (Spice, K2)	Cannabinoid

4.3.1 Alcohol Abuse and Increased Oxidative Stress

Although alcoholic beverages such as beer and wine contain antioxidants, when abused, alcohol becomes a pro-oxidant instead of an antioxidant. Moderate alcohol consumption has many health benefits, but all alcohol abuse is dangerous for health. When alcohol is metabolized by alternative pathways involving CYP2E1, significant reactive oxygen species are generated and the body experiences increased oxidative stress. Alcoholics experience increased oxidative stress for two reasons: increased oxidative stress induced by alcohol when consumed in excess and poor nutritional status of many alcoholics with a lower concentration of antioxidants in their blood. Many alcoholics are malnourished because they ingest fewer nutrients (carbohydrate, protein, vitamins, etc.) than do healthy individuals, or alcohol and its metabolism interfere with proper absorption and digestion of these nutrients. In general, alcoholics suffer from protein, mineral, and vitamin deficiencies, especially deficiency of vitamin A [44].

When consumed in excess, alcohol can generate oxygen free radicals, inhibit glutathione synthesis, and impair antioxidant mechanisms in humans as well as increase the malondialdehyde level, a biomarker for lipid peroxidation. In addition, acetaldehyde, an intermediate in alcohol metabolism, is a highly reactive molecule and has a high affinity for sulfhydryl groups. It can impair functions of several enzymes by forming Schiff base with sulfhydryl groups at the active sites of such enzymes. Acetaldehyde also enhances lipid peroxidation. Acetaldehyde and aldehydic products of lipid peroxidation can also cause damage to hepatocytes along with other tissues. Autoantibodies toward distinct types of such adducts have been documented in patients with severe alcohol-induced liver diseases. Elevated adduct levels can also occur in erythrocytes of alcoholics, which may be related to alcohol-induced morphological aberration in hematopoiesis [45]. Based on a study of 28 patients with alcohol dependence and 19 healthy control subjects, Peng *et al.* [46]

observed a significantly higher concentration of malondialdehyde in patients with alcohol dependence compared to controls. In addition, activities of the antioxidant enzymes superoxide dismutase and glutathione peroxidase were respectively 86 and 37% lower in alcohol-dependent patients compared to controls, indicating lower antioxidant defense in patients with alcohol dependence. However, serum malondialdehyde concentrations decreased when these patients practiced abstinence, indicating that alcohol was responsible for the increased oxidative stress experienced by these patients. Administration of a high dose of alcohol to rats aggravates systemic and local oxidative stress, leading to lung injury ranging from mild pulmonary dysfunction to severe lung injury. In addition, reactive oxygen species generated in the presence of excess alcohol may damage lipid membrane, enzymes carrying the sulfhydryl group, and DNA; such oxidative damage may be involved in the etiology of diverse human diseases caused by alcohol abuse, including coronary heart disease, acute ischemia of the extremities, infection, neurodegenerative diseases, and cancer [47]. Drinking during pregnancy can cause fetal abnormalities (fetal alcohol syndrome), premature delivery, and even fetal death. Therefore, pregnant women, and women who want to become pregnant, should not consume alcoholic beverages. Alcohol use can lead to the generation of reactive oxygen species that produce an imbalance in the intracellular redox state, leading to an overall increase in oxidative stress that may cause damage to the central nervous system of the fetus, causing fetal alcohol syndrome [48]. Chen *et al.* [49] demonstrated a higher level of 8-hydroxy-2′-deoxyguanosine concentration, a marker of oxidative damage to DNA, in alcohol-dependent patients compared to controls, and oxidative damage of DNA persisted after 1 week of detoxification.

4.3.2 Drug Abuse and Oxidative Stress

Although increased oxidative stress due to alcohol dependence has been well documented in the literature, fewer studies have described increased oxidative stress in drug addicts. Because many drug abusers also abuse alcohol, it is relatively difficult to establish that the cause of increased oxidative stress in these individuals is related to only drug abuse. However, Kovacic [50] commented that a unifying theme for toxicity and addiction related to certain drugs can be linked to electron transfer, reactive oxygen species, and oxidative stress. These drugs include nicotine, cocaine, alcohol, phencyclidine, amphetamines, 3,4-methyledioxyamphetamine, morphine, heroin, tetrahydrocannabinol, and certain therapeutic drugs such as benzodiazepines, phenytoin, phenobarbital, aspirin, and acetaminophen.

Amphetamine and its derivatives are potential neurotoxins due to production of free radicals after abusing these drugs. Govitrapong *et al.* [51] observed a 21% increase in plasma lipid peroxidation products as measured by thiobarbituric acid reactive substances and a 32% decrease in erythrocyte antioxidant enzyme glutathione peroxidase in amphetamine users compared to controls. In addition, activities of other antioxidant enzymes in erythrocytes such as superoxide dismutase and catalase were also reduced by 31 and 14%, respectively. Methamphetamine administration is also associated with excessive oxidative stress, and even recent abstainers from methamphetamine had higher serum malondialdehyde levels compared to controls [52]. Increasing evidence suggests that 3,4-methylenedioxymethamphetamine (MDMA and Ecstasy)-induced toxicity involves the production of reactive oxygen species and reactive nitrogen species. The free radicals can originate from several molecular pathways, including oxidative deamination of monoamine, metabolic pathways, catecholamine auto-oxidation, and hyperthermia. These free radicals may cause lipid peroxidation and cellular death [53].

Kovatsi *et al.* [54] reported elevated plasma 15-F(2t)-isoprostane, a marker of lipid peroxidation, in chronic heroin users compared to controls based on their study of 42 chronic heroin abusers and 42 healthy controls, indicating that chronic abuse of heroin is associated with increased oxidative stress. Cocaine-induced cardiovascular diseases, such as hypertension, thrombosis, myocardial dysfunction, cardiac arrhythmia, and endocarditis, may be linked to cardiac oxidative stress. Cerretani *et al.* [55] reported that cocaine administration can compromise the heart's antioxidant defense. In animal models, the reactive oxygen species generated as a result of cocaine administration trigger cardiac injury and apoptosis. Cocaine can also increase infiltration of inflammatory cells in the heart.

4.4 CONCLUSION

Oxidative stress induced by cigarette smoke is responsible for many health risks associated with smoking. In addition, smokers have a shorter life span than nonsmokers, although smokers who quit smoking by the age of 30 years (35 years in another study) may avoid this risk. Although nicotine is not a carcinogen, it may be a tumor promoter. Electronic cigarettes may cause less harm than convention cigarettes, but safety issues regarding the use of e-cigarettes are still under investigation. Alcohol and drug abuse are serious public health concerns. Both alcohol abuse and drug abuse increase oxidative stress in the human body, and many adverse health effects of alcohol and drug abuse are related to increased oxidative stress. In addition, many alcoholics and drug abusers are poorly nourished and often show nutritional deficiency, including deficiency of antioxidant vitamins and micronutrients.

REFERENCES

[1] Law MR, Eald NJ. Environmental tobacco smoke and ischemic heart disease. Prog Cardiovasc Dis 2003;46:31–8.

[2] McGinnis J, Foege WH. Actual causes of deaths in the United States. JAMA 1993;270:2207–12.

[3] Vogl M, Wenig CM, Leidl R, Pokhrel S. Smoking and health related quality of life in English population: implications for economic evaluations. BMC Public Health 2012;12:203.

[4] Doll R, Peto R, Boreham J, Sutherland I. Mortality in relation to smoking: 50 years of observations on male British doctors. BMJ 2004;328(7455):1519.

[5] Sakata R, McGale P, Grant EJ, Ozasa K, et al. Impact of smoking on mortality and life expectancy in Japanese smokers: a prospective cohort study. BMJ 2012;345:e7093.

[6] Hayashida K, Imanaka Y, Murakami G, Takahashi Y, et al. Difference in lifetime medical expenditures between male smokers and nonsmokers. Health Policy 2010;94:84–9.

[7] Barreiro E, Peinado VI, Galdiz JB, Ferrer E, et al. Cigarette smoke induced oxidative stress: a role in chronic obstructive pulmonary disease skeletal muscle dysfunction. Am J Respir Crit Care Med 2010;182:477–88.

[8] Norman V. An overview of the vapor phase, semivolatile and nonvolatile components of cigarette smoke. Rec Adv Tob Sci 1977;3:28–58.

[9] Baker RR. Smoke chemistry. In: Davis DL, Nielsen MT, editors. Tobacco: production, chemistry, and technology. Oxford: Blackwell; 1999. pp. 398–439.

[10] McRae DD. The physical and chemical nature of tobacco smoke. Rec Adv Tob Sci 1990;16:233–323.

[11] Hoffmann D, Hoffman I. Tobacco smoke components. Beitr Tabakforsch Int 1998;18:49–52.

[12] Hoffmann D, Hoffman I, El-Bayoumy K. The less harmful cigarette: a controversial issue. A tribute to Ernst L. Wynder. Chem Res Toxicol 2001;14:767–90.

[13] Baker RR. Smoke generation inside a burning cigarette: modifying combustion to develop cigarettes that may be less hazardous to health. Prog Ener Combust Sci 2006;32:373–85.

[14] Shin HJ, Sohn HU, Han JH, Park CH, et al. Effect of cigarette filters on the chemical composition and *in vitro* biological activity of cigarette mainstream smoke. Food Chem Toxicol 2009;47:192–7.

[15] Hughes JR. Tobacco withdrawal in self quitters. J Consult Clin Psychol 1992;60:689–97.

[16] Mannino DM. Why won't our patients stop smoking: the power of nicotine addiction. Diabetes Care 2009;32(Suppl. 2):5426–8.

[17] Lessov-Schlaggar CN, Pergadia M, Khroyan TV, Swan GE. Genetics of nicotine dependence and pharmacotherapy. Biochem Pharmacol 2008;75:178–95.

[18] Cardinale A, Nastrucci C, Cesaria A, Russo P. Nicotine: specific role in angiogenesis, proliferation and apoptosis. Crit Rev Toxicol 2012;42:68–89.

[19] D'Alessandro A, Boeckelmann I, Hammwhoner M, Goette A. Nicotine, cigarette smoking and cardiac arrhythmia: an overview. Eur J Prev Cardiol 2012;19:297–305.

[20] Swan GE, Lessov-Schlaggar CN. The effect of tobacco smoke and nicotine on cognition and the brain. Neurophysiol Rev 2007;17:259–73.

[21] Tostes RC, Carneiro FS, Lee AJ, Giachini FR, et al. Cigarette smoking and erectile dysfunction: focus on NO bioavailability and ROS generation. J Sex Med 2008;5:1284–95.

[22] Pryor WA. Cigarette smoke radicals and the role of free radicals in chemical carcinogenicity. Environ Health Perspect 1997;105:875–82.

[23] Evans MD, Church DF, Pryor WA. Aqueous cigarette tar extracts damage human alpha-1-proteinase inhibitor. Chem Biol Interact 1991;79:151–64.

[24] Spencer JP, Jenner A, Chimel K, et al. DNA damage by gas phase cigarette smoke apparently involves attack by reactive nitrogen species in addition to oxygen. FEBS Lett 1995;375:179–82.

[25] Seet R, Lee CYJ, Loke VM, Huang SH, et al. Biomarkers of oxidative damage in cigarette smoke: which biomarkers might reflect acute versus chronic oxidative stress? Free Rad Biol Med 2011;50:1787–93.

[26] Szumska M, Wielkosznski T, Tyrpien K. 3-Nitrotyrosine determination as nitrosative stress marker and health attitudes of medical students considering exposure to environmental tobacco smoke. Przegl Lek 2012;69:798–802.

[27] Padmavathi P, Reddy VD, Kavitha G, Paramahamsa M, et al. Chronic cigarette smoking alters erythrocyte membrane lipid composition and properties in male human volunteers. Nitric Oxide 2010;23:181–6.

[28] Alberg A. The influence of cigarette smoking on circulating concentrations of antioxidant micronutrients. Toxicology 2002;180:121–37.

[29] Bloomer RJ. Decreased blood antioxidant capacity and increased lipid peroxidation in young cigarette smokers compared to nonsmokers: impact of dietary intake. Nutr J 2007;6:39.

[30] Csordas A, Bernhard D. The biology behind the atherothrombotic effects of cigarette smoke. Nat Rev Cardiol 2013;10:219–30.

[31] Faux SP, Tai T, Thorne D, Xu D, et al. The role of oxidative stress in the biological responses of lung epithelial cells to cigarette smoke. Biomarkers 2009;14(Suppl. 1):90–6.

[32] Sangani R, Ghio A. Lung injury after cigarette smoking is particle related. Int J COPD 2011;6:191–8.

[33] Pryor WA. Cigarette smoke radicals and the role of free radicals in chemical carcinogenicity. Environ Health Perspect 1997;105(Suppl. 4):875–82.

[34] Shiverick KT, Salafia C. Cigarette smoking and pregnancy: I. Ovarian, uterine and placental effects. Placenta 1999;20:265–72.

[35] Dittrich R, Schibel A, Hoffmann I, Mueller A, et al. Influence of maternal smoking during pregnancy on oxidant status in amniotic fluid. In Vivo 2012;26:813–18.

[36] Cano M, Thimmalappula R, Fujihara M, Nagai N, et al. Cigarette smoking, oxidative stress, the anti-oxidant response through Nrf2 signaling, and age-related macular degeneration. Vision Res 2010;50:652–64.

[37] Wagener TL, Siegel M, Borrelli B. Electronic cigarette: achieving a balanced perspective. Addiction 2012;107:1545–8.

[38] Pellegrino RM, Tinghino B, Magiaracina G, Marani A, et al. Electronic cigarettes: an evaluation of exposure to chemicals and fine particulate matter (PM). Ann Ig 2012;24:279–88.

[39] US Department of Health and Human Services, Substance Abuse and Mental Health Services Administration (SAMHSA), Office of Applied Studies. Results from the 2010 National Survey on Drug Use and Health: National findings. Washington, DC: US Department of Health and Human Services.

[40] White AM, Hingson RW, Pan IJ, Yi HY. Hospitalization for alcohol and drug overdoses in young adults ages 18–24 in the United States, 1999–2008: results from the nationwide inpatient sample. J Stud Alc Drugs 2011;72:774–86.

[41] Fass JA, Fass AD, Garcia A. Synthetic cathinones (bath salts): legal status and patterns of abuse. Ann Pharmacother 2012;46:436–41.

[42] Spiller HA, Ryan ML, Weston RG, Jansen J. Clinical experience with and analytical confirmation of bath salts and legal high (synthetic cathinones) in the United states. Clin Toxicol 2011;49:499–505.

[43] Wells D, Ott CA. The new marijuana. Ann Pharmacother 2011;45:414–17.

[44] Lieber CS. Relationship between nutrition, alcohol use and liver disease. Alc Res Health 2003;27:220–31.

[45] Niemela O. Distribution of ethanol induced protein adducts in vivo: relationship to tissue injury. Free Rad Biol Med 2001;31:1533–8.

[46] Peng FC, Tang SH, Huang MC, Chen CC, et al. Oxidative status in patients with alcohol dependence: a clinical study in Taiwan. J Toxicol Environ Health A 2005;68:1497–509.

[47] Aytacoglu BN, Calikoglu M, Tamer L, Coskun B, et al. Alcohol induced lung damage and increased oxidative stress. Respiration 2006;73:100–4.

[48] Brocardo PS, Gil-Mohapel J, Christie BR. The role of oxidative stress in fetal alcohol spectrum disorders. Brain Res Rev 2011;67:209–25.

[49] Chen CH, Pan CH, Chen CC, Huang MC. Increased oxidative DNA damage in patients with alcohol dependence and its correlation with alcohol withdrawal severity. Alc Clin Exp Res 2011;35:338–44.

[50] Kovacic P. Unifying mechanism for addiction and toxicity of abused drugs with application to dopamine and glutamate mediators: electron transfer and reactive oxygen species. Med Hypotheses 2005;65:90–6.

[51] Govitrapong P, Boontem P, Kooncumchoo P, Pinweha S, et al. Increased blood oxidative stress in amphetamine users. Addict Biol 2010;15:100–2.

[52] Huang MC, Lin SK, Chen CH, Pan CH, et al. Oxidative stress status in recently abstinent methamphetamine abusers. Psychiatry Clin Neurosci 2013;67:92–100.

[53] Fiaschi AI, Cerretani D. Causes and effects of cellular oxidative stress as a result of MDMA abuse. Curr Pharm Biotechnol 2010;11:444–52.

[54] Kovatsi L, Njau S, Nikolaou K, Tsolakidou A, et al. Isoprostane as a marker of oxidative stress in chronic heroin users: correlation with duration of heroin use or concomitant hepatitis C infection. Am J Drug Alc Abuse 2010;36:13–7.

[55] Cerretani D, Fineschi V, Bello S, Riezzo I, et al. Role of oxidative stress in cocaine-induced cytotoxicity and cocaine-related death. Curr Med Chem 2012;19:5619–23.

CHAPTER 5

Oxidative Stress Induced by Household Chemicals

CONTENTS

5.1 INTRODUCTION

Many household chemicals not only are toxic but also can induce substantial oxidative stress. Moreover, household chemicals are a major cause of poisoning in children and pets. According to the 28th annual report of the American Association of Poison Control Centers, the three substance classes most frequently encountered in human exposures and poisoning were analgesics (11.5%), cosmetics/personal care products (7.7%), and household cleaning substances (7.3%). However, the three substance classes most frequently involved in poisoning of children ages 5 years or younger were cosmetic/personal care products (13.2%), analgesics (9.4%), and household cleaning substances (9.2%) [1]. Therefore, it is important to keep all these products out of reach of children. McGuigan [2] reported that in the United States, unintentional poisoning in children due to ingestion of household chemicals (perfume, cologne, aftershave, and household cleaning products) is common, and the majority of the cases studied by the author involved 1- and 2-year-old children. McKenzie *et al.* [3] reported that household bleach was the most common agent involved in poisoning of children aged 1–3 years, and the primary mechanism of exposure was ingestion. Even pet dogs and cats are exposed to these toxic household chemicals. One report noted that dogs were exposed more than cats to these household chemicals (84.7 vs. 15.3%), and oral exposure was most common for both dogs and cats (95.5% of toxicity cases) [4].

Ingestion of household chemicals by children can have serious consequences. Patra *et al.* [5] described a case of a near fatality in a child who ingested nail polish remover containing acetone. Even acetone-free nail polish remover containing γ-butyrolactone is toxic if ingested because this compound is metabolized into toxic γ-hydroxybutyric acid. A 15-month-old infant presented with coma and respiratory collapse in the hospital after sucking the contents of such nail polish remover container. The package insert provided

A. Dasgupta and K. Klein: Antioxidants in Food, Vitamins and Supplements. DOI: http://dx.doi.org/10.1016/B978-0-12-405872-9.00005-7

no information regarding toxicity of the product [6]. Sculptured nails are acrylic artificial fingernails that are glued onto real nails, and a special glue remover is needed to remove such nails. Some sculptured nail removers contain acetonitrile, which is very toxic if ingested because the human liver can transform acetonitrile into deadly cyanide. After swallowing a mouthful of sculptured nail polish remover at night, a 16-month-old boy was found dead in his bed the following morning. A second child who also ingested sculptured nail polish remover survived after intensive therapy. Both boys had blood cyanide levels in the potentially toxic range. The delay in onset of cyanide poisoning was due to the time needed for human liver enzymes to transform less toxic acetonitrile into cyanide [7].

Adults may also suffer severe poisoning and even death from ingestion of household chemicals. Cording *et al.* [8] described a case of an 89-year-old woman suffering from Alzheimer's disease who died after accidentally drinking Pine-Sol cleaning solution, which contains two toxic ingredients (isopropanol and 1-α-terpineol) [8]. Ethylene glycol, which is in antifreeze solution, and methanol, which is in windshield wiper fluid, can cause severe toxicity and may be fatal if ingested in sufficient quantities. In one study involving 119 cases of methanol poisoning and 88 cases of ethylene glycol poisoning, Coulter *et al.* reported 21 fatalities from methanol poisoning and 19 fatalities due to ethylene glycol poisoning [9]. In 1995, a police officer in Georgia who was married to a woman named Lynn Turner died suddenly. The cause of death was cardiac failure. However, in 2001, when a firefighter living with Lynn Turner died suddenly due to ethylene glycol poisoning, authorities exhumed the police officer's body for further investigation. It was established that ethylene glycol was also the cause of the police officer's death. In 2002, Lynn Turner was charged with murder using antifreeze (containing ethylene glycol). The local media provided extensive coverage of the event. Following that coverage, cases of poisoning with ethylene glycol increased in the state of Georgia (both self-poisoning and malicious poisoning). Lynn Turner was convicted for two murders and was sentenced to life in prison without parole [10].

Routes of exposure to household chemicals include the following:

- *Inhalation*: Most common route of exposure, which may cause coughing, a burning sensation, and shortness of breath
- *Skin contact*: Probably the next most common route of exposure, which may cause skin irritation, blisters, burning sensation and redness of the skin, and even burn
- *Eye contact*: Less common route of exposure except in cases of accidents or exposure to children who may play with these chemicals
- *Ingestion*: Uncommon route of exposure except in children.

5.2 VARIOUS HOUSEHOLD PRODUCTS AND THEIR ACTIVE INGREDIENTS

Household chemicals include auto products; chemicals inside the home (air fresheners, bleach, cleaning products, etc.); chemicals used for landscaping and gardening (including pesticides); personal care products such as antiperspirant, shampoo, and hair spray (which are relatively nontoxic except after ingestion of a large amount by a child); home maintenance products (paint, stain removers, etc.); and home office household products (adhesives, ink, correction fluid, electronic equipment cleaners, etc.). Although not all household products are toxic, some products contain acid, alkali, inorganic substances, and volatile organic substances that can be toxic. Chemicals found in various toxic household products are listed in Table 5.1.

Although less toxic than bleach and oven cleaners, some cosmetic products and office supply products contain volatile compounds that may cause toxicity if vapor is inhaled for a long time or if children ingest such products. Glue sniffing is an abusive behavior among some children and adolescents as well as among adults. This behavior is dangerous because solvent abuse and glue sniffing may result in severe toxicity, resulting in hospital admission and possibly death.

Many chemicals in household products can produce oxidative stress. In general, volatile organic compounds, which are found in many household chemicals, can easily vaporize, and inhaling such vapor may cause both oxidative stress and toxicity. In addition, glue sniffing is a major concern with regard to adolescents. Such practice can cause both increased oxidative stress and toxicity.

5.2.1 Methanol-Induced Oxidative Stress and Toxicity

Methanol is a toxic alcohol that should not be ingested or inhaled, and exposure to methanol vapor is a safety concern. Methanol is found not only in windshield wiper fluids but also in many other household products, including carburetor cleaner, paints, varnishes, paint thinners, and various cleaning products. Methanol is used in preparing denatured alcohol because the addition of methanol to ethanol makes it toxic and undrinkable (methylated spirit). In addition, denatured alcohol is cheap because it is exempted from the excise tax that is applicable to ethanol. Denatured alcohol is used as a fuel for spirit burners, camping stoves, and Canned Heat, which is designed to be burned directly from its can. Oxidative stress is produced during methanol metabolism. The metabolism of methanol is similar to that of alcohol (ethanol), where alcohol dehydrogenase and acetaldehyde dehydrogenase 2 are the major enzymes involved. This metabolism takes place mainly in the human liver.

Table 5.1 Household Chemicals Containing Toxic Ingredients

Household Chemical	Active Ingredient
In the kitchen	
Oven cleaners	Lye (sodium/potassium hydroxide)
Window/glass cleaner	Ammonia and isopropanol
All-purpose cleaners	Ammonia, phosphate
In the laundry room	
Bleach	Sodium hypochlorite
Spot and grease remover	Chlorinated hydrocarbon
Insect repellent	Organophosphorus or carbamate
In the bathroom	
Toilet bowl cleaner	Strong acid or acid + surfactant
Drain cleaner	Lye or sulfuric acid
Mold removers	Chlorine/ammonium chloride
Nail polish remover	Acetone or butyrolactone
Artificial nail remover	Acetonitrile
In the living room	
Rug/carpet cleaners	Naphthalene and perchloroethylene
Furniture polishes	Ammonia, phenol, nitrobenzene petroleum distillate
Air freshener	Formaldehyde, petroleum distillates, *p*-dichlorobenzene, aerosol propellants
In the garage	
Antifreeze	Ethylene glycol
Motor oil	Complex hydrocarbon, metal
Windshield washer	Methanol
Battery	Sulfuric acid and lead
Paint/paint thinner	Toluene
In the yard	
Insect repellents	Insecticides
Rodent killer	Warfarin or super warfarin
Swimming pool tablet	Sodium or calcium hypochlorite
Swimming pool cleaner	Chlorine, strong acid, hypochlorite
In the office/computer room	
Glue, rubber cement	Hexane
Permanent marker	Xylene
Typewriter correction fluid	Acetone

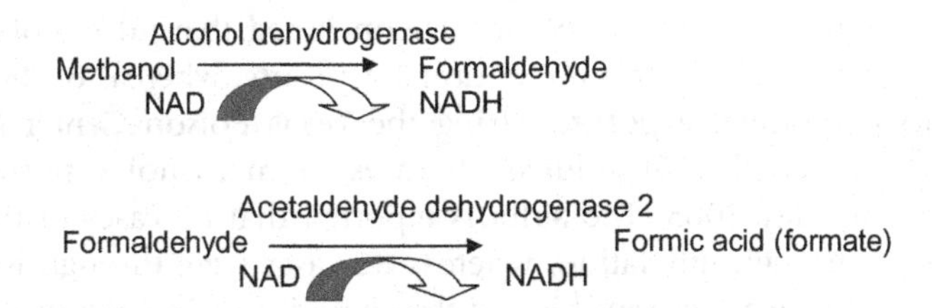

During methanol metabolism, there is an increase in NADH concentration, which is responsible for generation of superoxide anion due to increased mitochondrial electron flow along the respiratory electron transport chain. Therefore, free radicals generated during metabolism of methanol are responsible for increased oxidative stress following exposure to methanol. Using electron spin resonance, Skrzydlewska *et al.* [11] observed an increase in free radical signal in rats 6 and 12 h after intoxication with methanol. After 7 days of treatment with methanol, the malondialdehyde level was also significantly increased, indicating that methanol induced oxidative stress in the rat model. Aromatic amino acids such as tryptophan, tyrosine, and phenylalanine are very sensitive to oxidative damage by free radicals. The authors also observed a 30% decrease in tryptophan residue in proteins after 2 days of exposure to methanol. The level of bi-tyrosine doubled, which is a useful marker for protein modification by hydroxyl radicals, indicating oxidative damage. In another report, the authors observed an increased level of malondialdehyde not only in serum but also in liver and erythrocytes of rats after 7 days of treatment with methanol. In addition, the antioxidant enzyme superoxide dismutase and antioxidant α-tocopherol levels were diminished in erythrocytes. Moreover, levels of the antioxidants ascorbic acid and glutathione as well as antioxidant enzymes glutathione peroxidase and glutathione reductase were diminished in serum, liver, and erythrocytes [12]. Saoudi *et al.* [13] also observed significantly increased hepatic and renal lipid peroxidation evaluated by thiobarbituric acid reactive substances in rats treated with methanol compared to controls. They commented that the methanol-induced oxidative stress is due to the formation of superoxide anion and hydrogen peroxide during metabolism of methanol to formaldehyde and, finally, to formate. Metabolism of methanol by catalase also produces hydrogen peroxide, causing oxidative stress [14].

In addition to producing oxidative stress, methanol is a very toxic chemical. Human exposure to methanol may occur through different routes, including inhalation, cutaneous exposure, and ingestion. Methanol is well absorbed through all three routes and may cause toxicity regardless of the route of exposure. Davis *et al.* [15] noted that the average methanol exposure rate reported to the American Association of Poison Control Centers between 1993 and 1998 was 2254 cases annually, and one death occurred in every

183 exposures. In this report, the authors concluded that 90.3% of cases of methanol toxicity were due to unintentional exposure, whereas 8.3% of cases were due to intentional exposure. Using the Texas Poison Center Network database, Givens *et al.* [16] studied all cases of methanol exposure from January 2003 to May 2005. The authors reported that 87 cases of methanol exposure were through inhalation, whereas 81 cases were through ingestion. Carburetor cleaner was responsible for the majority of inhalation cases (79 of 87), whereas ingestion involved mostly windshield washer fluid (39 of 81) and carburetor cleaner (20 of 81). Whereas most cases of inhalation exposure to methanol (78%) were intentional, most cases of ingestion of methanol were either accidental (65%) or suicide attempts (24%). A majority of these patients (56% of patients in the inhalation group and 46% of patients in the ingestion group) were admitted to the hospital, and some patients in both groups experienced vision loss. This report indicates that exposure to methanol through inhalation may cause serious toxicity, although some previous reports have indicated that individuals who abuse methanol-containing products by inhalation are at low risk of developing methanol toxicity. Methanol is also absorbed from the skin and may cause methanol toxicity, as reported by Downie *et al.* [17]. A consultant supervising tank cleaning prior to methanol loading wore a positive pressure breathing apparatus but no protective clothing. After working for 2 or 3 h in the tank, he exited and worked on deck but was still wearing his methanol-soaked clothing, which eventually dried out because methanol is very volatile, with a low boiling point of only 65°C. However, he developed visual disturbances, a typical symptom of methanol toxicity, 8 h after exposure. Fortunately, he recovered completely in the hospital without loss of vision. Exposure to methanol is dangerous during pregnancy because these low-molecular-weight substances can cross the placenta and affect the fetus. Belson and Morgan [18] reported a case in which a 28-year-old pregnant woman was poisoned with methanol (route of exposure not reported) and was treated aggressively, including the use of hemodialysis. However, methanol was also detected in the blood of the newborn infant. Despite aggressive therapy, both the mother and her newborn died from methanol poisoning.

Methanol itself is relatively nontoxic. Therefore, methanol toxicity is a classic example of "lethal synthesis," in which metabolites of methanol—formaldehyde and formic acid—are the major cause of methanol toxicity. Formic acid is responsible for blindness and death because it is a transitory metabolite. Symptoms of methanol toxicity may not appear initially but, rather, may appear 12–24 h after exposure. Visual disturbances, including blurred vision, sensitivity to light, witnessing a virtual snowstorm, and, in some cases, partial or total loss of light perception, are commonly present in the individual presenting in the emergency room with suspected methanol toxicity. The major complication of methanol poisoning is loss of vision, including partial or total blindness. The lethal dose of methanol in humans is not fully established. Although it is assumed that ingestion of 30–100 mL of methanol may cause death, death from methanol may occur even after ingestion of 15 mL of 40% methanol. In contrast, there is a published report that neither death nor blindness occur after consumption of even 500 mL of methanol. However, cases of visual impairment and blindness are more common with methanol ingestion, and blindness may result from consuming at little as 4 mL of methanol [19]. The mechanism of methanol-induced blindness is not fully understood. Animal experiments indicate that formic acid, a metabolite of methanol, is responsible for most of the damage. Retinol dehydrogenase, an enzyme very similar to alcohol dehydrogenase, is present in the human retina, and metabolism of methanol inside the retina produces formic acid, which causes major methanol-induced retinal toxicity [20].

Ethanol is an effective antidote for methanol poisoning, and the sooner the therapy can be initiated, the better is the outcome. Ethanol is a preferred substrate for alcohol dehydrogenase, and in the presence of ethanol, metabolism of methanol to toxic formic acid metabolite is greatly reduced. Another effective therapy for methanol overdose is hemodialysis. Methanol is a small molecule with a molecular weight of only 32. Methanol can be effectively removed from circulation using hemodialysis. Usually, hemodialysis along with ethanol therapy for methanol poisoning should be initiated if the blood methanol level is 50 mg/dL or greater. Hemodialysis may also be initiated if an individual has ingested 30 mL or more of methanol or based on other clinical indications as determined by the physician treating such overdose [21]. Fomepizole (4-methylpyrazole), a potent competitive inhibitor of alcohol dehydrogenase, has recently been used as an antidote to treat methanol poisonings. This antidote can slow toxic formaldehyde metabolite formation from methanol, and small formaldehyde buildup can be secreted in urine. Bicarbonate therapy can also be used with fomepizole therapy to correct metabolic acidosis caused by methanol [22].

5.2.2 Ethylene Glycol-Induced Oxidative Stress and Toxicity

Ethylene glycol is found in automobile antifreeze (major use) and hydraulic brake fluid. A major commercial application of ethylene glycol is deicing fluid for aircrafts. Ethylene glycol is not volatile, and risk of inhalation exposure is minimal. Ethylene glycol is first metabolized into glycoaldehyde by alcohol dehydrogenase. Then glycoaldehyde is metabolized by acetaldehyde dehydrogenase, producing glycolic acid. However, glycoaldehyde is further metabolized by CYP2E1 to glyoxal. Glyoxal is then further metabolized into glycolic acid or glyoxylic acid. Glycolic acid can also be converted into glyoxylic acid by the action of either glycolate oxidase or lactate dehydrogenase. Glyoxylic acid is then converted into oxalic acid mainly by the action of lactate dehydrogenase. However, another enzyme, glycolate oxidase, is also capable of converting glyoxylic acid into oxalic acid. Oxalic acid can form crystals with calcium, producing calcium oxalate, which can deposit into kidneys, causing renal failure. Kruse [23] provides an in-depth review of methanol and ethylene glycol toxicity.

$$\underset{\text{Ethylene glycol}}{CH_2\text{–}OH\,|\,CH_2\text{–}OH} \xrightarrow{\text{Alcohol dehydrogenase}} \underset{\text{Glycoaldehyde}}{CH_2\text{–}OH\,|\,CHO}$$

$$CH_2\text{–}OH\,|\,CHO \xrightarrow{\text{Acetaldehyde dehydrogenase}} CH_2\text{–}OH\,|\,COOH \quad \text{(glycolic acid)}$$

Elevated lipid peroxidation product and reduced level of antioxidant enzymes have been reported in rat after exposure to ethylene glycol. Jurczyk *et al.* [24] reported that after 4 weeks of exposure to ethylene glycol, the malondialdehyde concentration in erythrocytes of rats increased from a pre-exposure value of 0.14 μmol/g of hemoglobin to 0.24 μmol/g of hemoglobin, indicating significant lipid peroxidation due to ethylene glycol-induced oxidative stress. Activities of the antioxidant enzymes catalase and superoxide dismutase were also diminished significantly due to exposure of ethylene glycol. Oxalate can induce lipid peroxidation *in vitro*. When rats were exposed to drinking water containing 0.75% ethylene glycol for 15 days, malondialdehyde concentrations were increased from a mean of 0.82 μmol in 24-h urine in control animals to a mean of 1.16 μmol in 24-h urine in experimental rats. Further increases in malondialdehyde concentrations were observed in rats treated for 30 and 60 days with ethylene glycol; a maximum mean malondialdehyde concentration of 2.06 μmol in 24-h urine was observed in rats exposed to ethylene glycol for 60 days. Lipid peroxidation products as measured as thiobarbituric acid reactive substances in kidney tissues increased from a mean of 0.48 nmol/mg of protein in control rats to 0.64 nmol/mg of protein in rats exposed to ethylene glycol for 15 days. In rats that were exposed to ethylene glycol for 60 days, the mean value of lipid peroxidation product was 1.61 nmol/mg of protein. Moreover, as expected, urinary secretion of oxalate was significantly increased in rats exposed to ethylene glycol. Renal tubular cell damage in these animals was also confirmed by an increase in urinary marker enzymes. The authors concluded that ethylene glycol-induced renal cell damage is associated with lipid peroxidation, indicating cell injury due to production of free radicals [25]. Studies have shown that renal cell exposure to oxalate or calcium oxalate leads to the formation of reactive oxygen species and the development of oxidative stress. Kamboj *et al.* [26] observed that lipid peroxidation measured as thiobarbituric acid reactive substances was increased by 59.4% in rats exposed to ethylene glycol for 7 weeks. In addition, exposure to ethylene glycol significantly reduced gene expression as well as activities of the antioxidant enzymes superoxide dismutase and glutathione transferase.

Ethylene glycol is very toxic, and ingestion may cause death. However, occupational exposure to ethylene glycol is not a health hazard as long as ethylene glycol does not come in contact with skin because it is absorbed from the skin. Airport workers involved in deicing aircraft during winter months using ethylene glycol-containing deicing fluid are not at risk for ethylene glycol toxicity. In one study, the authors measured ethylene glycol levels in 154 breathing zone air samples and 117 urine samples from 33 aviation workers exposed to deicing fluid (basket operators, deicing truck drivers, leads, and coordinators) during 42 work days over a winter period of 2 months at a Montreal airport. Ethylene glycol concentrations in air samples were relatively low, and measurable amounts of ethylene glycol in urine were only found in basket operators and coordinators, but some of them did not wear masks or were accidently sprayed with deicing fluid. However, acute or chronic renal toxicity, the major complication of ethylene glycol exposure, was not found in any of the aviation workers [27]. Absorption of ethylene glycol through the skin may cause serious toxicity, especially if skin lesions are present. Bouattar *et al.* [28] reported a case of toxicity in a 38-year-old man who presented at the hospital with nausea, vomiting, abdominal pain, and worsening mental status. The patient also experienced renal failure and was treated with hemodialysis. The renal biopsy revealed the presence of calcium oxalate crystals, a characteristic of ethylene glycol poisoning. It was later discovered that the patient worked in a cement factory and handled ethylene glycol without protective gloves. In addition, the patient had cutaneous psoriasis for 10 years. The authors concluded that cutaneous contact with ethylene glycol may cause poisoning in the presence of skin lesions.

The major route of exposure to ethylene glycol is ingestion of ethylene glycol-containing fluids either by accident or in intentional suicide. Ethylene glycol toxicity in humans occurs in several stages. The first stage is the neurological stage, in which mild euphoria-like ethanol poisoning may be observed within 30 min of ingestion of ethylene glycol. Other neurological symptoms include nystagmus, ataxia, seizure, and even coma, and these symptoms may be observed between 30 min and up to 12 h after ingestion of ethylene glycol. The next stage of ethylene glycol poisoning is cardiac symptoms, including mild hypertension and tachycardia. Finally, between 24 and 72 h after exposure, symptoms of renal failure may be observed, especially in patients who are not treated.

Major complications of ethylene glycol poisoning are metabolic acidosis and renal failure, and these complications may be fatal. The lethal dose of ethylene glycol is usually assumed to be 100 mL, but there are reports of fatalities from ethylene glycol poisoning with ingestion of only 30 mL [29]. If untreated, death may follow from ethylene glycol poisoning within 8–24 h after poisoning. On the other hand, the prognosis of ethylene glycol poisoning is good if it is treated in a timely manner. A 36-year-old man with a history of depression consumed a massive amount of ethylene glycol (3 L) in a suicide attempt. On admission, his blood ethylene glycol level was 1889 mg/dL, a potentially fatal level. Despite ingesting a lethal amount of ethylene glycol, the patient survived due to prompt medical attention and aggressive treatment using hemodialysis [30]. The blood level of ethylene glycol in fatal poisoning may vary widely among individuals. Rosano *et al.* [31] reviewed 12 medical examiners' cases in which fatality was due to ethylene glycol poisoning and observed that the ethylene glycol concentrations ranged widely from 5.8 to 779 mg/dL, with a mean of 183 mg/dL. The concentration of glycolic acid, a metabolite of ethylene glycol, varied from 81 to 177 mg/dL. Calcium oxalate crystals were detected in renal tissues.

Toxicity from ethylene glycol is mostly due to ingestion. Children and animals often consume large amounts of ethylene glycol because of its sweet taste. In an attempt to prevent ethylene glycol poisoning, denatonium benzoate may be added to ethylene glycol because this agent has a bitter taste. In 10 states, the addition of a bittering agent to antifreeze formulations containing ethylene glycol is compulsory (Box 5.1). Ethylene glycol poisoning is

Box 5.1 STATES IN WHICH ADDITION OF A BITTERING AGENT IS COMPULSORY IN ANTIFREEZE CONTAINING ETHYLENE GLYCOL

- Alabama
- Arizona
- California
- Maine
- New Jersey
- New Mexico
- Oregon
- Tennessee
- Virginia
- Washington

treated using bicarbonate, ethanol, fomepizole, and hemodialysis. If treatment can be started soon enough after ingestion, simple intravenous administration of ethanol may be sufficient for full recovery from ethylene glycol poisoning. Hemodialysis, along with ethanol or fomepizole infusion and bicarbonate therapy to correct metabolic acidosis, may also be needed when treating patients with more serious ethylene glycol toxicity.

5.2.3 Oxidative Stress and Toxicity Caused by Volatile Organic Compounds in Household Products

Various household products contain volatile organic compounds (Table 5.2). However, benzene, toluene, ethylbenzene, and xylene are most commonly found in non-occupational and occupational environments. Benzene is a known carcinogen linked to leukemia, and both toluene and xylene are known to increase the risk of cancer in paint industry workers [32]. Oxidative stress has been implicated as these volatile organic compounds. Kim *et al.* [33] reported that following 1 week of exposure of shipbuilding workers to volatile organic compounds (who are exposed to these chemicals due to their occupation), the urinary levels of hippuric acid (metabolite of toluene) and methyl hippuric acid (metabolite of xylene) were significantly elevated compared to pre-exposure levels. Moreover, hippuric acid both before and after exposure was associated with increased malondialdehyde (a marker of lipid peroxidation) levels before and after exposure. Evaluation of gene expression associated with hippuric acid levels showed significant alteration after exposure. The authors concluded that occupational exposure

Table 5.2 Volatile Organic Compounds in Various Household Products

Volatile Organic Compound	Examples of Household Products
Acetone	Nail polish remover, markers, typewriter correction fluid
Benzene[a]	Paint and varnish remover, varnish, airplane glue
Ethylbenzene	Ink
Chlorobenzene/dichlorobenzene	Spot and grease remover
Fluorocarbons	Hair spray, analgesic spray, refrigerator coolant
Hexane	Glue, rubber cement
Propane/butane gas	Lighter fluid, gas to light grill, spray paint
Styrene	Paint and adhesives
Toluene	Paint, spray paint, paint thinner, shoe polish, adhesives
Xylene	Permanent marker

[a]*Highly volatile and toxic; in 1978, the Consumer Protection Safety Commission banned the use of benzene in many household products.*

to toluene induces oxidative stress and various gene expression changes, some of which may be related to oxidative stress. Halifeoglu *et al.* [34] reported that plasma malondialdehyde levels were significantly higher in 18 employees who worked with paint thinner compared to 18 controls (mean malondialdehyde value of 2.0 nmol/mL in workers vs. mean value of 1.0 nmol/mL in controls), indicating that paint thinner inhalation may increase lipid peroxidation products due to increased oxidative stress. Paint thinners contain mainly toluene and also other volatile organic compounds, such as acetone, isobutyl acetate, and isobutanol. Unlike other components of paint thinner, toluene can diffuse into body fluids and may cause toxicity to intestine, kidneys, liver, spleen, and brain through generation of reactive oxygen species. Toxicities of various volatile compounds found in household products are summarized in Table 5.3.

Individuals occupationally exposed to other volatile compounds are also subjected to increased oxidative stress and are prone to asthma and various lung dysfunctions. Hairdressers are exposed to ammonia during hair bleaching and a variety of volatile organic compounds during their work hours. In one study, the authors observed increased serum levels of an oxidative marker of DNA damage, 8-hydroxy-2′-deoxyguanosine, in hairdressers, indicating that these individuals are subjected to increased oxidative stress due to exposure to volatile organic compounds at hair salons [35]. Exhaled breath is a good method for assessing lung function in workers exposed to volatile organic compounds and other pollutants in the air of work environments.

Table 5.3 Toxicity of Various Volatile Organic Compounds in Household Products

Volatile Compound	Oxidative Stress?	Toxicity
Acetone	Yes	Moderately toxic but breathing acetone vapor for a long time may cause burning sensation to nose and throat; toxic to small children if ingested
Benzene	Yes	Prolonged exposure linked to cancer and leukemia
Chlorobenzene	Yes	Eye and skin irritation
Dichlorobenzene	Yes	Eye and skin irritation; very high exposure linked to cancer
Fluorocarbons	Yes	Toxic to immune system and liver
Hexane	Yes	Prolonged exposure may cause nerve damage; may also be linked to cancer
Propane/butane gas	Yes	Highly flammable but low toxicity
Toluene	Yes	Prolonged exposure linked to cancer
Xylene	Yes	Prolonged exposure linked to cancer

Professional cleaners working in hospitals exhibited higher levels of markers of oxidative stress as well as higher amounts of ammonia and other chemicals in their exhaled breath, indicating that they suffer from increased oxidative stress due to the use of cleaning chemicals, including detergents, that contain volatile compounds [36]. Even photocopying can increase oxidative stress in the human body because ozone is emitted during the photocopying process. In one study involving 80 photocopy operators and 80 healthy volunteers, the authors found that markers of oxidative stress were elevated in the blood of photocopy operators, indicating that these operators experienced a higher level of oxidative stress than healthy volunteers [37].

Volatile organic compounds being emitted from various household chemicals can enter the body through breathing if a person is exposed directly to these chemicals for a prolonged time. However, when these household chemicals are kept in a closed container and used only as needed, they are fairly safe. Nevertheless, some volatile organic substances may still escape into indoor air, causing air pollution and impaired pulmonary function, especially in the elderly, by inducing oxidative stress. Moreover, oxidative stress can be generated due to passive inhalation of volatile organic compounds. In a study of 154 elderly people (aged 60 years or older), the authors found not only compromised pulmonary function but also elevated concentrations of markers of oxidative stress (8-oxo-2′-deoxyguanosine and malondialdehyde–thiobarbituric acid adduct) in urine specimens of the participants. In general, two organic compounds—toluene and xylene—were responsible for producing most significant oxidative stress in these elderly subjects. The authors concluded that exposure to toluene and xylene may exert harmful effects on lung function, and oxidative stress could be involved in the pathogenesis of these elderly people because there was a significant correlation between markers of oxidative stress and diminished lung function [38].

Acetone, which is in many household chemicals, is also volatile. Using a rat model, Mathias *et al.* [39] demonstrated that acute administration of acetone produced oxidative stress as evident by alteration of hepatic oxidative parameters. Styrene, which is in household products such as paints and adhesives, is also volatile and can impair lung function through inflammation of lung cells mediated by oxidative stress [40]. Chlorinated benzene (chlorobenzene and dichlorobenzene), which is in various paints, can also cause oxidative stress and damage to lung tissue [41]. Incense burning is a popular practice in many homes and also in temples, but it increased the risk of asthma attack in a child who was genetically deficient in activity of antioxidant enzyme glutathione S-transferase [42]. Chuang *et al.* [43] studied particulate matter generated during burning of candles and incense during church services and observed that the levels of PM_{10} and $PM_{2.5}$ reached 9.6 and 38.9 $\mu g/m^3$, respectively, during candle/incense burning; such elevated levels exceed the

European Union air quality guidelines. Therefore, these increased levels of particulate matter could increase oxidative stress in attendees breathing such air. The authors concluded that generation and subsequent inhalation of such particulate matter during church activities may pose significant risks in terms of respiratory health. Reactive oxygen species capable of oxidation of DNA are generated during incense burning, and such species are associated with incense particulate matter in the form of both nano-soot and micro-soot. Therefore, breathing such particulate matter poses a health risk [44].

5.3 GLUE AND SOLVENT ABUSE: A SIGNIFICANT PROBLEM WITH ADOLESCENTS

Solvent (inhalant) abuse is common among adolescents in the United States, but this problem is often overlooked. Various readily available household products are abused, including glue, adhesives, nail polish, nail polish remover, cigarette lighter fluid, butane gas, gas (petrol), air fresheners, deodorant, hairspray, pain-relieving spray, typewriter correction fluid, paint thinners, paint removers, and a variety of other agents. These household and office products contain toxic solvent such as toluene, acetone, hexane, chlorinated hydrocarbon, xylene, propane gas (grills, spray paint), butane gas (lighter fluid, spray paint), and fluorocarbons. Abuse of these household chemicals produces euphoria, but such abuse is also dangerous because prolonged abused of toluene and chlorinated hydrocarbon can cause permanent damage to the nervous system, lungs, heart, and kidneys. However, the major risk of solvent abuse is sudden death from cardiac arrest following arrhythmia [45]. Steffee *et al.* [46] reported two cases of fatal volatile solvent inhalation abuse: gasoline sniffing in a 20-year-old man and aerosol air freshener inhalation in a 16-year-old girl.

In addition to organ toxicity and sudden death, glue and solvent abuse increases oxidative stress due to inhalation of volatile organic compounds present in these household products. Oxidative metabolism of toluene, benzene, and acetone generates reactive oxygen species. In addition, quinones are generated during metabolism of toluene and benzene, which in turn can disrupt the redox system, generating more oxygen free radicals. Metabolism of these organic compounds may also stimulate CYP2E1, which is prone to radical formation. In addition, inhalation of volatile organic compounds can trigger inflammatory responses in the lungs, and inflammatory mediators released by phagocytic leukocytes and infiltrating macrophages as a result are known to generate reactive oxygen and reactive nitrogen species in the lungs. These processes can induce oxidative stress. Toluene, a common solvent in paint thinner, is also known to produce oxidative stress in the brain, causing elevated lipid peroxidation and neurological damage. Rats exposed to toluene also exhibited elevated lipid peroxidation products and evidence of oxidative damage to the DNA. Toluene exposure also increases lipid peroxidation products (both malondialdehyde and protein carbonyl levels) in human erythrocytes [47]. Dundarz *et al.* [48] reported elevated levels of malondialdehyde in both erythrocytes and plasma of 37 adolescents with a history of inhalant abuse compared to 27 well-matched healthy adolescents, indicating increased lipid peroxidation in adolescents who abused solvent. In addition, the activity of glutathione peroxidase was reduced in both erythrocytes and plasma of solvent abusers, providing further evidence of elevated oxidative stress in adolescents who abused solvents.

5.4 SKIN CONTACT WITH HOUSEHOLD CHEMICALS AND OXIDATIVE STRESS

Phenol is an organic compound used not only in industry for manufacturing of various chemicals but also in paint and household cleaning products. Phenol is toxic to the skin, causing rash, depigmentation, and dermatitis. Dermal exposure to phenol produces electron spin resonance-detectable PBN (*N-t*-butyl-α-phenylnitrone) spin-trapped signals of lipid-derived radicals, which supports the hypothesis that phenol-induced radical formation is responsible for phenol-induced dermal toxicity. Although phenol is less volatile than benzene and toluene, breathing phenol vapor may generate further oxidative stress [49]. Sodium or potassium hydroxide (strong alkali, caustic, lye) -containing household cleaning products can cause skin burn (alkali burn) after contact with skin, and oxidative stress is one of the mechanisms by which strong alkali causes skin damage. Ingestion of strong alkali-containing products can cause severe burns in the esophagus. Using a rat model, Gunel *et al.* [50] demonstrated elevated malondialdehyde concentration in esophageal tissue of rats exposed to 10% sodium hydroxide solution, indicative of lipid peroxidation. In addition, the level of antioxidant glutathione was significantly reduced in these tissue samples. The authors concluded that reactive oxygen species may play an important role in the early phase of caustic esophageal burn by increasing tissue damage. Brankov *et al.* [51] studied 17 children with severe caustic injuries of the esophagus and stomach who eventually developed acute pancreatitis. These patients had elevated amylase and C-reactive proteins. The authors noted that ingestion of corrosive substances caused increased production of oxygen free radicals, which cause lipid peroxidation and tissue damage. The authors further commented that "corrosive oxidative stress" is the cause of the acute pancreatitis seen in these children after ingestion of corrosive agents.

Strong acid, such as sulfuric acid and hydrochloric acid, can also cause significant skin damage upon skin contact and even severe chemical burn. Sulfuric acid is one of the most widely used industrial chemicals; it is used in the electrical industry, chemical industry, and jewelry industry, and also in many household products. Vapors are also irritants to mucus, and ingestion of even a small amount of sulfuric acid may be fatal. Accidental exposure to sulfuric acid-containing compounds is especially dangerous in children, but adults may also suffer serious consequences. Aound *et al.* [52] reported a case of a 26-year-old jeweler who died after accidental ingestion of sulfuric acid. Benomran *et al.* [53] reported a case of a man who died after inhalation of sulfuric acid fumes in an attempt to clear a blockage in a drain pipe using strong sulfuric acid. Autopsy revealed severe pulmonary edema, and traces of acid and sulfate compounds were also detected in his upper respiratory passage. Exposure of alkali or acid to eyes is dangerous, and washing of eyes with water should be initiated immediately. Chemical eye injury due to exposure to acid or alkali is a medical emergency requiring immediate transfer of the victim to the nearest emergency medical center for immediate therapy, beginning with effective eye irrigation [54].

In addition to strong acid and alkali, large numbers of chemicals, including agricultural chemicals, industrial chemicals, and household chemicals, are known to produce diverse types of skin injury on contact. Kovacic and Somanathan [55] reported that oxidative stress and electron transfer comprise a portion of the key mechanism in producing dermal toxicity by these agents. Pesticides are used in insect repellent, and warfarin or super warfarin is used in rodent killer. Major insecticides in the United States are organophosphates and carbamates (carbamates are less toxic than organophosphates). These compounds can cause toxicity and even death if ingested in large amounts, mostly during suicide attempts. These insecticides can cause life-threatening poisoning if ingested by children. Both organophosphorus and carbamate insecticides are known to produce significant oxidative stress. Ogut *et al.* [56] studied a group of 94 pesticide-formulating workers (with at least 5 years of experience in pest control in apple and cherry production) and 45 controls. They observed that chronic exposure to organophosphorus, synthetic pyrethroid, and carbamate pesticides was associated with increased lipid peroxidation products in erythrocytes.

5.5 CONCLUSION

Many household compounds are toxic and may cause oxidative stress. In addition, accidental poisoning of toddlers is a common problem but a preventable cause of child death. Adolescents should be counseled not only about the dangers of alcohol and drug abuse but also about the dangers of abusing household products, including glue sniffing and solvent abuse. It is important to store all household products, including cosmetic products, completely out of reach of children. Moreover, to avoid health hazards and increased oxidative stress burden of the body during house cleaning, it is important to wear gloves and a face mask to avoid skin contact with household cleaning products as well as to avoid breathing organic vapor in air.

REFERENCES

[1] Bronstein AC, Spyker DA, Cantilena Jr LR, Green JL, et al. 2010 annual report of the American Association of Poison Control Center's national poison data system (NPDS): 28th annual report. Clin Toxicol 2011;49:910–41.

[2] McGuigan MA. Common culprits in childhood poisoning: epidemiology, treatment and parental advice for prevention. Pediatr Drugs 1999;1:313–24.

[3] McKenzie LB, Ahir N, Stolz U, Nelson NG. Household cleaning product-related injuries treated in US emergency departments in 1990–2006. Pediatrics 2010;126:509–16.

[4] Mahdi A, Van der Merwe D. Dog and cat exposure to hazardous substances reported to the Kansas State veterinary diagnostic laboratory 2009–2012. J Med Toxicol 2013;9:207–11.

[5] Patra S, Sikka G, Khaowas AK, Kumar V. Successful intervention in a child with toxic methemoglobin due to nail polish remover poisoning. Ind J Occup Med 2011;15:137–8.

[6] Brown JJ, Nanayakkara CS. Acetone free nail polish removers: are they safe? Clin Toxicol 2005;43:297–9.

[7] Caravati EM, Litovitz TL. Pediatric cyanide intoxication and death from an acetonitrile containing cosmetic. JAMA 1988;260:4370–3.

[8] Cording CJ, Vallaro GM, Deluca R, Camporese T, et al. A fatality due to accidental Pine Sol ingestion. J Anal Toxicol 2000;24:664–7.

[9] Coulter CV, Farquhar SE, McSherry CM, Isbister GK, et al. Methanol and ethylene glycol acute poisonings—predictors of mortality. Clin Toxicol 2011;49:900–6.

[10] Morgan BW, Geller RJ, Kazzi ZN. Intentional ethylene glycol poisoning increases after media coverage of antifreeze murders. West J Med 2011;12:296–9.

[11] Skrzydlewska E, Elas M, Farbiszewski R, Roszkowska A. Effect of methanol intoxication on free radical induced protein oxidation. J Appl Toxicol 2000;20:239–43.

[12] Skrzydlewska E, Farbiszewski R. Lipid peroxidation and antioxidant status in the liver, erythrocytes and serum of rats after methanol intoxication. J Toxicol Environ Health A 1998;53:637–49.

[13] Saoudi M, Ben Hsouna A, Trigui M, Jamoussi K, et al. Differential oxidative stress responses to methanol in intraperitoneally exposed rats: ameliorative effects of *Opunita vulgaris* fruit extract. Toxicol Ind Health 2012;28:549–59.

[14] Bailey LA, Prueitt RL, Rhomberg LOP. Hypothesis based weight of evidence evaluation of methanol as a human carcinogen. Regul Toxicol Pharmacol 2012;62:278–91.

[15] Davis LE, Hudson D, Benson BE, Jones Easom LA, et al. Methanol poisoning exposures in the United States: 1993–1998. J Toxicol Clin Toxicol 2002;40:499–505.

[16] Givens M, Kalbfleisch K, Bryson S, Carl R. Comparison of methanol exposure routes reported to Texas poison control centers. West J Emerg Med 2008;9:150–3.

[17] Downie A, Khattab TM, Malik MI, Samara IN. A case of percutaneous industrial methanol toxicity. Occup Med (London) 1992;42:47–9.

[18] Belson M, Morgan BW. Methanol toxicity in a newborn. J Toxicol Clin Toxicol 2004;42:673–7.

[19] Kruse JA. Methanol poisoning. Intens Care Med 1992;18:391–7.

[20] Garner CD, Lee EW, Terzo TS, Louis-Ferdinard RT. Role of retinal metabolism in methanol induced retinal toxicity. J Toxicol Environ Health 1995;44:43–56.

[21] Gonda A, Gault H, Churchill D, Hollomby D. Hemodialysis for methanol intoxication. Am J Med 1978;64:749–58.

[22] Hovda KE, Anderson KS, Urdal P, Jacobsen D. Methanol and formate kinetics during treatment with fomepizole. Clin Toxicol 2005;43:221–7.

[23] Kruse JA. Methanol and ethylene glycol intoxication. Crit Care Clin 2012;28:661–711.

[24] Jurczyk AP, Galecki P, Jankowska B, Meissner E, et al. Effect of ethylene glycol on antioxidative enzymes and lipid peroxidation activity in erythrocytes [in Polish]. Arch Med Sadowej Kryminol 2002;52:147–53.

[25] Thamilselvan S, Hackett RL, Khan SR. Lipid peroxidation in ethylene glycol induced hyperoxaluria and calcium oxalate nephrolithiasis. J Urol 1997;157:1059–63.

[26] Kamboj P, Aggarwal S, Puri S, Singla SK. Effects of aqueous extract of *Tribulus terrestris* on oxalate induced oxidative stress sin rats. Indian J Nephrol 2011;21:154–9.

[27] Gerin M, Patrice S, Begin D, Goldberg MS, et al. A study of ethylene glycol exposure and kidney function of aircraft de-icing workers. Int Arch Occup Environ Health 1997;69: 255–65.

[28] Bouattar T, Madani N, Hamzaqui H, Alhamany Z, et al. Severe ethylene glycol intoxication by skin absorption. Nephrol Ther 2009;5:205–9 [in French].

[29] Walder AD, Tyler CKG. Ethylene glycol antifreeze poisoning: three case reports and a review of treatment. Anesthesia 1994;49:964–7.

[30] Johnson B, Meggs WJ, Bentzel CJ. Emergency department hemodialysis in a case of severe ethylene glycol poisoning. Ann Emerg Med 1999;33:108–10.

[31] Rosano TG, Swift TA, Kranick CJ, Sikirica M. Ethylene glycol and glycolic acid in postmortem blood from fatal poisonings. J Anal Toxicol 2009;33:508–13.

[32] Lundberg I, Milatou-Smith R. Mortality and cancer inducing among Swedish paint industry workers with long-term exposure to organic solvents. Scand J Work Environ Health 1998;24:270–5.

[33] Kim JH, Moon JY, Park EY, Lee KH, et al. Changes in oxidative stress biomarkers and gene expression levels in workers exposed to volatile organic compounds. Ind Health 2011;49:8–17.

[34] Halifeoglu I, Canatan H, Ustundag B, Illhan N, et al. Effect of thinner inhalation on lipid peroxidation and some antioxidant enzymes in people working with paint thinner. Cell Biochem Funct 2000;18:263–7.

[35] Ma CM, Lin LY, Chem HW, Huang LC, et al. Volatile organic compounds exposure and cardiovascular effects in hair salons. Occup Med (London) 2010;60:624–30.

[36] Corradi M, Gergelove P, Di Pilato E, Folesani G, et al. Effect of exposure to detergents and other chemicals on biomarkers of pulmonary response in exhaled breath from hospital cleaners: a pilot study. Int Arch Occup Environ Health 2012;85:389–96.

[37] Zhou JF, Cai D, Tong GZ. Oxidative stress and potential free radical damage associated with photocopying. A role of ozone? Free Rad Res 2003;37:137–43.

[38] Yoon HI, Hong YC, Cho SH, Kim H, et al. Exposure to volatile organic compounds and loss of pulmonary function in the elderly. Eur Respir J 2010;36:1270–6.

[39] Mathias MG, Almeida BB, Buerno JE, Portari GV, et al. Lipid peroxidation and antioxidant system in rats acutely treated with acetone. Exp Clin Endocrinol Diabetes 2010;118:368–70.

[40] Mogel I, Baumann S, Bohme A, Kohajda T, et al. The aromatic volatile organic compounds toluene, benzene and styrene induce COX-2 and prostaglandins in human lung epithelial cells via oxidative stress and p38 MEPK activation. Toxicology 2011;289:28–37.

[41] Feltens R, Mogel I, Roder-Stolinski C, Simon JC, et al. Chlorobenzene induces oxidative stress in human lung epithelial cells *in vitro*. Toxicol Appl Pharmacol 2010;242:100–8.

[42] Wang IJ, Tsai CH, Chen CH, Tung KY, et al. Glutathione S-transferase, incense burning and asthma in children. Eur Respir J 2011;37:1371–7.

[43] Chuang HC, Jones T, BeruBe K. Combustion particles emitted during church services: implications for human respiratory health. Environ Int 2012;40:137–42.

[44] Chuang HC, Jones TP, Lung SC, BeruBe K. Soot-driven reactive oxygen species formation from incense burning. Sci Total Environ 2011;409:4781–7.

[45] Flanagan RJ, Ruprah M, Meredith TJ, Ramsey JD. An introduction to the clinical toxicology of volatile substances. Drug Saf 1990;5:359–83.

[46] Steffee CH, Davis GJ, Nicol KK. A whiff of death: fatal volatile solvent inhalation abuse. South Med J 1996;89:879–84.

[47] Martinex-Alfaro M, Alcaraz-Contreras Y, Carabez-Trejo A, Leo-Amador GE. Oxidative stress effects of thinner inhalation. Indian J Occup Environ Med 2011;15:87–92.

[48] Dundarz MR, Turkbay T, Akay C, Sarici SU, et al. Antioxidant enzymes and lipid peroxidation in adolescents with inhalant abuse. Turk J Pediatr 2003;45:43–5.

[49] Murrat YR, Kisin E, Castronova V, Kommineni C, et al. Phenol induced *in vivo* oxidative stress in skin: evidence for enhanced free radical generation, thiol oxidation and antioxidant depletion. Chem Res Toxicol 2007;20:1769–77.

[50] Gunel E, Caglayan E, Caglayan O, Akilloglu I. Reactive oxygen radical levels in caustic esophageal burns. J Pediatr Surg 1999;34:405–7.

[51] Brankov O, Shivachev KH, Drebov R, Dumanov K. Clinical manifestation of acute pancreatitis in children with caustic ingestion injury: the role of oxidative stress. Khirurgiia (Sofia) 2007;3:5–8 [in Bulgarian].

[52] Aound R, Matar N, Melkane A, Nassar M, et al. Sulfuric acid ingestion. J Trauma 2011;70:E98.

[53] Benomran FA, Hassan AL, Masood SS. Accidental fatal inhalation of sulfuric acid fumes. J Forensic Leg Med 2008;15:56–8.

[54] Rodrigues Z. Irrigation of the eye after alkaline and acid burn. Emerg Nurse 2009;17:26–9.

[55] Kovacic P, Somanathan R. Dermal toxicity and environmental contamination: electron transfer, reactive oxygen species, oxidative stress, cell signaling, and protection by antioxidants. Rev Environ Contam Toxicol 2010;203:119–38.

[56] Ogut S, Gultekin F, Kisioglu AN, Kucukoner E. Oxidative stress in the blood of farm workers following intensive pesticide exposure. Toxicol Ind Health 2011;27:820–5.

CHAPTER 6

Psychological Stress-Induced Oxidative Stress

Differences Between Personality Types, Gender, and Race

CONTENTS

6.1 INTRODUCTION

Normal human metabolism produces free radicals. However, psychological stress can produce increased oxidative stress in the human body. Psychological stress is a part of life, but the magnitude of stress and subsequent oxidative stress varies between personality types and between gender. In this chapter, the emphasis is on psychological stress due to life events, anxiety, frustration, or loneliness and how such stress can produce oxidative stress, causing damage to the human body. *Stress* can be defined as a demand (psychological or physical) that is outside the norm of an individual, and it is manifested by biochemical change in the body. Stress is an adaptive mechanism of humans and animals, but the level of stress experienced by different individuals in a similar situation may vary widely. For example, a well-prepared student may feel less stress when taking a test than an ill-prepared student. Common symptoms of psychological stress are summarized in Box 6.1.

Leeming [1] proposed an interesting hypothesis suggesting that the human race carries an inherited stress response from early human evolution in the hot tropical climate of Africa. Early human diet in Africa was rich in vitamin C but minimal in sodium, whereas present-day intake of high sodium is incompatible with homeostasis of the human body, especially when subjected to any form of stress. The climatic conditions of Europe and North America are ill suited to experience stress because it appears that a constituent part of the bodily stress response is designed to be effective in a hot tropical climate in which stress can be accompanied by sweating of the overheated body through thermoregulation. Without low sodium intake and overheating of the body as a response to stress followed by sweating, in modern-day life, stress can result in lower immune function and reduced ability of the body to resist disease. It is well documented in the medical literature that prolonged psychological stress is linked to many diseases

A. Dasgupta and K. Klein: Antioxidants in Food, Vitamins and Supplements. DOI: http://dx.doi.org/10.1016/B978-0-12-405872-9.00006-9

Box 6.1 SIGNS AND SYMPTOMS OF SIGNIFICANT STRESS

Short-Term Stress

- Increased blood pressure
- Increased heart rate
- Tension headache
- Anxiety attack
- Back pain
- Diarrhea
- Upset stomach

Long-Term Stress

- Unhealthy eating pattern and weight gain/weight loss
- Trouble sleeping, including onset of insomnia
- Fatigue
- Difficulty with relationships
- Onset of depression, which may lead to major depression or psychiatric illness

Box 6.2 DISEASES ASSOCIATED WITH CHRONIC PSYCHOLOGICAL STRESS

- Asthma
- Cancer
- Cardiovascular diseases
- Immune dysfunction
- Hypertension
- Metabolic syndrome
- Stroke
- Substance abuse
- Type 2 diabetes
- Premature aging
- Rheumatoid arthritis
- Various psychological disorders

(Box 6.2). The central link between prolonged psychological stress and various diseases, including premature aging, is probably related to chronic elevation of cortisol and other stress hormones. Chronic stress also increases oxidative stress in the human body, which plays an important role in the pathogenesis of many diseases linked to chronic psychological stress. In one study, it was observed that plasma malondialdehyde (lipid peroxidation marker) was directly correlated with the severity of emotional stress experienced by healthy subjects [2]. Sivonova *et al.* [3] investigated whether psychological stress could induce oxidative stress in medical students by measuring various parameters on the day of an examination and between examination periods. The results demonstrated that on the day of an examination, the oxidative damage to the DNA and sensitivity to the lipid oxidation were significantly increased compared to the same parameters during nonstress periods (between examinations). A significant decrease in plasma antioxidant activity in students who were under oxidative stress on the examination day was also observed. The authors concluded that psychological stress induced by examination increased oxidative stress in medical students.

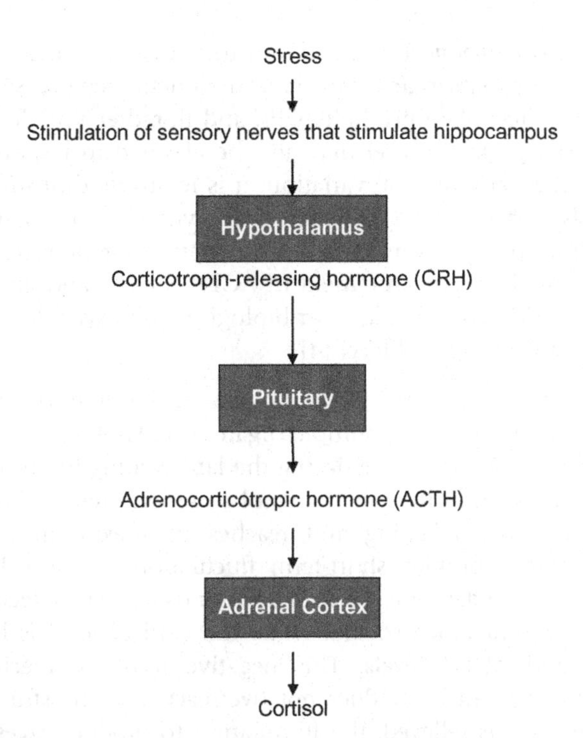

FIGURE 6.1 Mechanism of cortisol release in response to stress.

6.2 EFFECT OF STRESS ON HPA AXIS, CORTISOL, AND OTHER BIOMOLECULES

The biochemical mechanism of stress is exerted through the hypothalamus–pituitary–adrenal cortex (HPA) axis. Psychological events initiate HPA response predominantly by activating specific neuroendocrine neurons in the paraventricular nucleus of the hypothalamus. The magnitude of activation depends on the psychological component of the situation, such as unpredictability, uncontrollability, anticipation, ego involvement, and habituation. As a result of stimulation of the hypothalamus by stress, corticotropin-releasing hormone (CRH) is secreted. In addition, arginine vasopressin (ARV) may be released. CRH and ARV are transported to the pituitary gland (situated just beneath the hypothalamus), stimulating release of adrenocorticotropic hormone (ACTH). ACTH is released in the bloodstream and eventually activates the adrenal cortex, the outer portion of the adrenal gland located at the top each kidney, to release cortisol (Figure 6.1). Cortisol is a glucocorticoid hormone that has many important metabolic functions. In addition, the adrenal cortex releases mineralocorticoid (e.g., aldosterone) and androgen (e.g., testosterone) hormones. These hormones are biosynthesized from cholesterol and are called *steroid hormones*. In addition to release of cortisol, the perception of a stressful situation activates the sympathetic nervous system, which in turn releases catecholamines (adrenalin and noradrenalin) from the adrenal medulla. The prolactin level may also be elevated in response to stress, but due to wide interindividual variation, it is relatively difficult to correlate prolactin levels with psychological stress. However, cortisol is considered the major stress hormone. Salivary cortisol is a useful biomarker for measuring psychological stress, but measurement of ACTH, blood, and urinary cortisol must also be considered to account for biological and psychological variables that affect the salivary cortisol level [4].

After cortisol is released by the adrenal cortex, it travels through the bloodstream and exerts its effect on multiple organs. The HPA axis demonstrates a circadian pattern, with low levels during the late evening hours and the lowest level after midnight in humans. Cortisol secretion then begins to rise several hours prior to awakening and reaches its maximum level by late morning. However, complex short-term fluctuation of cortisol levels also takes place during the daytime. Cortisol also sends a negative feedback mechanism to the hypothalamus so that when the cortisol level is high, it suppresses CRH and ACTH levels. This negative feedback mechanism also ensures that the human body does not overreact in a stressful situation or that when the stress is relieved, the stimulation to produce excess cortisol is no longer present. This constant adjustment and readjustment of the cortisol level is termed the *allostatic load*. A person with a high allostatic load may feel anxious all the time [5].

Cortisol is the primary glucocorticoid hormone and has the opposite effect of insulin because it increases blood sugar. Cortisol also affects immune function, blood pressure, and binds with receptors in the brain affecting neurological function, thus preparing the human body to deal with stress. In addition to increased blood pressure and increased glucose levels in blood, heart rate is increased so that there is better blood flow to all organs, including the brain and heart muscle, as well as other muscles. Cortisol also increases oxidative stress, can cause damage to nerve cells (neurons), and reduces concentrations of antioxidant enzymes [6].

Dehydroepiandrosterone (DHEA) is a steroid hormone found in higher concentrations in younger adults and in lower concentrations in the elderly. DHEA can antagonize the effect of cortisol, and it is found in lower amounts in response to stress. In addition, levels of DHEA are lowered in depression and certain mental disorders. It has been well documented that chronic stress can cause depression. Within the brain, the serotonergic system is also activated in response to stress. However, the role of serotonin in the adaptive mechanism of stress is not fully understood. A lower serotonin level is associated with depression, and serotonin function is reduced in the presence of elevated cortisol [7]. Stress can alter gene expression in the brain, and certain genetic polymorphisms can increase the risk of an individual experiencing higher stress levels. Variations of genes involved in regulating the sympathetic system or in the HPA axis are associated with altered stress response [8].

The endocannabinoid system, which exists in both the brain and periphery, has been implicated as both a regulator and an effector of the stress response. The endocannabinoid system is composed of two G-proteins coupled to cannabinoid receptors (CB1 and CB2) as well as two arachidonate-derived ligands—arachidonylethanolamide (anandamide) and 2-arachidonylglycerol. Anandamide is a member of the *N*-acetyl ethanolamide family; other members of the family are palmitoylethanolamide (PEA) and oleoylethanolamide (OEA). The cannabinoid system has a protective role in stress and anxiety because it can counteract the effect of stress by participating in a cortisol-mediated negative feedback mechanism in order to inhibit hyperactivity of the HPA axis in response to stress. In a study of 71 adult volunteers, Dlugos *et al.* [9] demonstrated that psychosocial stress increased the serum concentration of anandamide and other members of the *N*-acetyl ethanolamide family. Interestingly, increased serum PEA levels were also positively correlated with cortisol levels, and these findings support a protective role of the endocannabinoid system in stress and anxiety.

6.3 PSYCHOLOGICAL STRESS INDUCES OXIDATIVE STRESS: ROLE OF ELEVATED CORTISOL LEVEL

As mentioned previously, psychological stress can induce oxidative stress. However, the molecular mechanism regarding the HPA axis and increased oxidative stress has not been completely elucidated. Severe life stress such as early maternal separation, maternal neglect of a child, prolonged loneliness, social isolation, and divorce might lead to severe and prolonged depression and potentially psychiatric illness, and the link between oxidative stress and many psychiatric disorders has been well documented. Severe stress may induce changes in neurons located in the prefrontal cortex, amygdala, insula, basal ganglia, and anterior cingulate cortex. Magnetic resonance imaging demonstrated smaller frontal lobes in subjects with a history of traumatic childhood experience who developed bipolar disorders later in life compared to normal controls [10]. The developing HPA axis is under strong social regulation in infancy and early childhood, and it is vulnerable to the quality of child caregiving. Child maltreatment has a complex and long-term influence on basal cortisol levels and HPA responsiveness to stress, which are influenced by many factors including genetic makeup. Interestingly, maltreated children with internalization problems demonstrate elevated basal cortisol levels in the early morning, whereas adults maltreated as children may exhibit low basal cortisol levels but elevated ACTH levels in response to psychological stress [11]. Animal experiments have linked early maternal separation with lifelong changes in HPA axis activity. Pesonen *et al.* [12] studied 282 subjects, of whom 85 were nonseparated, 129 were separated from their father, and 68 were separated from both parents during World War II. The authors observed that whereas subjects who were separated from their father did not differ significantly from nonseparated subjects, those separated from both parents had higher average salivary cortisol and plasma ACTH levels across all time points compared to the nonseparate group. Women showed higher baseline plasma cortisol and ACTH compared to men. The authors concluded that separation from both parents during childhood may alter the stress physiology of individuals much later in life.

Divorce is a very stressful life event that affects both the divorcing couple and their children. Women are particularly susceptible to negative effects associated with divorce, such as depression, insomnia, overeating, excessive drinking, or even substance abuse. Children whose parents are going through a divorce can be engaged in anxious behavior, including anger and hostility. There may be an association between severe mental consequences of divorce and increased oxidative stress. Divorced men showed a lower antioxidant defense system in the central nervous system (CNS) and decreased resistance of the brain to oxidative damage compared to nondivorced men. Children experiencing divorce of parents or adults who experienced divorce of parents during childhood showed decreased antioxidant resistance against reactive oxygen species in several tissues and also in the brain and lungs. In addition, children who were exposed to severe psychological stress during divorce of their parents may be more susceptible to brain cancer as well as elevated reactive oxygen species and reactive nitrogen species in the CNS. Research also indicates that neurophysiological and psychological alterations induced by war experiences are related to brain oxidative stress. In addition, posttraumatic stress disorder observed in US soldiers returning from the Iraq War was associated with reactive oxygen species-mediated changes in the brain [13].

Epel *et al.* [14] examined the stress level of healthy premenopausal women who were the biological mother of either a healthy child (19 women) or a chronically ill child (30 women). They observed that women who cared for chronically ill children experienced a much higher level of psychological stress than did women who cared for healthy children. As expected, women who experienced a higher level of psychological stress also demonstrated a higher amount of oxidative stress than the women in the control group, and the magnitude of oxidative stress also increased with the number of years these mothers cared for their sick children. In addition, women who experienced psychological stress also showed accelerated cell aging, as demonstrated by shortening of telomeres compared to that of the control group as well as lower telomerase activity. Moreover, women who experienced the highest stress demonstrated 10 years of additional aging of cells, as evidenced by shortening of telomeres, compared to women who did not experience such stress. Telomeres are strips of DNA–protein complex at the end of chromosomes that protect and stabilize the chromosomes. When a cell divides, the telomere is shortened until a critical length is achieved and the cell enters senescence and eventually apoptosis. Telomeres are therefore considered the biological clock of the cell, and telomere length can be used as a marker of the cell's aging process. Oxidative stress shortens telomere length in cell culture models. Epel *et al.* concluded that psychological stress experienced by these women resulted in significant increases in oxidative stress that caused shorter telomere length. These authors also reported that women who experienced stress also showed an elevated concentration of 8-hydroxy-2′-deoxyguanosine (also known as 8-oxo-7,8-dihydro-2′-deoxyguanosine), the marker of oxidative damage of DNA.

Increased oxidative stress and lower antioxidant defense in individuals suffering from chronic psychological stress may also be linked to chronic elevation of cortisol, a well-documented biomarker for psychological stress. Although acute elevation of cortisol in response to stress is necessary for survival, chronic psychological stress associated with prolonged hypercortisolism leads to a dysregulation of the HPA axis that is detrimental to health. Cortisol released due to activation of the HPA axis is known to bind with intracellular glucocorticoid receptors, and it exerts its effect by altering gene expression. Therefore, it can be speculated that oxidative stress induced by high cortisol is mediated by transcriptionally determined changes in DNA/RNA repair, the antioxidant defense system, as well as the formation of reactive oxygen species. Elevated cortisol level may also lead to increased oxidative stress where the number of reactive oxygen species produced by mitochondria exceeds antioxidant defense of the body. Excessive reactive species can damage DNA, lipids, and proteins. Damage of DNA in individuals suffering from prolonged stress is possibly linked to premature aging. In animal experiments, it has been demonstrated that both psychological stress and administration of corticosterone (cortisol equivalent in animal) lead to lower antioxidant defense in the brain, liver, and heart of rats as evidenced by lower activities of free radical scavenging enzymes such as superoxide dismutase, catalase, glutathione transferase, and glutathione reductase as well as the nonenzymatic antioxidants glutathione and serum urate. In addition, markers of lipid peroxidation and other oxidative markers were elevated in brain, liver, and heart. The authors concluded that oxidative stress is a major mechanism in the maladaptation to chronic stress, and stress hormones have a role in increasing oxidative stress [15]. Joergensen *et al.* [16] studied the correlation between urinary excretion of cortisol and markers of oxidatively damaged DNA and RNA in a sample of 220 elderly men and women. They observed a significant positive correlation (using regression analysis) between 24-h urinary cortisol and 8-oxo-7,8-dihydro-2-deoxyguanosine, the marker of total oxidative damage of DNA. Individuals in the highest quartile of 24-h cortisol excretion had a 57% higher level of oxidative marker of DNA damage compared to individuals in the lowest quartile. The authors concluded that their findings support the hypothesis of cortisol-induced oxidative damage to DNA as a mechanism of complex association between stress and various diseases as well as aging.

6.4 EFFECT OF PERSONALITY TRAITS ON EXPERIENCING PSYCHOLOGICAL STRESS

Although factors such as a death in the family, divorce, caring for a sick child, discrimination, loss of job, physical and mental abuse, and a high level of stress in a job can initiate chronic stress, people with certain personality traits may also be more prone to stress. Interestingly, one study observed that female but not male caregivers whose parents needed care experienced stress and adverse mental health consequences from the stress [17]. In general, personality types can be classified as type A and type B personality, although many individuals may have a mixture of both traits or show a more complex personality type such as types C and D. It has been speculated that people with type A personality are more prone to stress and stress-related illness as well as higher blood pressure and higher cholesterol levels than people with type B personality [18]. People with type A personality are also more prone to coronary heart disease [19]. Sirri *et al.* [20] observed in relation to type A personality a significant difference in the prevalence of cardiovascular diseases (36.1%) compared to other diseases (10.8%). Type A behavior was also associated with irritable mood, but, in general among cardiac patients, those with type A personality were less depressed, demoralized, and worried about illness than other personality types. However, these findings have been disputed in the medical literature. Mitaishvili and Danelia [21] found no significant correlation between type A personality and coronary heart disease. Traits expected in type A and type B personality are listed in Box 6.3.

Type C personality, also known as cancer-prone personality (first described by Lydia Temoshol in cancer patients), is indicative of a person who represses his or her emotions. Studies have confirmed that having this type of personality may increase the risk of cancer through immune and

Box 6.3 CHARACTERISTICS OF TYPE A AND TYPE B PERSONALITIES

Type A Personality

- Highly competitive and accept challenges
- Aggressive and try to complete project before deadlines
- Impatient of often intolerant of others
- Not a good listener and may interrupt others when they are talking
- Strong desire to be in charge of the situation/task
- Strong desire to acquire things/objects or assets
- Do not like to wait
- Few interests outside work
- Little or no physical activity
- Smoking or drinking habit may present
- Irregular or excessive eating pattern
- Hold back feelings/few social activities
- Often take work home
- Not satisfied with work/life

Type B Personality

- Less competitive and may not mind leaving things unfinished for a while
- Calm and can wait
- Capacity to tolerate others and can be a good listener
- Easygoing
- Interested in activities other than work
- Enjoy social networking
- Less prone to alcohol or tobacco abuse
- Enjoy physical activities
- Good eating habits
- Not a perfectionist
- Can express feelings
- Satisfied with work and life
- Prefer not to take work home

hormonal pathways [22]. In addition, people with type C personality are not risk takers but have excellent organization skills, and they keep their workplace neat and clean. Although they live inside a shell and do not share their emotions and problems with others, when speaking, they speak less and slowly. Because individuals with type C personality deny their feelings and cannot stand up for their rights, they suffer from stress and depression more than people with any other personality type.

Individuals with type D personality have a tendency to experience negative emotions and to inhibit these emotions by avoiding social contact with others. They are not risk takers. Moreover, they are afraid of rejection and prefer to stick to a routine rather than the uncertainty of change, even if such change could be beneficial to them. This is the reason why this personality type is also termed *distress personality*. People with type D personality experience a high amount of stress and are prone to major depression. These individuals tend to have higher cortisol, which may be a mediating factor for oxidative stress and increased risk of heart disease. Patients with heart disease who have type D personality are at higher risk of morbidity and mortality [23]. Kupper *et al.* [24] observed that type D personality patients suffering

Box 6.4 GENDER DIFFERENCE IN RISK OF DEVELOPING VARIOUS DISEASES

- For similar age, men are prone to
  - Cardiovascular diseases, including myocardial infarction
  - Infectious diseases
- For similar age, women are more prone to rheumatoid arthritis
  - Systemic lupus erythematosus
  - Multiple sclerosis
  - Stress-related fibromyalgia
  - Chronic pain

from congestive heart failure are characterized by increased oxidative stress burden and decreased antioxidant level compared to non-type D personality.

6.5 GENDER DIFFERENCE IN EXPERIENCING PSYCHOLOGICAL STRESS

In general, women are more prone to stress than men, although women generally have better coping skills than men to deal with the stress. For similar age, men are prone to diseases different from those that women experience, and this gender difference may be indirectly related to how men and women cope with stress and major life events. Gender differences in risk of various diseases are listed in Box 6.4. In addition, women and men may respond to the same "stressor" differently; for example, a long-distance relationship may produce more stress in a woman than in a man. In general, during stress men may prefer the "fight-or-flight" approach, whereas women prefer the "tend-and-befriend" approach. *Tending* involves nurturant activities designed to protect the woman and her offspring and that promote safety and reduce stress. *Befriending* is the creation and maintenance of a social network that may provide a safety net and reduce stress. Both animal and human studies indicate that the hormone oxytocin, which is secreted from the brain in conjunction with female reproductive hormones, may be related to different patterns of response of women to stress [25]. During stress, cortisol and epinephrine increase blood pressure, but oxytocin can counteract the action of cortisol. Women show a more complex reaction to stress because the menstrual cycle, pregnancy, and menopause necessarily affect only women.

For premenopausal women, the response to stress may vary during the menstrual cycle. During the luteal phase, the concentration of salivary and blood cortisol increases more in response to stress than during the follicular phase, indicating that female reproductive hormones also play a role in response to stress. It is important to effectively manage stress during pregnancy because prenatal stress has been linked to long-lasting effects on the behavioral and physiological development of offspring [26]. A higher stress level during pregnancy may lead to a shorter pregnancy and smaller infant. One study observed that infants born to mothers with higher cortisol levels near the end of pregnancy showed more crying, fussing, and negative facial expression [27]. Based on a study of 158 pregnant women, Baibazarova *et al.* [28] observed that plasma cortisol concentration was correlated with amniotic fluid cortisol, which in turn was correlated with lower birth weight. The authors concluded that fetal cortisol may be an important predictor of infant outcome, which shed light on the mechanism through which prenatal stress may affect birth outcome and temperament of the infant. Another complicated factor during pregnancy is the ability of the placenta to synthesize CRH. Desensitization of the maternal pituitary to CRH and resetting after birth may be a factor in postpartum depression [29].

A significant difference in response to psychological stress exists in premenopausal and most menopausal women. Research studies using stressors such as public speaking and mathematical tasks to evaluate stress response indicated that blood pressure was increased more in postmenopausal women compared to premenopausal women in response to the stress. Estrogen significantly reduced the response to stress in premenopausal women. Postmenopausal women who received estrogen appeared to experience blunting of the effect of stress [30]. Based on a study of 81 women, Glover *et al.* [31] reported that women with low estrogen levels were more vulnerable to posttraumatic stress disorder than were women with normal estrogen levels. Low estrogen in women with normal menstrual cycles was also associated with an inability to suppress fear responses even in safe conditions. Estrogen also markedly increases binding of oxytocin in the hypothalamus area of the brain, thus increasing its action. As mentioned previously, oxytocin can counteract stress and, along with estrogen, can modulate nurturing maternal behavior after childbirth (oxytocin level increases after childbirth). Oxytocin also has an anxiolytic effect and reduces inhibition, thus facilitating social networking [32]. Although oxytocin is also produced in men, women generally benefit from the calming effect of oxytocin more than men because release of oxytocin in response to stress in general is greater in women than in men. In addition, estrogen also facilitates the calming effect of oxytocin [33].

Childs *et al.* [34] compared Tier Social Stress Test (TSST; a reliable laboratory procedure to induce stress in human subjects) responses in healthy men and women. Women subjects were tested twice, once during the follicular phase and once during the luteal phase of the menstrual cycle. The authors observed that men exhibited greater cortisol response to stress than did women. In addition, women in the luteal phase exhibited the greatest subjective (greater feeling of anger and depression than experienced by men and also women in the follicular phase) and allopregnanolone (a neurosteroid that is secreted in response to stress and has a calming effect) response to stress, whereas women in the follicular stage exhibited blunted noradrenaline response. However, women who were most sensitive to subjective stress also exhibited the highest concentrations of salivary cortisol but the smallest progesterone response to stress. These findings indicate the presence of gender differences in terms of response to stress.

Sex differences in physiological and subjective responses to acute stress may contribute to sex differences in the prevalence of gender-related disorders. For example, women are more likely than men to be diagnosed with anxiety and related disorders. It has been hypothesized that this difference is related to gender differences in responding to stress. Women typically report higher levels of negative effect to stress than do men. Following TSST, women reported more fear, irritability, and confusion and less happiness compared to men. However, the authors found no significant difference in plasma cortisol levels between men and women [35]. Sex differences in cortisol reactivity to stress may be modulated by hormonal variation in women during various phases of the menstrual cycle. Irie *et al.* [36] reported that the marker of oxidative damage to DNA, 8-hydroxydeoxyguanosine, correlated with perceived workload and perceived psychological stress in female workers only; no such association was found in male workers. The authors concluded that psychological stress and perceived overwork appear to be related to the pathogenesis of cancer via oxidative damage particularly in female workers, indicating gender differences in response to stress.

6.6 CORTISOL AND ETHNICITY: EFFECT OF PSYCHOLOGICAL STRESS

In general, African-Americans show flatter morning-to-evening cortisol slopes than do Caucasians, and Hispanics exhibit significantly lower evening cortisol levels than do Caucasians. Greater psychosocial risk and less parental monitoring were associated with flatter cortisol slopes [37]. In a study of 98 subjects, Chong *et al.* [38] observed that whites had a 36% higher relative mean cortisol response to the TSST compared to blacks. Whites also showed higher mean ACTH levels compared to blacks 25 min after the start of the test. However, there was no difference in prolactin levels between whites and blacks. The authors concluded that whites may have a more robust HPA axis response to stress than that in blacks. Cohen *et al.* [39] reported that lower socioeconomic status (education and income) and being black were associated with higher evening levels of cortisols and flatter diurnal rhythm.

Experiencing racial discrimination is a major psychological stress that may cause higher blood pressure, increased obesity, cardiovascular reactivity, worse self-reported health, and even earlier morbidity. Tull *et al.* [40] hypothesized that chronic defeat response to such social stressors may increase the likelihood of HPA axis dysfunction, leading to dysregulation of cortisol, accumulation of abdominal fat, and glucose intolerance. African-Caribbean women who have experienced a high level of internalized racism are at higher risk of developing abdominal obesity and glucose intolerance. The authors demonstrated that a higher mean perceived stress score and a greater tendency to defeat response (restraint, denial, and behavioral disengagement) among African-Caribbean women (experiencing a higher level of internalized racism) were significantly correlated with dysregulation of cortisol [40]. Szanton *et al.* [41] reported that in a multivariable linear regression model, racial discrimination (based on self-reported racial discrimination) was significantly associated with increased red blood cell oxidative stress (measured by fluorescent heme degradation product). The authors used a biracial sample of 3271 subjects between the ages of 30 and 64 years. When stratified by race, perception of any discrimination was not associated with red blood cell oxidative stress in whites but there was a significant association regarding this parameter in African-Americans.

6.7 CONCLUSION

Stress is a part of life, and there are individual and gender differences in response to stress. It has been well-established that glucocorticoid hormones, especially cortisol, are secreted in response to stress stimuli that activate the HPA axis. The endocannabinoid system has also been implicated in the adoptive mechanism of stress. Glucocorticoid-induced fast feedback inhibition of HPA axis is mediated by a nongenomic signaling mechanism that involves endocannabinoid signaling at the level of the paraventricular nucleus of the hypothalamus [42]. Chronic stress is detrimental to human health because it results in glucocorticoid receptor resistance that results in failure to downregulate the inflammatory response, which plays an important role in the pathogenesis of many diseases [43]. It is important to cope with stress adequately by a adopting a healthy lifestyle (see Chapter 17).

REFERENCES

[1] Leeming RA. From the common cold to cancer: how evolution and the modern lifestyle appear to have contributed to such eventualities. Med Hypotheses 1994;43:145–50.

[2] Aleksandrovskii IA, Poiurovskii MV, Neznamov GG, Seredeniia SB, et al. Lipid peroxidation in emotional stress and neurotic disorders [in Russian]. Zh Nevropatol Psikhiatr Im S S Korsakova 1988;88:95–101.

[3] Sivonova M, Zitnanova I, Hlincikova L, Skodacek I, et al. Oxidative stress in university students during examination. Stress 2004;7:183–8.

[4] Hellhammer DH, Wust S, Kudielka BM. Salivary cortisol as a biomarker in stress research. Psychoneuroendocrinology 2009;43:163–71.

[5] Adinoff B, Iranmanesh A, Veldhuis J, Fischer L. Disturbances of the stress response: the role of HPA axis during alcohol withdrawal and abstinence. Alc Health Res World 1998;22:67–72.

[6] McIntosh LJ, Sapolsky RM. Glucocorticoids may enhance oxygen radical mediated neurotoxicity. Neurotoxicology 1996;17:873–82.

[7] Herbert J. Stress, the brain and mental illness. BMJ 1997;315:530–5.

[8] Ising M, Holsboer F. Genetics of stress response and stress related disorders. Dialogues Clin Neurol 2006;8:433–44.

[9] Dlugos A, Childs E, Stuhr KL, Hillard CJ, et al. Acute stress increases circulating anandamide and other *N*-acetyl ethanolamines in healthy humans. Neuropsychopharmacology 2012;37:2416–27.

[10] Lyoo IK, Han MH, Cho DY. A brain MRI study in subjects with borderline personality disorder. J Affect Disord 1998;50:235–43.

[11] Tarullo AR, Gunnar MR. Child maltreatment and the developing HPA axis. Horm Behav 2006;50:632–9.

[12] Pesonen AK, Raikkonen K, Feldt K, Heinonen K, et al. Childhood separation experience predicts HPA axis hormonal responses in late adulthood: a natural experiment of World War II. Psychoneuroendocrinology 2010;35:758–67.

[13] Schiavone S, Jaquet V, Trabace L, Krause KH. Severe life stress and oxidative stress in the brain: from animal models to human pathology. Antioxid Redox Signal 2013;18:1475–90.

[14] Epel ES, Blackburn EH, Lin J, Dhabhar D, et al. Accelerate telomere shortening in response to life stress. Proc Natl Acad Sci USA 2004;101:17312–15.

[15] Zafir A, Banu N. Modulation of *in vivo* oxidative status by exogenous corticosterone and restraint stress in rats. Stress 2009;12:167–77.

[16] Joergensen A, Broedbaek K, Weimann A, Semba R, et al. Association between urinary excretion of cortisol and markers of oxidatively damaged DNA and RNA in humans. PLoS One 2011;6:e20795.

[17] Amirkhanyan AA, Wolk DA. Parent care and the stress process: findings from panel data. J Gerontol B Psychol Sci Soc Sci 2006;61:S248–55.

[18] Smyth KA, Yarandi HN. A path model of type A and type B responses to coping and stress in employed black woman. Nurs Res 1992;41:260–5.

[19] Abbott AV, Snyder C, Vogel ME. Type A behavior and coronary heart disease. Am Fam Physician 1984;30:209–17.

[20] Sirri L, Fava GA, Guidi J, Porcelli P, et al. Type A behavior: a reappraisal of its characteristics in cardiovascular disease. Int J Clin Pract 2012;66:854–61.

[21] Mitaishvili N, Danelia M. Personality type and coronary heart disease. Georgian Med News 2006;134:58–60.

[22] Eskelinen M, Ollonen P. Assessment of cancer prone personality characteristics in healthy study objects and in patients with breast disease and breast cancer: a prospective case–control study in Finland. Anticancer Res 2011;31:4013–17.

[23] Sher L. Type D personality: the heart, stress and cortisol. QJM 2005;98:323–9.

[24] Kupper N, Girdon Y, Winter J, Denollet J. Association between type D personality, depression, and oxidative stress in patients with chronic heart failure. Psychosom Med 2009;71:973–80.

[25] Taylor SE, Klein LC, Lewis BP, Gruenewald TL, et al. Biobehavioral response to stress in females: tend and befriend, not fight or flight. Psychol Rev 2000;107:411–29.

[26] De Weerth C, Buitelaar JK. Physiological stress reactivity in human pregnancy—a review. Neurosci Biobehav Rev 2005;29:295–312.

[27] De Weerth C, van Hess Y, Buitelaar JK. Prenatal maternal cortisol levels and infant behavior during first 5 months. Early Hum Dev 2003;74:139–51.

[28] Baibazarova E, van de Beek C, Cohen-Kettenis PT, Buitelaar J, et al. Influence of prenatal maternal stress, maternal plasma cortisol and cortisol in the amniotic fluid on birth outcomes and child temperament at 3 months. Psychoneuroendocrinology 2013;38:907–15.

[29] Thomson M. The physiological roles of placental corticotropin releasing hormone in pregnancy and childbirth. J Physiol Biochem 2013;69:559–73.

[30] Lindheim SR, Legro RS, Bernstein L, Stanczyk FZ, et al. Behavioral stress responses in premenopausal and postmenopausal women and the effects of estrogen. Am J Obstet Gynecol 1992;167:1831–6.

[31] Glover EM, Jovanovic T, Mercer KB, Kerley K, et al. Estrogen levels are associated with extinction deficits in women with posttraumatic stress disorder. Biol Psychiatry 2012;72:19–24.

[32] McCarthy MM. Estrogen modulation of oxytocin and its relation to behavior. Adv Exp Med Biol 1995;395:235–45.

[33] Jezova D, Jurankova E, Mosnarova A, Kriska M, et al. Neuroendocrine responses during stress with relation to gender differences. Acat Neurobiol Exp (Warsaw) 1996;56:779–85.

[34] Childs E, Dlugos A, De Wit H. Cardiovascular, hormonal and emotional response to the TSST in relation to sex and menstrual cycle phase. Psychophysiology 2010;47:550–9.

[35] Kelly MM, Tyrka AR, Anderson GM, Price LH, et al. Sex differences in emotional and physiological responses to the Trier Social Stress Test. J Behav Ther Exp Psychiatry 2008;39:87–98.

[36] Irie M, Asami S, Nagata S, Miyata M, et al. Relationship between perceived workload, stress and oxidative DNA damage. Int Arch Occup Environ Health 2001;74:153–7.

[37] Martin CG, Bruce J, Fischer PA. Racial and ethnic differences in diurnal cortisol rhythms in preadolescents: the role of parental psychosocial risk and monitoring. Horm Behav 2012;61:661–8.

[38] Chong RY, Uhart M, McCaul ME, Johnson E, et al. Whites have a more robust hypothalamic pituitary adrenal axis response to a psychological stressor than blacks. Psychoneuroendocrinology 2008;33:246–54.

[39] Cohen S, Schwartz JE, Epel E, Kirschbaum C, et al. Socioeconomic status, race and diurnal cortisol decline in the Coronary Artery Risk Development in Young Adults (CARDIA) study. Psychosom Med 2006;68:41–50.

[40] Tull ES, Sheu YT, Butler C, Cornelious K. Relationship between perceived stress, coping behavior and cortisol secretion in women with high and low levels of internalized racism. J Natl Med Assoc 2005;97:206–12.

[41] Szanton SL, Rifkind JM, Mohanty JG, Miller 3rd ER, et al. Racial discrimination is associated with a measure of red blood cell oxidative stress: a potential pathway for racial health disparities. Int J Behav Med 2012;19:489–95.

[42] Evanson NK, Tasker JG, Hill MN, Hillard CJ, et al. Fast feedback inhibition of HPA axis by glucocorticoids is mediated by endocannabinoid signaling. Endocrinology 2010;151:4811–19.

[43] Cohen S, Janicki-Deverts D, Doyle WJ, Miller GE, et al. Chronic stress, glucocorticoid receptor resistance, inflammation and disease risk. Proc Natl Acad Sci USA 2012;109:5995–9.

CHAPTER 7

Oxidative Stress and Cardiovascular Diseases

CONTENTS

7.1 INTRODUCTION

Cardiovascular disease is a broad term that includes several diseases that affect the human heart and that can be broadly classified as diseases of the coronary arteries and diseases of the heart including congenital defect. In the United States, cardiovascular disease is the main cause of mortality, accounting for approximately 37% of all deaths. *Acute coronary syndrome* is a broad term that covers unstable angina, non-ST segment elevated myocardial infarction (NSTEMI), and ST segment elevated myocardial infarction (STEMI). Every year, several million individuals are admitted to the emergency department with symptoms suggestive of acute coronary syndrome, but only an estimated 1.4 million are admitted to the hospital with acute coronary syndrome and approximately 70% of these patients suffer from unstable angina or NSTEMI. In patients with NSTEMI, assessment of clinical symptoms and measurement of cardiac biomarkers are critical for proper diagnosis. Myocardial infarction is responsible for approximately 500,000 deaths per year in the United States [1].

The relation between plasma cholesterol and risk for atherosclerosis was extensively investigated in the Framingham Heart Study, which was initiated in 1948 in Framingham, Massachusetts. A total of 5209 men and women were enrolled, to study risk factors for heart diseases. The study was conducted under the direction of the National Heart Institute, which is now the National Heart, Lung, and Blood Institute. Several important guidelines regarding risk of cardiovascular diseases emerged from the Framingham Heart Study. In addition to lipid profile, there are other important risk factors for cardiovascular diseases, which are summarized in Box 7.1.

According to the World Health Organization (WHO), the majority of cardiovascular diseases can be prevented by risk factor modification and change in lifestyle. Approximately 70% of Americans are overweight, and

A. Dasgupta and K. Klein: Antioxidants in Food, Vitamins and Supplements. DOI: http://dx.doi.org/10.1016/B978-0-12-405872-9.00007-0

Box 7.1 RISK FACTORS FOR CARDIOVASCULAR DISEASES OTHER THAN ABNORMAL LIPID PROFILE

Unmodifiable Risk Factors

- Male sex
- Advanced age (male > 45 years, female > 55 years)
- Postmenopausal
- Family history (myocardial infarction or sudden death younger than 55 years of age in father or other male first-degree relative or younger than 65 years of age in mother or other first-degree female relative)
- Genetic factors (African-Americans, Mexican-Americans, Native American Indians, and people from the Indian subcontinent have a higher risk of heart diseases than Caucasians)

Modifiable Risk Factors

- Abnormal lipid profile (also can be genetic)
- Hypertension
- Diabetes
- Smoking
- Obesity (> 20% of ideal body weight)
- Physical inactivity
- Excessive use of alcohol (moderate drinking protects against cardiovascular diseases and stroke)
- Poor diet (no or few fruits and vegetables and high in carbohydrates)
- Excessive stress

fewer than 15% of children and adults exercise sufficiently. Among American adults, 11–13% have diabetes and 34% have hypertension, indicating the severity of the problem and risk of cardiovascular diseases in Americans [2]. Initial results of the Framingham study established the link between a high concentration of total cholesterol and a low concentration of high-density lipoprotein (HDL) cholesterol and risk for cardiovascular diseases, but elevated triglyceride was thought to play only a minor role in increasing the risk of cardiovascular diseases [3]. However, later reports observed the link between elevated triglycerides as a risk factor for heart diseases. Currently, treatment of lipid disorders is based on the third report of the expert panel of the National Cholesterol Education program, and a desirable total cholesterol level of less than 200 mg/dL has been universally accepted. Desirable and elevated lipid parameters related to risk of cardiovascular diseases are listed in Table 7.1. In the United States, approximately 16.3% of the population suffers from high cholesterol levels of 240 mg/dL or higher. This population has a cardiovascular risk factor twice as high as that of people with an optimal cholesterol level of 200 mg/dL or lower [4]. Another lipid parameter, lipoprotein(a), may also be used along with other lipid parameters to investigate risk of cardiovascular disease in a patient. A value greater than 30 mg/dL is undesirable. In addition to lipid profile, other nonlipid parameters, such as homocysteine and C-reactive protein, also indicate risk of cardiovascular diseases. In patients with suspected myocardial infarction, troponin I, troponin T, and total creatinine kinase and creatinine kinase isoenzyme (CK-MB) are used because these parameters are established cardiac markers.

Table 7.1 Lipid Profile and Risk of Cardiovascular Diseases

Analyte	Cholesterol Level (mg/dL)	Comment
Total cholesterol	<200	Desirable
	200–239	Borderline high
	>240	High
Low-density lipoprotein (LDL) cholesterol	<100	Optimal
	100–129	Near optimal
	130–159	Borderline high
	>160	Highly elevated
High-density lipoprotein (HDL) cholesterol	<40	Low
	≥60[a]	High (desirable)
Triglycerides	<150	Desirable
	150–199	Borderline high
	200–499	High
	>500	Very high

[a]*High HDL cholesterol is not a risk factor for cardiovascular disease; only low HDL cholesterol (<40 mg/dL) is a risk factor for both men and women.*

7.2 OXIDATIVE STRESS AND MECHANISM OF ATHEROSCLEROSIS

Involvement of both hyperlipidemia and oxidative stress in the pathogenesis of atherosclerosis is well-established [5]. Endothelial dysfunction that leads to atherosclerosis is associated with increased vessel wall thickness caused by fatty buildup (fatty streak) in areas inside the coronary artery wall. Increased vascular production of reactive oxygen species is responsible for oxidation of low-density lipoprotein (LDL), forming oxidized LDL (ox-LDL). Sources of reactive oxygen species include xanthine oxidase, lipoxygenase, nicotinamide adenine dinucleotide phosphate (NADPH) oxidase, and uncoupling of nitric oxide synthase [6]. It has been demonstrated that transport of ox-LDL across the endothelium into the arterial wall is necessary for the development of fatty streaks. Therefore, ox-LDL is believed to play a role in the initiation of the atherosclerotic process [7]. Moreover, ox-LDL affects other potentially atherogenic events, including cytotoxicity, increased expression of adhesion molecules with enhanced monocyte and leukocyte binding to endothelial cells, and foam cell formation. In addition, ox-LDL stimulates the release of interleukin-1 from endothelial cells and macrophages [8].

However, ox-LDL does not explain the entire complex mechanism of involvement of oxidative stress in the pathogenesis of atherosclerosis. A mechanism involving both ox-LDL and nitric oxide (NO) has also been proposed because endothelial dysfunction is characterized by loss of NO bioactivity [9]. NO promotes vasodilation and inhibits the effect of cytokines and the adhesion molecule-related atherosclerotic process. Oxygen free radicals may directly reduce the bioavailability of NO because superoxide reacts with NO forming peroxynitrite, another reactive species. Reactive oxygen species and peroxynitrite can oxidize tetrahydrobiopterin, a critical co-factor for endothelial NO synthase [10]. There is increasing clinical evidence that phospholipid oxidation products (ox-PLs) also play a role in the pathogenesis of atherosclerosis because ox-PLs interact with endothelial cells, monocytes/macrophages, platelets, smooth muscle cells, and HDL to promote atherogenesis. Current research indicates that the oxidized fatty acid moiety of ox-PL can protrude in the aqueous phase and can interact with receptors. However, the effects of ox-PL are also mediated by gene regulation and transcription. Although ox-PL affects multiple genes and pathways, some of which are proatherogenic and some are protective, concentrations at which ox-PL is present in vessel wall of atherogenic lesions, its effect is mostly to promote atherogenesis [11]. Oxidative stress can also directly activate genes responsible for inflammation of blood vessels, including coronary arteries. Oxidative stress may also directly alter the signaling mechanism of cells on arterial wall that regulate proper function of the arteries. Interestingly, estrogen protects against such harmful effects of oxidative stress and protects women from myocardial infarction before the menopause [12]. However, heart attack may also occur in women before the menopause due to many other factors. microRNAs have emerged as important regulators of gene expression that post-transcriptionally modify cellular function. Studies have indicated that there may a direct link between altered microRNA expression and the pathogenesis of atherosclerosis [13]. Atherosclerosis, the main cause of cardiovascular diseases, is a chronic inflammatory condition with immunocompetent cells in lesions that predominantly produce proinflammatory cytokines. In addition, dead cells and ox-LDL are abundant in atherosclerotic plaques, and rupture of plaques causes the disease. It has been established that ox-LDL has proinflammatory and immunostimulatory properties and is capable of causing cell death at a higher rate than normal [14].

7.3 ELEVATED LEVELS OF MARKERS OF OXIDATIVE STRESS IN PATIENTS WITH CARDIOVASCULAR DISEASES

Many investigators have demonstrated elevated levels of various markers of oxidative stress in patients with cardiovascular diseases. In a study including 38 patients with acute myocardial infarction, Levy *et al.* [15] demonstrated that all vitamins, including antioxidant vitamins, were lower in these patients than in controls. The authors observed a 33% decrease in vitamin A concentration, a 10% decrease in vitamin E concentration, and a 37% decrease in carotenoid level in patients compared to controls. In addition, oxidation of plasma using a water-soluble azo compound resulted in a 165% elevation in the concentration of malondialdehyde (as measured by thiobarbituric acid reactive substances) and 81% elevation in the concentration of conjugated dienes in patients with acute myocardial infarction compared to controls. Increased lipid peroxidation also occurred in patients who underwent successful thrombolytic therapy. In a study of 34 patients with myocardial infarction, 19 with unstable angina, and 40 healthy controls, Loeper *et al.* [16] observed that malondialdehyde concentrations were elevated in patients with myocardial infarction during the initial 48 h after an attack, and the magnitude of the increase correlated with the severity of the illness. Although after 48 h malondialdehyde concentrations were decreased, levels were still higher than those in controls even after 12 days. The activity of superoxide dismutase, an important antioxidant enzyme, was also reduced during the initial 48 h after myocardial infarction. Similar but less remarkable changes were observed in patients suffering from unstable angina. Weitz *et al.* [17] demonstrated an elevated concentration of pentane (a direct index of lipid peroxidation) in patients with acute myocardial infarction, indicating increased oxidative stress experienced by these patients. Using rat and rabbit models, Baker *et al.* [18] observed free radical generation during myocardial ischemia and reperfusion using electron spin spectroscopy.

Kotur-Stevuljevic *et al.* [19] measured oxidative stress status parameters, inflammation markers, and lipid status parameters in 188 coronary heart disease patients and 197 cardiovascular disease-free age-matched controls. The authors observed that plasma malondialdehyde (measured by thiobarbituric acid reactive substances) and superoxide anion (measured by the rate of nitroblue tetrazolium reduction) concentrations were significantly higher in patients with cardiovascular diseases compared to controls. For example, mean malondialdehyde level was 3.43 μmol/L in patients with coronary heart diseases and 2.91 μmol/L in controls. Concentrations of plasma superoxide dismutase were significantly reduced in patients compared to controls, indicating that patients experienced significantly increased oxidative stress compared to controls. Using multiple stepwise regression analysis, the authors showed that both fibrinogen and high sensitivity C-reactive protein independently correlated with plasma malondialdehyde levels in patients with various cardiovascular diseases. The relationship between oxidative stress parameters and inflammatory markers indicates strong mutual involvement of both factors in the pathogenesis of atherosclerosis.

Increased oxidative stress is also produced during reperfusion of the heart after myocardial infarction due to concurrent rapid release of free radicals.

In a study of 45 patients with acute myocardial infarction and 24 healthy controls, Simic *et al.* [20] observed that at the time of admission, the activities of antioxidant enzymes superoxide dismutase and glutathione peroxidase in erythrocytes were significantly decreased compared to those in controls. In patients with successful reperfusion, 90 min after initiation of thrombolytic therapy, elevated levels of both antioxidant enzymes were observed; however, after 90 min, levels of both enzymes started to decline, with superoxide dismutase reaching its lowest level on the second day and glutathione peroxidase reaching its lowest level on the first day after acute myocardial infarction. After that period, levels of both enzymes in erythrocytes gradually increased (between days 3 and 8) and finally reached levels observed in patients treated with streptokinase. In contrast, antioxidant enzyme levels remained low during a period of 7 days in patients without successful reperfusion. Plasma levels of malondialdehyde coincided with the time course of antioxidant enzymes observed in the study. The authors concluded that prolonged ischemic heart disease reduces antioxidant enzyme levels due to increased oxidative stress. In contrast, in patients with successful preperfusion, antioxidant enzymes levels returned to normal.

Reintroduction of oxygen during reperfusion salvage of myocardium that produces reactive oxygen species can damage not only lipids and proteins but also DNA. In a study of 28 patients, Himmetoglu *et al.* [21] observed that the plasma levels of 8-hydroxydeoxyguanosine (a marker of oxidative damage of DNA) were significantly elevated 3 h after initiation of therapy but subsequently started to decline on the third day and on the day patients were discharged. However, levels were still higher at the time of discharge compared to those of controls. Interestingly, the level of 8-hydroxydeoxyguanosine was slightly decreased in patients treated with thrombolytic agent but there was a more significant decrease in patients treated with primary percutaneous transluminal coronary angioplasty. In a study involving 15 patients with STEMI and 10 patients with stable ischemic heart disease, the authors observed that patients with STEMI had significantly higher concentrations of malondialdehyde and 4-hydroxy-2-nonenal (another marker of lipid peroxidation) before reperfusion compared to patients with stable ischemic heart disease [22]. After reperfusion, concentrations of 4-hydroxy-2-nonenal in erythrocytes in patients with STEMI were higher than those in stable ischemic heart disease patients. At the same time, the activity of superoxide dismutase significantly decreased in patients with impaired tissue perfusion (myocardial blush grade <2), indicating that diminished antioxidant defense after reperfusion was associated with impaired myocardial perfusion.

In chronic heart failure, the heart shows abnormalities in both structure and function that impair its ability to fill with or eject blood and/or it has reduced capacity to meet the necessary cardiac output. Heart failure may be related to an inherited abnormality or other factors, such as chronically increased cardiac workload or loss of muscle mass due to myocardial infarction. Cardiac stressors that may demand higher cardiac workload include cardiomyopathy, hypertrophy, and abnormalities in electrical function, calcium handling, cell variability, and other factors. Heart failure and many other processes that predispose to heart failure are associated with increased oxidative stress. Furthermore, oxidative stress is increased in the failing heart, which contributes to the pathogenesis of myocardial remodeling that eventually results in heart failure [23]. In one study, the authors measured markers of oxidative stress and antioxidant enzyme activities in 120 patients with chronic heart failure and 69 healthy controls [24]. The authors observed that lipid peroxidation products as measured by plasma malondialdehyde concentration and oxidation of plasma protein (measured as protein thiol groups and reactive carbonyl derivatives) were increased in patients with chronic heart failure compared to controls. In addition, the activity of the antioxidant enzyme glutathione peroxidase was significantly lower in patients with congestive heart failure compared to controls, indicating that patients with ischemic heart failure are in a state of oxidative stress and carbonyl stress. Interestingly, plasma malondialdehyde was an independent predictor of mortality in these patients. The authors observed that a plasma malondialdehyde level greater than 8.0 μmol/L was associated with an eightfold higher mortality risk. Mondal *et al.* [25] studied 10 heart failure patients supported by continuous-flow left ventricular assist devices (LVADs) and 10 age- and sex-matched normal controls. They observed that patients supported by LVADs showed significantly higher levels of reactive oxygen species in blood leukocytes (measured by flow cytometry) and higher levels of ox-LDL compared to healthy controls. In addition, superoxide dismutase activities decreased with increasing postoperative period. Moreover, there was a significantly higher number of γ-H2AX (a member of the histone protein family 2A, member X) foci in lymphocytes, indicating DNA damage (DNA double-strand breaks induce phosphorylation of H2AX at serine 139, known as γ-H2AX), in heart failure patients. In addition, various markers of DNA repair (expression of DNA ligase IV, Ku70, and Ku80 proteins quantitated in blood lymphocytes by immunocytochemistry) were also high after 1 week of intervention. Another marker of DNA repair, Mre11(meiotic recombination 11 homolog A, a repair protein for DNA with double-strand break), was elevated after 3 months of LVAD transplantation, indicating abnormal DNA repair. The authors concluded that continuous-flow LVAD-implant patients not only demonstrated elevated oxidative stress and DNA damage in blood leukocytes but also showed abnormalities in DNA repair pathways.

7.4 EXERCISE, YOGA, OXIDATIVE STRESS, AND CARDIOVASCULAR DISEASES

The value of exercise and yoga in reducing the risk of cardiovascular diseases has been studied. The American College of Cardiology/American Heart Association recommends at least 30 min of moderate (at 50–70% of maximal predicted heart rate) exercise on most days to reduce the risk of cardiovascular events [26]. Studies of human subjects indicate that regular exercise can improve cardiovascular function not only in healthy young subjects but also in older people and people with underlying risk factors such as patients with metabolic syndrome, hypertension, type 2 diabetes, stable cardiovascular diseases, and congestive heart failure, and survivors of myocardial infarction. Although acute exercise increases oxidative stress, mild oxidative stress can act as a stimulus of the physiological antioxidant system and triggers various physiological adaptations. Regular physical exercise, in contrast to acute exercise, delays oxidative stress-induced cell damage by improving the antioxidant protective mechanism in the myocardium. This cardioprotective mechanism of exercise is evidenced by increased myocardial manganese–superoxide dismutase activity. Aerobic exercise also reduces ventricular epicardial fat thickness. Exercise also increases the expression of cardiac heat shock proteins that can combat oxidative stress. In addition, endoplasmic reticulum stress protein levels are increased, and these proteins are known to have a cardioprotective role under stress. Moreover, regular exercise can also result in cardiac mitochondrial adaptations so that mitochondria generate fewer reactive oxygen species. Modulation of the sarcolemmal and mitochondrial potassium channel by regular exercise also provides further cardioprotection. Exercise-induced upregulation of endothelial nitric oxide synthase (eNOS) expression results in increased nitric oxide synthesis, which also provides cardioprotection [27]. It has been postulated that exercise augments glutathione replenishment and protects myocardium from electromechanical dysfunction and cell death. Adaptive cardioprotective signaling is triggered by reactive oxygen species from NADPH oxidase, and it leads to improved glutathione replenishment through redox-dependent modification in glutathione reductase [28].

Many clinical studies have shown benefits of exercise in reducing not only the risk of cardiovascular events but also all-cause mortality (Box 7.2). Whereas most studies have shown a 20–35% reduction in relative risk of death in both men and women who exercise regularly, Myers *et al.* [29] reported that exercise reduces all-cause mortality and death from cardiovascular diseases by up to 50%. Regular exercise increases plasma antioxidant defense and decreases platelet sensitivity and aggregability, thus lowering the risk of adverse thrombotic events. Regular exercise increases the plasma nitric

Box 7.2 BENEFITS OF EXERCISE TO REDUCE RISK OF CARDIOVASCULAR DISEASE

- Boosts antioxidant levels
- Reduces inflammation by decreasing epicardial fat
- Increases expression of cardiac heat shock proteins
- Increases vascular expression of eNOS and increases nitric oxide production
- Exercise-induced blood vessel remodeling
- Increases expression of endoplasmic reticulum stress proteins that provide cardioprotection
- Exercise-induced mitochondrial adoption to produce less reactive oxygen species
- Modulation of sarcolemmal and mitochondrial potassium channels
- Produces a short-term proinflammatory response followed by long-term anti-inflammatory effect
- Hematological benefits

oxide level, which may be linked to diminished platelet activity. Di Massimo *et al.* [30] studied the effect of 20 weeks of exercise in sedentary individuals on plasma antioxidant defense and platelet responsiveness. They observed that regular exercise significantly reduced platelet aggregation and increased total plasma antioxidant capacity as well as decreased lipid peroxidation products in plasma measured as thiobarbituric acid reactive substances. Plasma superoxide dismutase activity was also increased after 20 weeks of exercise, further demonstrating better antioxidant defense of plasma against oxidative stress. In addition, an increase in lag time in the oxidation of LDL was observed. The level of HDL cholesterol that provides cardioprotection was increased and correlated with both total antioxidant capacity of plasma and nitrite/nitrate content of both plasma and platelet cytosol. These findings indicate that regular moderate exercise leads to better antioxidant defense of the body against oxidative stress as well as inhibition of oxidation of LDL.

Mind–body therapy that promotes spiritual well-being may be associated with improved regression of coronary artery plaques, leading to decreased risk of cardiovascular diseases. Yoga may be helpful in reducing the risk of cardiovascular diseases. Manchanda *et al.* [31] performed a randomized prospective controlled trial involving 42 men with angiographically proven coronary artery disease. They divided patients into two groups: One group of patients received standard care (control group), whereas the other group received a lifestyle modification program involving yoga. At the end of 1 year, patients in the yoga group showed a significant reduction in the number of anginal episodes per week, improved exercise capacity, and decreased body weight compared to patients in the control group. In addition, patients in the yoga group showed a greater reduction of serum total cholesterol, LDL cholesterol, and triglyceride levels compared to patients in the control group. Coronary angiograms repeated after 1 year showed that patients in the yoga group had significantly more lesions regressed compared to the control group (20% in the yoga group vs. 2% in the control group), and fewer lesions progressed (5% in the yoga group vs. 37% in the control group). Interestingly, only 1 patient in the yoga group required revascularization procedure (coronary angioplasty or bypass) during the study period, whereas 8 patients in the control group required such intervention. Yogendra *et al.* [32] reported similar findings in a study that included 71 patients in a yoga study group and 42 patients in a standard care group in which all patients had angiographically proven coronary artery diseases. The authors reported that after 1 year, serum cholesterol was reduced by 23.3% in the yoga group versus 4.4% in the control group, and LDL cholesterol reduction was 26% in the yoga group versus 2.6% in the control group. Regression of disease was also more significant in the yoga group than in the control group.

Lakkireddy *et al.* [33] observed that in patients with paroxysmal atrial fibrillation, the practice of yoga had many beneficial effects, including improved heart rate, blood pressure, and reduced symptomatic atrial fibrillation episodes. In addition, yoga reduced depression and anxiety and improved the quality of life in these patients. The beneficial effects of yoga in relation to cardiovascular diseases may be related to reducing oxidative stress in these patients. In a study of 104 subjects, Yadav *et al.* [34] observed that the practice of yoga reduced oxidative stress in the subjects as evidenced by a significant reduction in lipid peroxidation products in blood (measured as thiobarbituric acid reactive substances) after yoga intervention. However, Cheema *et al.* [35] observed no improvement in heart rate variability or other related parameters in 37 adults employed in university-based office positions and who participated in a 10-week worksite-based yoga program.

7.5 DIET, OXIDATIVE STRESS, AND CARDIOVASCULAR DISEASES

Results from observational studies suggest that a high fruit and vegetable intake reduces the risk of coronary heart disease, and in 2003 WHO recommended a daily consumption of 400–500 g of fruits and vegetables. Eating fruits and vegetables that can provide enough antioxidant defenses is very useful in lowering the risk of cardiovascular diseases. In one large study involving 313,074 men and women without previous history of ischemic heart disease, the authors observed that people who consumed eight servings of fruits and vegetables per day (a total of 640 g of fruits and vegetables per day) had a 22% lower risk of fatal ischemic heart disease than people who consumed less than three servings of fruits and vegetables per day (<240 g of fruits and vegetables) [36]. Another large study involving participants from 52 countries also concluded that high intake of fruits and vegetables can reduce the risk of acute myocardial infarction. Whereas an Oriental diet (high intake of tofu, soy, and other sauces) does not increase the risk of acute myocardial infarction, a Western diet, which is rich in fried foods and salty snacks, increases the risk of acute myocardial infarction because fried foods are high in calories and salt. Moreover, fried foods may contain higher amounts of saturated fatty acids that may subsequently increase the risk of cardiovascular disease. However, a prudent diet (high in fruits and vegetables) reduces the risk. The authors also commented that unhealthy dietary intake increases the risk of acute myocardial infarction globally and accounts for approximately 30% of the population-attributable risk [37]. Studies suggest that fruits containing relatively high concentrations of antioxidant flavonols, anthocyanins, and procyanidins, such as pomegranate, purple grapes, and berries, can lower the risk of cardiovascular diseases, particularly due to their anti-hypertensive effect, inhibition of platelet aggregation, and increased endothelial vasodilation [38].

Eating fish at least twice a week can reduce the risk of death from sudden acute myocardial infarction by up to 50%. Fish is a rich source of omega-3 fatty acids, which have many beneficial effects, including reducing the risk of heart attack and preventing clot formation [39]. However, consuming large amounts of fish may have adverse effects due to environmental contaminants such as organic and inorganic mercury (particularly methylmercury), dioxins, and other compounds that may also be present in fish, especially predatory fish. Pregnant women, women planning to become pregnant, and children should not eat predator fish, including shark, swordfish, king mackerel, golden bass, and golden snapper, in which high amounts of mercury may accumulate. Other people can eat up to 7 ounces of these fish per week. Salmon, cod, flounder, catfish, and other seafood such as crabs and scallops may also contain mercury but in much lower amounts.

Mediterranean diet is beneficial for reducing the risk of cardiovascular diseases. Mediterranean diet uses olive oil as the principal source of fat, and it uses an abundance of plant foods (vegetables, fresh and dried fruits, whole grain cereals, and nuts and legumes). It includes low to moderate amounts of fish, eggs, cheese, yogurt, and poultry and very low amounts of red meat, saturated fats, and margarines, as well as moderate consumption of wine. Mediterranean diet is also rich in α-linolenic acid (ALA). The Lyon diet study involving patients who survived first myocardial infarction showed that Mediterranean diet can provide a striking protective effect in these patients, as evidenced by a 40–70% reduction in the risk of recurrence after a 4-year follow-up. ALA may also have antiarrhythmic properties [40]. In addition to reducing the risk of cardiovascular diseases, Mediterranean diet also reduces the risk of cancer and Parkinson's and Alzheimer's disease—all of which are linked to oxidative stress. The mechanism of protection from these diseases in the Mediterranean diet may be partly related to the antioxidative effect of such diet [41]. Moderate consumption of alcohol, especially red wine, also has a cardioprotective effect because red wine is rich in antioxidants.

Anxiety can produce oxidative stress and increases the concentration of cortisol, which may increase the risk of various diseases including cardiovascular diseases. Depression is also an independent predictor of mortality and cardiac morbidity in patients with coronary heart disease [42]. Depression can also induce oxidative stress. Therefore, depression must be treated to reduce the risk of coronary heart disease and heart attack. In one report, the authors demonstrated the usefulness of selective serotonin reuptake inhibitors in treating depressed patients who also had coronary artery disease [43]. Emotionally stressful events, particularly those that cause anger, are capable of triggering the onset of acute myocardial infarction [44].

7.6 DIETARY SUPPLEMENTS AND CARDIOVASCULAR DISEASES

Omega-3 polyunsaturated fatty acids have beneficial effects in reducing the risk of cardiovascular diseases (Box 7.3). The human body cannot synthesize these fatty acids; they must be obtained from diet. Nutritionally important omega-3 polyunsaturated fatty acids include ALA (18-carbon chain), eicosapentaenoic acid (EPA; 20-carbon chain), and docosahexaenoic acid (DHA; 22-carbon chain). ALA is found in vegetable oils, whereas a major source of DHA and EPA is fish oil. However, ALA is less effective at reducing cardiovascular events compared to long-chain omega-3 polyunsaturated fatty acids such as DHA and EPA. Fish oil supplementation has a cardioprotective effect and may lower the incidence of heart failure. The beneficial effects of fish oil include inhibition of thromboxane production, increased production of prostacyclin, increased fibrinolytic activity of plasma, modification of leukotriene and cytokine production, reduction in vasospastic response to catecholamines, reduction in blood viscosity, and decreased platelet activity factor and platelet-derived growth factor, as well as reduced oxygen radical generation.

Box 7.3 DIETARY SUPPLEMENTS AND FUNCTIONAL FOODS THAT MAY HAVE SOME CARDIOPROTECTIVE EFFECT

- Omega-3 fatty acid (fish oil supplement)
- Coenzyme Q10
- Carnitine
- Garlic
- Hawthorn
- Resveratrol (abundant in red wine)

Therefore, supplementation with omega-3 polyunsaturated fatty acids provides overall health benefits with minimal risk [45].

Coenzyme Q10 (CoQ10) is a naturally occurring substance that is synthesized from tyrosine in the human body and also found in the diet (fish, organ meat (offal), germs of whole grain, etc.). CoQ10 is involved in the mitochondrial transport chain and plays a critical role in energy production. Low serum CoQ10 may increase the risk of heart failure, and oral CoQ10 supplement improves left ventricular contractility in patients with congestive heart failure without any adverse effects. An arrhythmia-suppressing effect of CoQ10 has also been shown. In one clinical trial using 144 patients with acute myocardial infarction, 2320 mg per day of CoQ10 for 28 days reduced angina and total arrhythmia, and it improved vascular function [46]. In a New York Heart Association study, the authors observed that CoQ10 significantly improved the clinical outcome in patients with stage III or IV heart failure [47]. CoQ10 likely exerts its beneficial effects by inhibiting LDL oxidation and improving mitochondrial function in the myocardium. L-Carnitine is a naturally occurring substance that is synthesized from tyrosine and involved in mitochondrial energy production. L-Carnitine is a positive inotropic agent, and supplementation with propionyl-L-carnitine is beneficial in the prevention and treatment of ischemic heart disease, congestive heart failure, hypertrophic heart disease, and peripheral arterial disease. Garlic supplementation may lower the risk of cardiovascular disease. Hawthorn, a herbal supplement, may improve exercise tolerance, shortness of breath, and fatigue. Resveratrol, a component of wine, may favorably alter the progression of atherosclerosis. Nuts are rich in antioxidant polyphenols and are beneficial for health; there is an inverse relationship between nut intake and cholesterol level [48].

Vitamin E is an excellent antioxidant, but Chung [49] commented that clinical trials on the ability of vitamin E to prevent cardiovascular disease provided disappointing results because it did not reduce mortality. However, consumption of food rich in vitamin E has been associated with lower risk of coronary heart disease in middle-aged to older men and women. The American Heart Association does not support the use of vitamin E supplementation to prevent cardiovascular disease but recommends the consumption of foods abundant in antioxidant vitamins and other nutrients [50].

7.7 CONCLUSION

Cardiovascular diseases are caused by a complex interaction between gene and environment, but the role of oxidative stress in increasing the risk of these diseases has been well-established. Oxidation of LDL plays an important role in plaque formation, and reducing oxidative stress may be beneficial in lowering the risk of cardiovascular disease. Evidence indicates that a diet rich in fruits and vegetables, which are rich in antioxidants, is beneficial in lowering the risk of cardiovascular disease. Functional foods such as nuts may also have a beneficial effect. Moderate consumption of alcohol, especially red wine, may be beneficial in lowering the risk of cardiovascular disease and other illness. Lifestyle modifications that include regular exercise and yoga also have beneficial effects.

REFERENCES

[1] Movahed MR, John J, Hashemzadeh M, Hashemzadeh M. Mortality trends in non-ST-segment elevation myocardial infarction (NSTEMI) in the United States from 1998 to 2004. Clin Cardiol 2011;34:689–92.

[2] Kones R. Primary prevention of coronary heart disease: integration of new data, evolving views, revised goals, and role of rosuvastatin in management: a comprehensive survey. Drug Des Dev Ther 2011;5:325–80.

[3] Gordon T, Kannel WB, Castelli WP, Dawber TR. Lipoproteins, cardiovascular disease and death: the Framingham study. Arch Intern Med 1981;141:1128–31.

[4] Vijayakrishnan R, Kalyatanda G, Srinivasan I, Abraham GM. Compliance with the adult treatment panel III guidelines for hyperlipidemia in a resident run ambulatory clinic: a retrospective study. J Clin Lipidol 2013;7:43–7.

[5] Peluso I, Morabito G, Urban L, Loannone F, et al. Oxidative stress in atherosclerosis development: the central role of LDL and oxidative burst. Endocr Metab Immune Disord Drug Targets 2012;12:351–60.

[6] Victor VM, Apostolova N, Herance R, Hernadez-Mijares A, et al. Oxidative stress and mitochondrial dysfunction in atherosclerosis: mitochondria-targeted antioxidants as potential therapy. Curr Med Chem 2009;16:4654–67.

[7] Yoshida H. Front line of oxidized lipoproteins: role of oxidized lipoproteins in atherogenic and cardiovascular disease risk. Rinsho Byori 2010;58:622–30.

[8] Thomas CE, Jackson RL, Ohlweiler DF, Ku G. Multiple lipid oxidation products in low density lipoprotein induce interleukin-1 beta release from human blood mononuclear cells. J Lipid Res 1994;35:417–27.

[9] Victor VM, Rocha M, Sola E, Banuls C, et al. Oxidative stress endothelial dysfunction and atherosclerosis. Curr Pharm Des 2009;15:2988–3002.

[10] Landmesser U, Harrison DG. Oxidant stress as a marker for cardiovascular events: Ox marks the spot. Circulation 2001;104:2638–40.

[11] Lee S, Birukov KG, Romanoski CE, Springstead JR, et al. Role of phospholipid oxidation products in atherosclerosis. Circ Res 2012;111:778–99.

[12] Kondo T, Hirose M, Kageyama K. Roles of oxidative stress and redox regulation in atherosclerosis. J Atheroscler Thromb 2009;16:523–38.

[13] Zampetaki A, Dudek K, Mayr M. Oxidative stress in atherosclerosis: the role of microRNAs in arterial remodelling. Free Rad Biol Med 2013;64:69–77.

[14] Frostegard J. Immunity, atherosclerosis and cardiovascular disease. BMC Med 2013;11:117.

[15] Levy Y, Bartha P, Ben-Amotz A, Brook G, et al. Plasma antioxidants and lipid peroxidation in acute myocardial infarction and thrombolysis. J Am Coll Nutr 1998;17:337–41.

[16] Loeper J, Goy J, Rozensztajn L, Bedu O, et al. Lipid peroxidation and protective enzymes during myocardial infarction. Clin Chim Acta 1991;196:119–25.

[17] Weitz ZW, Brinbaum AJ, Sobotka PA, Zarling EJ, et al. High breath pentane concentrations during acute myocardial infarction. Lancet 1991;337:933–5.

[18] Baker JE, Felix CC, Olinger GN, Ksalyanaraman B. Myocardial ischemic and reperfusion: direct evidence for free radical generation by electron spin resonance spectroscopy. Proc Natl Acad Sci USA 1988;85:2786–9.

[19] Kotur-Stevuljevic J, Memon L, Stefanovic A, Spasic S, et al. Correlation of oxidative stress parameters and inflammatory markers in coronary artery disease patients. Clin Biochem 2007;40:181–7.

[20] Simic D, Mimic-Oka J, Pljesa M, Milanovic D, et al. Time course of erythrocyte antioxidant activity in patients treated by thrombolysis for acute myocardial infarction. Jpn Heart J 2003;44:823–32.

[21] Himmetoglu S, Dincer Y, Bozcali E, Ali Vural V, et al. Oxidative DNA damage and antioxidant defense after reperfusion in acute myocardial infarction. J Investig Med 2009;57:595–9.

[22] Kaminski K, Bonda T, Wojtkowska I, Dobrzycki S, et al. Oxidative stress and antioxidative defense parameters early after reperfusion therapy for acute myocardial infarction. Acute Card Care 2008;10:121–6.

[23] Hafstad AD, Nabeebaccus AA, Shah AM. Novel aspect of ROS signalling in heart failure. Basic Res Cardiol 2013;108:359.

[24] Radovanovic S, Savic-Radojevic A, Pljesa-Ercegovac M, Djukic T, et al. Markers of oxidative damage and antioxidant enzyme activities as predictors of morbidity and mortality in patients with heart failure. J Card Fail 2012;18:493–501.

[25] Mondal NK, Sorensen E, Hivala N, Feller E, et al. Oxidative stress, DNA damage and repair in heart failure patients after implantation of continuous flow left ventricular assist devices. Int J Med Sci 2013;10:883–93.

[26] Third Report of the National Cholesterol Education Program (NCEP). Expert panel on detection, evaluation and treatment of high blood cholesterol in adults (Adult Treatment Panel III) final report. Circulation 2002;106:3121–43.

[27] Golbidi S, Laher I. Exercise and the cardiovascular system. Cardiol Res Pract 2012; 2012:210852.

[28] Frasier CR, Moukdar F, Patel H, Sloan RC, et al. Redox-dependent increase in glutathione reductase and exercise preconditioning: role of NADPH oxidase and mitochondria. Cardiovasc Res 2013;98:47–55.

[29] Myers J, Kaykha A, George S, Abella J, et al. Fitness versus physical activity pattern in predicting mortality in men. Am J Med 2004;117:912–18.

[30] Di Massimo C, Scarpelli P, Penco M, Tozzi-Cinacarelli MG. Possible involvement of plasma antioxidant defenses in training associated decrease of platelet responsiveness in humans. Eur J Appl Physiol 2004;91:406–12.

[31] Manchanda SC, Narang R, Reddy KS, Sachdeva U, et al. Retardation of coronary atherosclerosis with yoga lifestyle intervention. J Assoc Physicians India 2000;48:687–94.

[32] Yogendra J, Yogendra HJ, Ambardekar S, Lele RD, et al. Beneficial effects of yoga lifestyle on reversibility of ischemic heart disease: Caring Heart Project of International Board of Yoga. J Assoc Physicians India 2004;52:283–9.

[33] Lakkireddy D, Atkins D, Pillarisetti J, Ryschon K, et al. Effect of yoga on arrhythmia burden, anxiety, depression and quality of life in paroxysmal atrial fibrillation: the YOGA group My Heart Study. J Am Coll Cardiol 2013;61:1177–82.

[34] Yadav RK, Ray BB, Vempati R, Bijlani RL. Effect of a comprehensive yoga-based lifestyle modification program on lipid peroxidation. Indian J Physiol Pharmacol 2005;49:358–62.

[35] Cheema BS, Houridis A, Busche L, Raschke-Cheema V, et al. Effect of office worksite-based yoga program on heart variability outcome of a randomized controlled trial. BMC Complement Alt Med 2013;13:82.

[36] Crowe FL, Roddam AW, Key TJ, Appleby PN, et al. Fruits and vegetable intake and mortality from ischemic heart disease: results from the European Prospective Investigation into Cancer and Nutrition (EPIC) heart study. Eur Heart J 2011;32:1235–43.

[37] Iqbal R, Anand S, Ounpuu S, Islam S, et al. Dietary patterns and the risk of acute myocardial infarction in 52 countries: results of the INTERHEART study. Circulation 2008; 118:1929–37.

[38] Chong M, Macdonald R, Lovegrove JA. Fruit polyphenols and CVD risk: a review of human intervention studies. Br J Nutr 2010;4:S28–39.

[39] Kris-Etherton PM, Harris WS, Apple LJ. Fish consumption, fish oil, omega-3-fatty acids and cardiovascular disease. Circulation 2002;106:2747–57.

[40] De Lorgeril M, Salen P. The Mediterranean-style diet for the prevention of cardiovascular diseases. Public Health Nutr 2006;9:118–23.

[41] Hadziabdic MO, Bozikov V, Pavic E, Romic Z. The antioxidative protecting role of the Mediterranean diet. Coll Antropol 2012;36:1427–34.

[42] Freedland KE, Carney RM. Depression as a risk factor for adverse outcomes in coronary heart disease. BMC Med 2013;11:131.

[43] Paraskevaidis I, Palios J, Parissis J, Filippatos G, et al. Treating depression in coronary artery disease and chronic heart failure: what's new in using selective serotonin re-uptake inhibitors? Cardiovasc Hematol Agents Med Chem 2012;10:109–15.

[44] Mittleman MA, Maclure M, Sherwood JB, Multry RP, et al. Triggering of acute myocardial infarction onset by episodes of anger: Determinants of Myocardial Infarction Onset Study Investigators. Circulation 1995;92:1720–5.

[45] Jarreau RK, Lee JH, Lavie CJ. Should we start prescribing omega-3-polyunsaturated fatty acids in chronic heart failure? Curr Heart Fail Rep 2012;9:8–13.

[46] Singh RB, Wander GS, Rastogi A, Shukla PA, et al. Randomized double blind placebo controlled trial of coenzyme Q10 in patients with acute myocardial infarction. Cardiovasc Drugs Ther 1998;12:347–53.

[47] Pepe S, Marasco SF, Haas SJ, Sheeran FL, et al. Coenzyme Q10 in cardiovascular disease. Mitochondrion 2007;7(Suppl):S154–67.

[48] Mullin GE. The heart speaks II: embracing integrative medicine for heart health. Nutr Clin Pract 2012;27:568–71.

[49] Chung MK. Vitamins, supplements, herbal medicines and arrhythmias. Cardiol Rev 2004;12:73–84.

[50] Saremi A, Arora R. Vitamin E and cardiovascular disease. Am J Ther 2010;17:e56–65.

CHAPTER 8

Oxidative Stress and Cancer

CONTENTS

8.1 INTRODUCTION

The impact of oxidative stress has been recognized since the 1950s [1]. According to Benz and Yau [2], oxidative stress has been implicated in the pathophysiology of many diseases, including cancer. Cancer is the second most common cause of mortality in the United States after cardiovascular diseases, including myocardial infarction. According to the American Cancer Society, approximately 1,660,290 new cases of cancer were expected to be diagnosed in 2013. Moreover, an estimated 580,350 Americans were projected to die of cancer. Fortunately, the death rate from all types of cancer, including lung, colorectum, breast, and prostate cancer—the four most common types of cancer in the United States— is on the decline. However, obesity and lack of sufficient physical activity may contribute to an increased rate of death due to certain types of cancer, such as cancer of the pancreas and kidney as well as adenocarcinoma of the esophagus, which are associated with excess weight. Obesity and lack of sufficient physical activity may also increase the incidence of certain types of cancer and may adversely affect cancer survivors' quality of life [3]. Siegel *et al.* [4] commented that although a reduction in overall death from cancer since 1990 in men and since 1991 in women translated into avoidance of approximately 898,000 deaths, individuals with a low level of education are twice as likely to die of cancer compared to most educated individuals. The prevalence of common types of cancer according to data provided by the National Cancer Institute is summarized in Table 8.1. Reactive species have been shown to induce lipoperoxidation and alter DNA and proteins. These alterations may be responsible for triggering tumorigenesis. Dietary modifications and implementation of exercise regimens have proven to be beneficial in the prevention and treatment of cancer. Various types of cancer that can be linked to oxidative stress are listed in Box 8.1.

A. Dasgupta and K. Klein: Antioxidants in Food, Vitamins and Supplements. DOI: http://dx.doi.org/10.1016/B978-0-12-405872-9.00008-2

Table 8.1 Prevalence of Common Types of Cancer in the United States

Cancer Type	Prevalence in Men (per 100,000 Men; All Races)	Prevalence in Women (per 100,000 Women; All Races)
Most common cancer in women		
Breast cancer		128.3
Most common cancer in men		
Prostate cancer[a]	152.0	
Other common cancer types		
Bladder cancer	36.6	8.9
Colon and rectal cancer	52.2	39.3
Endometrial cancer		24.3
Leukemia (all types)	16.3	10.0
Melanoma[b]	27.4	16.7
Non-Hodgkin lymphoma	23.9	16.4
Pancreatic cancer	13.9	10.9
Thyroid cancer	6.1	18.2

[a]*Overall rate of prostate cancer (all races) is 152 per 100,000 men, but in the black population the prevalence is 228.5 per 100,000 men.*
[b]*Melanoma is predominantly observed in the white population, with a prevalence of 31.9 per 100,000 men and 20.0 per 100,000 women.*
Source: *Data from the National Cancer Institute.*

Box 8.1 VARIOUS TYPES OF CANCER ASSOCIATED WITH OXIDATIVE STRESS

- Bladder cancer
- Breast cancer
- Colon and rectal cancer
- Endometrial cancer
- Kidney cancer
- Leukemia
- Liver cancer
- Lung cancer
- Melanoma
- Oral cancer
- Ovarian cancer
- Pancreatic cancer
- Prostate cancer
- Thyroid cancer

8.2 LINK BETWEEN OXIDATIVE STRESS AND CANCER

Oxidative stress occurs when reactive oxygen and/or nitrogen species are generated in excess. Although the body is equipped with resilient antioxidant defense systems, oxidative stress can sometimes become overwhelming [1]. Several proposed etiological mechanisms exist, but the general consensus is that excessive production of free radicals negatively impacts macromolecules, leading to oncogenesis [5]. Studies have shown that when polyunsaturated fatty acids (the lipids in cell membranes) are exposed to reactive oxygen species, lipid peroxidation can occur. Ultimately, this process changes the fluidity of the cell membranes and leads to the generation of fatty acid free radicals and eventually lipid peroxidation breakdown products such as malondialdehyde, conjugated dienes, and hydroperoxides. Reactive oxygen species also attack DNA, giving rise to mutations that may involve tumor suppressor genes or oncogenes [6]. Collectively, these alterations cause chemotaxis of inflammatory cells and may modulate gene expression and cell proliferation [7,8]. Reactive oxygen species may also alter structural proteins via oxidation. Normally, when proteins are damaged, proteasomes break them down into respective mixture of amino acids for recycling [9]. However, if a protein becomes severely damaged, it begins to accumulate within cells. This triggers further inhibition of proteasomes, accelerated accumulation of damaged proteins, decreased protein turnover, and alteration of cellular organelles [10]. Oxidative stress can also activate various transcription factors, including NF-κB (nuclear factor-κB), AP-1 (activator protein-1), p53 (tumor suppressor protein known as tumor protein 53), HIF-1α (hypoxia-inducible factor-1α), PPAR-γ (peroxisome proliferator-activated receptor-γ), β-catenin/Wnt, and Nrf2 (nuclear factor E2-related factor 2). Activation of these transcription factors can affect the expression of more than 500 genes, including genes that regulate growth factors, inflammatory cytokines, chemokines, cell cycle regulator molecules, and anti-inflammatory molecules, indicating that oxidative stress, chronic inflammation, and cancer are closely related [11].

Oxidative damage to DNA is considered to be an important promoter of cancer as well as aging and neurological diseases. Highly reactive hydroxyl radicals can cause significant damage to DNA by promoting DNA oxidation. In addition, other oxidants including peroxynitrite, reactive intermediates generated by eosinophil peroxidase, and nitryl chloride, as well as oxidizing nitrating and chlorinating species generated through hydrogen peroxide-mediated reactions of myeloperoxidase with chloride and nitric oxide, can damage biomolecules including DNA. Although base excision repair pathways initiated by DNA glycosylase, nucleotide excision repair, and mismatch repair mechanisms are capable of oxidatively damaging DNA, highly damaged DNA cannot be repaired through these mechanisms. Oxidation of DNA may result in nucleobase modifications. Alteration of the nucleobase sequence may lead to carcinogenesis due to the mutagenic potential of the mispaired bases. In addition, DNA methylation can be affected by oxidation. Studies have shown that oxidized DNA tends to be hypomethylated. DNA methylation plays an important role in gene regulation, and hypomethylated DNA is found in many cancers [12]. Scott *et al.* [13] noted that oxidative DNA damage including 7,8-dihydro-8-oxoguanine plays an important role during cellular transformation. In addition, unusual activity and subcellular distribution of apurinic/apyrimidinic endonuclease 1, an essential DNA repair factor as well as a redox sensor influencing cell proliferation, is an important link between oxidative stress and the pathophysiology of cancer.

Many studies have established an association between cancer incidence and various disorders linked to glutathione peroxidase-related enzyme functions, most commonly glutathione *S*-transferase. This class of enzyme utilizes glutathione in a series of reactions transforming a wide range of compounds, including products of oxidative stress, carcinogens, and some drugs. Increased levels of oxidized glutathione have been found in advanced stages of cancer progression. In addition, elevated levels of thioredoxin have been reported in a wide range of human carcinomas, including cervical carcinoma, hepatoma, lung carcinoma, colorectal carcinoma, and hepatoma. Moreover, many cancer cells can secrete thioredoxin, and elevated levels of thioredoxin in human tumors may cause resistance to certain chemotherapeutic agents, including cisplatin and doxorubicin [14]. Tumor suppressor p53 plays an important role in regulation of angiogenesis. Nitric oxide production via inducible nitric oxide synthase (iNOS), an isoform of nitric oxide synthases, may directly induce mutation in p53, which may contribute to loss of its repressor activity. Moreover, nitric oxide has been reported to exert a dichotomous effect within a multistage model of cancer. Interestingly, nitric oxide also has tumoricidal effects [15]. Dual roles of nitric oxide in promoting cancer and its tumoricidal effects are summarized in Table 8.2.

8.3 OXIDATIVE STRESS GENERATED BY CHEMOTHERAPY AND RADIATION THERAPY IN CANCER PATIENTS

It has been well-established that some chemotherapy and radiation therapy also generate oxidative stress during cancer treatment. In general, chemotherapeutic agents can be classified into several categories, including alkylating agents (e.g., cyclophosphamide and ifosfamide), anthracyclines (e.g., doxorubicin and

Table 8.2 Dual Role of Nitric Oxide in Promoting and Preventing Cancer

Role of Nitric Oxide	Comments
Tumor-promoting role	
Genotoxicity	Formation of reactive nitrogen species that may destroy a variety of biomolecules, including DNA (oxidative damage, strand break, and nucleic acid modification), and inhibit damaged DNA repair
Induction and promotion of angiogenesis	Stimulating permeability of vascular epithelium; activation of cyclooxygenase resulting in increased production of prostaglandins and other mechanisms
Antiapoptotic effect	Mutation of p53 and subsequent reduction or loss of its tumor-suppressor property; suppression of ceramide generation and increase in Bcl-2 (B-cell lymphoma 2) expression
Promotion of metastasis	Promotes lymphangiogenesis and spread of cancer cells to lymph nodes as well as other mechanisms
Limiting host immune response	Suppression of activity of leukocytes in immune host response to tumor
Tumoricidal role	
Cytotoxic effect on tumor cells	Suppression of cellular respiration and shifts iron metabolism; suppression of DNA synthesis; activation of the proapoptotic caspase family of proteases

bleomycin), platinum compounds, mitotic inhibitors (e.g., vincristine), antimetabolites (e.g., 5-fluorouracil), camptothecin derivatives (e.g., topotecan), hormone therapies (e.g., tamoxifen), and biological response modifiers such as interferon. Chemotherapeutic agents that generate high amounts of reactive oxygen species include anthracyclines, platinum compound (cisplatin), alkylating agents, and campothecins. In general, anthracyclines generate the highest levels of reactive oxygen species. Doxorubicin, which is used for treatment of a wide variety of cancers, including non-Hodgkin's and Hodgkin's lymphoma, multiple myeloma, and lung, ovarian, gastric, breast, and thyroid cancer, and sarcoma, can induce significant oxidative stress in patients receiving doxorubicin therapy. Several mechanisms have been proposed for the action of doxorubicin, including intercalation of DNA leading to inhibition of DNA synthesis or inhibition of topoisomerase II and generation of free radicals that damage DNA and the cell membrane. However, severe cardiotoxicity of doxorubicin limits its use in cancer chemotherapy, and free radical generation leading to excessive oxidative stress is partly responsible for its cardiotoxicity. Many studies have demonstrated that several chemotherapeutic agents induce apoptosis in cancer cells, and acute generation of reactive oxygen species or transient reactive oxygen species burst is frequently observed in apoptosis induced by various anticancer drugs.

Moreover, toxicities of many anticancer agents are also related to elevated oxidative stress produced by these agents. Radiotherapy using X-rays and γ-rays as well as heavy particle radiations (protons or neutrons) eradicates cancer cells through free radical formation [16].

8.4 BLADDER CANCER AND OXIDATIVE STRESS

Urothelial carcinoma is the most common malignant tumor of the bladder. It is the eighth leading cancer-related cause of death in men and the fourth most common cancer diagnosed in the United States [4]. Smoking is the main risk factor, but occupational hazards, infections, and exposure to toxic chemicals have also been cited as potential risk factors. Several studies have evaluated the role of oxidative stress in urothelial carcinoma. The incidence of bladder cancer increases with age, possibly due to the accumulative effect of free radicals. In addition, urothelium is one of the slowest cycling epithelia in the body. Perse *et al.* [17] examined the association between oxidative stress and urothelial cancer. The authors compared the oxidative status of urothelial cells of young and aged mice and observed that the total antioxidant capacity of the urothelial cells decreased with age. In addition, they noted a significant increase in markers of oxidative stress such as lipid peroxides (malondialdehyde) and iNOS in urothelial cells of aged mice compared to young mice.

In a study of 35 patients with bladder cancer and 32 controls, Gecit *et al.* [18] demonstrated that serum malondialdehyde and nitric oxide levels were significantly higher in patients with bladder cancer compared to controls. In addition, levels of serum prolidase, a member of the matrix metalloproteinase family that plays a major role in collagen turnover, matrix remodeling, and cell growth, were also higher in patients with bladder cancer compared to controls. The total antioxidant capacity of serum was decreased in patients with bladder cancer compared to controls. The authors concluded that increased prolidase seems to be associated with increased nitric oxide and oxidative stress with decreased total antioxidant capacity of sera in patients with bladder cancer. Epidemiologic studies have also supported the fact that oxidative stress is involved in the pathophysiology of bladder cancer, and reducing the impact of oxidative stress by balanced nutrition can be preventative for urothelial carcinoma [19].

8.5 BREAST CANCER AND OXIDATIVE STRESS

Breast cancer is the most common cancer among women worldwide, and increased oxidative stress is associated with a higher risk of developing breast carcinoma. Lin *et al.* [20] observed that insulin-like growth factor-1

(IGF-1) interacts with 17β-estradiol via reactive oxygen species-dependent pathways involving mitogen-activated protein kinase (MAPK) activation. Consuming alcohol, especially heavy drinking (more than three drinks per day), may increase the risk of breast cancer in women. Although the precise mechanism by which alcohol consumption is linked to increased risk of breast cancer is not known, alcohol-induced oxidative stress may play a role [21]. Kedzierska *et al.* [22] reported elevated levels of thiobarbituric acid reactive substances in plasma and elevated isoprostane levels in urine of patients with invasive breast cancer compared to a control group. They concluded that free radical-induced peroxidation of unsaturated fatty acids in patients with breast cancer may be involved in the pathogenesis of breast cancer. However, treatment of breast cancer, including surgical intervention, chemotherapy, and/or radiation therapy, may also increase the oxidative stress. Tamoxifen is a widely used antiestrogen agent for treating patients with breast cancer. It has been speculated that tamoxifen is able to generate oxidative stress, thereby causing thiol depletion and activation of transcriptional factor NF-κB and thus aiding in apoptosis of cancer cells [23]. In addition, agents such as surfactin and rotenone have been shown to induce apoptosis of breast cancer cells via reactive oxygen species-dependent pathways [24,25].

8.6 COLON AND RECTAL CANCER AND OXIDATIVE STRESS

Colorectal carcinoma, another common cancer worldwide, is an epithelial derived carcinoma occurring sporadically or developing as a result of inherited risk factors. Epidemiological studies have linked the occurrence of sporadic colorectal carcinoma to a variety of environmental and lifestyle risk factors, including oxidative stress [26]. There is evidence for an adverse effect of obesity and protective effect of high physical activity against colon but not rectal cancer. Diet may have a protective effect, whereas there are also modest adverse effects of consumption of red and processed meat. In addition, polyunsaturated fatty acids in fish may also have a protective effect [27]. Oxidative stress seems to play a role in the pathophysiology of colorectal cancer due to rapid colonic mucosal turnover. Some investigators have suggested that fast cellular division is a potential risk factor for increased DNA oxidation due to the associated high cellular metabolic rate. This theory is supported by studies performed using rat colonocytes that demonstrated that lower crypt cells are more sensitive to oxidative damage than apical ones [28,29]. Additional evidence indicating the important role of oxidative stress in the pathophysiology of colorectal carcinoma was obtained from studying tissue samples derived from colorectal lesions. Haklar *et al.* [30] observed significantly elevated levels of reactive oxygen species in cancerous colon tissues, with hypochlorite making the most important contribution of reactive species, thus suggesting the role of inflammatory cells in the pathogenesis of the disease. Nitric oxide and peroxynitrite concentrations were also higher in cancerous cells [30]. In a study involving the examination of cancer tissue specimens collected from 55 patients with colorectal cancer, the authors observed higher levels of malondialdehyde and 4-hydroxy-2-nonenal in cancer tissue, whereas the activity of antioxidant enzyme catalase was significantly reduced [31]. The authors concluded that significant changes in antioxidant capacity of colorectal cancer tissues lead to enhanced action of oxygen radicals, resulting in lipid peroxidation. In a study of 39 patients with colon cancer and 105 healthy subjects, Guz *et al.* [32] observed elevated concentrations of 8-oxo-7,8-dihydroguanine (8-oxoGua) in leukocyte DNA of patients with colon adenomas compared to healthy individuals. In addition, levels of 5-methylcytosine were lower in these patients compared to healthy individuals. The authors concluded that the increased level of 8-oxoGua may lead to carcinogenesis not only via mispair/mutagenic potential of the modified base but also through its ability to influence gene expression by affecting DNA methylation. In another study, the authors observed significantly increased lipid peroxidation products in plasma of patients with colorectal cancer at various stages compared to controls. Total superoxide dismutase activities and glutathione reductase were lower in all colorectal carcinoma groups compared to controls. The authors concluded that colorectal carcinoma is characterized by increased oxidative stress and antioxidant imbalance. Moreover, progression of disease is followed by an increase in redox imbalance [33].

8.7 KIDNEY CANCER AND OXIDATIVE STRESS

Kidney cancer (renal cell carcinoma) is a malignant neoplasm that originates from the proximal tubular epithelium and is highly lethal. Oxidative stress is implicated in the pathogenesis of renal cell carcinoma. Ganesamoni *et al.* [34] assessed the role of oxidative stress in renal cell carcinoma by studying 68 patients and 30 healthy volunteers. The authors observed increased levels of reactive oxygen species in erythrocytes and white blood cells of patients with renal cell carcinoma compared to controls. Moreover, nitric oxide levels were increased in plasma of these patients compared to controls. In addition, the authors noted lower levels of glutathione and glutathione transferase in sera of patients with renal cell carcinoma. As expected, lipid peroxidation products were elevated in tumor tissues compared to normal tissues. Oxidative stress correlated with renal cell carcinoma grade and stage but decreased after curative resection.

8.8 LIVER CANCER AND OXIDATIVE STRESS

Malignant transformation of the liver parenchyma can result in the development of several types of cancer. Hepatocellular carcinoma (HCC) is the most common malignant neoplasm of the liver and represents approximately 85% of the primary malignant neoplasms occurring in the liver. Annually, 750,000 new diagnoses of HCC are made, and nearly 700,000 deaths worldwide are due to HCC [35]. Numerous risks factors have been identified for HCC, including persistent viral infections with hepatitis B or C, alcohol consumption, obesity, cigarette smoking, and diabetes mellitus [36]. In addition, the liver is a major site of metabolism that results in generation of reactive oxygen species. Therefore, increased oxidative stress may play a role in the pathogenesis of various liver cancers. During cellular respiration, high levels of reactive oxygen species are also generated. Because reactive oxygen species are known to cause alterations in DNA, it has been postulated that the liver is the primary location for oncogenesis [37]. Studies have shown that oxidative stress activates hepatic stellate cells and signaling pathways (particularly MAPKs), which collectively may promote cellular transformation [38,39].

Although the liver has a robust antioxidant defense system to neutralize free radicals, when the balance is shifted to elevated oxidative stress, hepatocarcinogenesis is likely to occur. Li *et al.* [40] reported that glutathione *S*-transferase P1, an important enzyme that can protect cells from oxidative stress in various human cancers, was significantly decreased (determined as glutathione *S*-transferase P1 protein expression in peripheral blood mononuclear cells) in patients with HCC compared to patients with chronic hepatitis B infection. The mRNA expression of glutathione *S*-transferase P1 was also reduced in patients with HCC, whereas plasma malondialdehyde and xanthine oxidase levels were significantly elevated, indicating that patients with HCC suffered from elevated oxidative stress compared to patients with chronic hepatitis B infection. Kinase inhibitors have been used in the treatment of HCC to alleviate oxidative stress. Published reports indicated that the overall survival rate of patients on sorafenib (a kinase inhibitor) was improved compared to that of the placebo group [41,42]. This clearly indicates that oxidative stress is paramount in hepatocellular carcinoma.

8.9 LEUKEMIA AND OXIDATIVE STRESS

Oxidative stress has been implicated in several oncohematologic diseases, including chronic lymphocytic leukemia, Hodgkin's lymphoma, multiple myeloma, and chronic myeloid leukemia (CML) [43]. CML is a clonal myeloproliferative disorder caused by neoplastic transformation of primitive hematopoietic cells. It accounts for approximately 14% of all leukemias, with an annual incidence of 1 or 2 cases per 100,000 people. In a study of 34 women and 30 men (ages 20–80 years; median, 44 years) with CML, Petrola *et al.* [44] observed significantly higher concentrations of nitrite and malondialdehyde (measured as thiobarbituric acid reactive substances) in the plasma of patients with CML who were being treated with tyrosine kinase inhibitors compared to imatinib. Elevated lipid peroxidation products in the plasma of patients with leukemia have also been reported by various investigators. In a study of 10 patients with acute myeloid leukemia (AML), 15 patients with CML, 5 patients with acute lymphocytic leukemia (ALL), and 20 healthy subjects, Ghalaut *et al.* [45] observed significantly elevated malondialdehyde (measured as thiobarbituric acid reactive substances) levels in sera of patients with AML and ALL compared to controls. Increased malondialdehyde levels were observed in sera of patients with CML, especially in those with blast crisis. However, CML patients with blast crisis who achieved remission showed lower levels of malondialdehyde compared to patients without remission. In a study of 128 CML patients and 50 age- and gender-matched controls, Ahmad *et al.* [46] observed significantly increased lipid peroxidation products in the plasma of CML patients compared to controls. The plasma levels of lipid peroxidation products continued to increase along with reduction of nonenzymatic antioxidant status in CML patients as the disease progressed. Akinlolu *et al.* [47] reported increased plasma malondialdehyde levels in patients with CML and non-Hodgkin's lymphoma compared to healthy controls.

8.10 LUNG CANCER AND OXIDATIVE STRESS

Worldwide, lung cancer accounts for 12% of total cancer diagnoses. Although there are several risk factors for developing lung cancer, smoking has been implicated in 90% of all cases [35]. Studies have shown that per puff, cigarette smoke contains 10^{14} free radicals [48]. Lung cancer and chronic obstructive pulmonary disease (COPD) may coexist in smokers because cigarette smoke induces oxidative stress and an inflammatory response in lung cells. Pastor *et al.* [49] assessed the protein content of bronchoalveolar lavage of patients with COPD, lung cancer, or both COPD and lung cancer using two-dimensional polyacrylamide gel electrophoresis (for protein separation) followed by analysis using matrix-assisted laser desorption/ionization time-of-flight mass spectrometry (MALDI-TOF). The authors observed that 16 oxidative stress regulatory proteins were differentially expressed in bronchoalveolar lavage of patients with COPD and lung cancer compared to controls, and a distinct proteomic reactive oxygen species protein signature emerged that was characteristic of lung cancer and COPD. Cobanoglu *et al.* [50] studied the plasma concentration of coenzyme Q10 and malondialdehyde (as thiobarbituric acid reactive substances) as well as whole blood concentration of 8-hydroxy-2-deoxyguanosine, a marker of oxidative DNA damage. The authors observed significantly elevated malondialdehyde level in patients with lung cancer (mean, 10.4 ng/mL) compared to healthy controls (mean, 4.07 ng/mL). The whole blood level of 8-hydroxy-2-deoxyguanosine was also significantly elevated in patients with lung cancer compared to controls. Coenzyme Q10 is known to inhibit lipid peroxidation, and levels of coenzyme Q10 in plasma lung cancer patients were significantly lower than those of controls. These data suggested that patients with lung cancer experienced significantly elevated oxidative stress compared to healthy controls.

8.11 ORAL (HEAD AND NECK) CANCER AND OXIDATIVE STRESS

Head and neck cancer refers to a group of biologically similar cancers that start in the lip, oral cavity, nasal cavity, paranasal sinuses, pharynx, and larynx, and 90% of such cancers are squamous cell carcinoma. Oral cancer is the sixth most common cancer worldwide with a global annual incidence of 640,000 cases, of which 62% occur in India and Southeast Asia. As with lung cancer, smoking is a major risk for the development of oral cancer because cigarette smoke contains numerous carcinogenic compounds, including *N*-nitrosamine/benzopyrene-like classical carcinogenic compounds that cause leukoplakia and submucous fibrosis, which may eventually lead to oral squamous cell carcinoma. Oxidative stress induced by cigarette smoke has been shown to induce DNA damage and cause widespread genomic instability. Although DNA repair systems exist, studies have shown that certain polymorphisms in hMLH1, a mismatch repair gene that plays a crucial role in correcting replication error, cell cycle arrest, apoptosis, and oxidative stress, may increase the risk of oxidative DNA damage in smokers [51]. Studies have also shown that tumor cells can generate reactive oxygen species. In a study of 36 patients with head and neck squamous cell carcinoma, Dequanter *et al.* [52] observed that the oxidized:reduced glutathione ratio in tumor tissue was lower compared to that in adjacent normal tissue in 38% of patients. Gupta *et al.* [53] reported elevated levels of malondialdehyde and depleted activity of antioxidant enzyme superoxide dismutase in the plasma of patients with head and neck squamous cell carcinoma compared to controls, indicating that oxidative stress was significantly elevated in these patients. In another study, the authors detected a higher concentration of nitric oxide in sera of patients with stage IV squamous cell carcinoma of the oral cavity, whereas higher levels of malondialdehyde and nitrotyrosine were detected in sera of patients with all stages of oral cancer compared to healthy controls [54].

8.12 MELANOMA AND OXIDATIVE STRESS

Histological examination of the skin reveals that the skin is composed of several different types of cells. Thus, numerous types of skin cancer are known to arise from the skin. Exposure of the skin to solar ultraviolet (UV) light is the main cause of skin cancer. The most important defense mechanism that protects skin against UV radiation is melanin synthesis. Caucasians are at higher risk of developing skin cancer including melanoma, a malignant skin lesion that arises from melanocytes due to low pigmentation of the skin. Although melanoma is not the most common form of skin cancer, it is one of the most deadly. DNA is the major target of direct or indirect UV-induced skin damage. In squamous cell carcinoma, UV-induced signature mutation in the p513 tumor suppressor gene is a common event [55]. In malignant melanoma, high levels of reactive oxygen species in the primary melanoma environment can also influence the behavior of tumor-associated macrophages. In addition, upregulation of nucleus-to-cytoplasm translocation of PPAR-γ may be responsible for the pro-invasiveness capacity of melanoma tumor-associated macrophages [56]. Brar *et al.* [57] noted that malignant melanoma cells spontaneously generate reactive oxygen species that promote constitutive activation of the transcription factor NF-κB. The authors also reported that reactive oxygen species were generated by isolated cytosol-free melanoma plasma membranes, which seem to be associated with NADPH oxidase production, and such reactive oxygen species may play a role in signaling malignant melanoma growth. Sander *et al.* [58] observed elevated levels of malondialdehyde in melanoma tissue of cancer patients compared to healthy controls. Malondialdehyde levels were elevated not only in melanoma tissues but also in surrounding keratinocytes. In contrast, a severely disturbed antioxidant balance with diminished antioxidant enzymes (superoxide dismutases and catalase) was observed in nonmelanoma tumors, whereas malondialdehyde was elevated in squamous cell carcinoma. The authors commented that these findings indicated that oxidative stress may play different roles in the pathogenesis of human skin cancer. In nonmelanoma skin cancer, a diminished antioxidant defense caused by chronic UV exposure may contribute to carcinogenesis, whereas melanoma cells exhibit elevated oxidative stress that may damage surrounding tissue, resulting in progression of metastasis.

8.13 OVARIAN CANCER AND OXIDATIVE STRESS

Ovarian carcinoma is the fifth most common cancer affecting women. The majority of ovarian tumors are derived from the epithelial lining of the ovary. Often, women experience vague, nonspecific symptoms. As a result, diagnosis is often delayed, resulting in a poor prognosis. The microRNA-200-dependent oxidative stress and fibrosis signatures are promising in patient stratification for developing therapeutic strategies [59]. Xia *et al.* [60] showed that ovarian cancer cells spontaneously produced reactive oxygen species. In addition, they demonstrated that reactive oxygen species regulated hypoxia-inducible factor-1 (HIF-1) and vascular endothelial growth factor (VEGF) expression in ovarian cancer cells, and elevated levels of reactive oxygen species were required for inducing angiogenesis and tumor growth. Chan *et al.* [61] observed that accumulation of reactive oxygen species during progression of ovarian cancer may cause the degradation of mitogen-activated protein kinase phosphatase (MKP3), which in turn leads to aberrant extracellular signal-regulated kinase (ERK1/2) activation and contributes to tumorigenicity and chemoresistance in human ovarian cancer cells. Mateescu *et al.* [62] reported that cross talk between oxidative stress and the miR-200 family of microRNA affects tumorigenesis and chemosensitivity in ovarian cancer.

Evidence of increased oxidative stress in patients with ovarian cancer was provided by the observation of elevated levels of lipid peroxidation products (measured as thiobarbituric acid reactive substances) and conjugated dienes in plasma of patients with ovarian cancer compared to controls. Lower levels of antioxidant enzymes such as superoxide dismutase and catalase as well as lower levels of antioxidants such as vitamin C and E were also observed in plasma of ovarian cancer patients compared to controls [63]. Sanchez *et al.* [64] reported elevated levels of malondialdehyde and oxidatively damaged DNA product 8-oxo-2′-deoxyguanosine in ovarian cancer biopsy tissues compared to normal biopsy tissues. Endometrial cancer is the most common form of gynecological malignancy. Patients with endometrial cancer also experience elevated oxidative stress, as evidenced by the observation of elevated levels of conjugated dienes in erythrocytes of female patients with malignant tumor of endometrium [65].

8.14 PANCREATIC CANCER AND OXIDATIVE STRESS

Pancreatic carcinoma is an epithelial malignant neoplasm of acinic cell origin with a low incidence of 1%. Risk factors include smoking, alcohol abuse, diabetes, and chronic pancreatitis. Interestingly, each of these risk factors is associated with increased oxidative stress [66]. Pancreatic cancer is aggressive and usually does not respond to therapy due to resistance to apoptosis. Reactive oxygen species have been implicated as a prosurvival antiapoptotic factor in pancreatic cancer cells. Human pancreatic adenocarcinoma cells such as PANC-1 can generate reactive oxygen species that are stimulated by growth factors such as insulin-like growth factor-1 or fibroblast growth factor-2. Growth factors can also stimulate NAD(P)H (nicotinamide adenine dinucleotide phosphate) oxidase activity in these cells, and Vaquero *et al.* [67] showed that nonmitochondrial NAD(P)H oxidase is a major source of growth factor-induced reactive oxygen species production by pancreatic cancer cells. Mochizuki *et al.* [68] demonstrated that inhibition of NAD(P)H leads to apoptosis, thus further demonstrating the role of reactive oxygen species in the pathophysiology of pancreatic cancer. Kodydkova *et al.* [69] reported reduced antioxidant defense capacity and increased markers of oxidative stress in patients with pancreatic cancer and chronic pancreatitis compared to controls.

8.15 PROSTATE CANCER AND OXIDATIVE STRESS

Prostate cancer is an important worldwide health problem occurring most commonly in men between ages 54 and 75 years. NAD(P)H activity is also noted to be increased in prostatic cancer cells. Reactive oxygen species play an important role in regulating growth and survival of prostate cancer. Brar *et al.* [70] studied NAD(P)H-regulated cell growth and concluded that inhibition of reactive oxygen species generation by either antioxidant or NAD(P)H oxidase inhibitors increased apoptosis of these prostate cancer cells. Kumar *et al.* [71] also demonstrated that NAD(P)H activity generates reactive oxygen species, which leads to oncogenesis by altering the redox homeostasis and directly affecting regulation of various cellular functions in prostate cancer. In a study of 55 men with prostate cancer and 55 age-matched healthy men, Battisti *et al.* [72] observed higher levels of lipid peroxidation products (measured as thiobarbituric acid reactive substances) and protein carbonylation products in sera of patients with prostate cancer compared to health controls. In addition, antioxidant enzyme catalase levels and levels of antioxidant vitamins (vitamins C and E) were reduced in sera of patients with prostate cancer compared to healthy controls, indicating that patients with prostate cancer experienced increased oxidative stress compared to healthy controls. Moreover, oxidative stress profile appeared to be modified by bone metastasis, treatment, and Gleason score.

8.16 THYROID CANCER AND OXIDATIVE STRESS

Oxidative stress has also been implicated in carcinoma of the thyroid gland. Superoxide dismutase, an antioxidant enzyme, has been shown to be decreased in numerous disorders of the thyroid, including carcinoma. Likewise, other antioxidant defenses have also been shown to be impaired. In a study of 43 patients with thyroid cancer and 43 healthy controls, Akinci *et al.* [73] observed that before thyroidectomy in thyroid cancer patients, malondialdehyde concentrations were significantly elevated compared to those of controls, but 20 days after thyroidectomy, malondialdehyde levels were decreased significantly compared to pre-surgery levels. Levels of glutathione peroxidase, an important antioxidant enzyme, were also lower before thyroidectomy, but levels were increased significantly post-surgery. However, superoxide dismutase levels did not change pre- and post-surgery. The authors concluded that oxidant/antioxidant balance was associated with thyroid cancer.

8.17 DIET/DIETARY SUPPLEMENTATION, CANCER, AND OXIDATIVE STRESS

As discussed previously, excess oxidative stress plays an important role in the development of various types of cancer. Because oxidative stress is caused by an imbalance between the production of reactive oxygen species and the body's antioxidant defenses, attempts have been made to use diet and dietary supplements to reduce the risk of various cancers by modifying oxidant/antioxidant imbalance. Epidemiological studies indicate that consumption of dietary phytochemicals found in fruits and vegetables can decrease the risk of various cancers. Among various vegetables, cruciferous species such as broccoli and related vegetables are associated with reduced cancer risks in organs such as colorectum, lung, prostate, and breast. Such protection may be related to glucosinolates, a class of sulfur-containing glycosides present in high amounts in cruciferous vegetables. These glycosides break down to isothiocyanates, which also provide health benefits [74]. Glucosinolates prevent cancer by inducing antioxidant and detoxifying enzymes such as glutathione S-transferase and uridine-5′-diphosphoglucuronosyltransferase and by inhibition of carcinogen-activating enzymes such as cytochrome P450 1A1 [75]. The Mediterranean diet, which is based on abundant and variable plant food, olive oil, low intake of red meat, and moderate consumption of wine, is associated with lower risk of cardiovascular diseases and cancer. The mechanism of such protection is probably related to the favorable effect of a balanced ratio of omega-6 to omega-3 fatty acids and high amounts of fiber, antioxidants, and other polyphenols found in fruits, vegetables, and wine [76].

Tea and coffee are rich in antioxidants. The prophylactic and therapeutic properties of tea have been attributed to tea polyphenolic compounds, particularly catechins in green tea and theaflavins in black tea, which are also excellent antioxidants. Tea polyphenols are excellent anticancer agents and can reduce the risk of various cancers, including HCC [77]. Coffee intake is also associated with reduced risk of liver cancer and chronic liver diseases [78]. Chapter 13 provides an in-depth discussion of antioxidant effect of tea and coffee.

Dietary augmentation has also been shown to be beneficial in renal cell carcinoma. A bioactive proteoglucan in maitake mushrooms (D-Fraction) has been shown to have immunomodulatory and anticancer properties. Moreover, vitamin C potentiates the effect [79]. The therapeutic benefit of dietary supplementation has also been evaluated in colorectal cancer (CRC). Probiotics may have preventive and therapeutic benefits because they may inhibit tumor progression, reduce oxidative stress, and stimulate the host immune system [80]. Dietary fibers also have a protective effect in CRC because they capture fats and biliary acids, thereby preventing their procarcinogenic activities. However, high consumption of red meat, smoked foods, or cold cuts, is believed to contribute to carcinogenesis because these foods directly affect epithelial turnover and induce metabolism of bile acids. In addition, excess consumption of alcohol in association with nicotine addiction increases the risk of CRC [81].

Studies have shown that dietary fat stimulates lipid peroxidation, and this may part of the reason why increased consumption of dietary fat may increase the risk of breast cancer. Vieira *et al.* [82] observed that intake of chicken and high-fat dairy products was associated with increased plasma lipid hydroperoxides in women with newly diagnosed breast cancer. In addition, intake of animal fat, dairy products, and sweet foods was associated with low levels of serum antioxidant capacity. Consumption of alcohol may increase the risk of breast cancer. Based on a literature review, Seitz *et al.* [83] concluded that the risk of breast cancer may increase by up to 4% in women who consume up to one drink per day. In addition, they observed that the risk increased by 40–50% with heavy alcohol consumption (defined as three or more drinks per day). Based on these findings, the authors concluded that women should not exceed one drink per day, and those with an increased risk for developing breast cancer should abstain from alcohol. Interestingly, women may receive all the benefits of moderate alcohol consumption with just one or two drinks per week. Although light consumption of alcohol can prevent certain types of cancer, heavy drinking is associated with increased risk of cancer. Jung *et al.* [84] observed a J-shaped association between alcohol consumption and risk of all types of cancer death. Relative to light drinkers (up to 90 g of alcohol per week, approximately up to 7 drinks per week), nondrinkers had a slightly higher risk of cancer (hazard ratio, 1.18). However, heavy drinkers (>504 g per week, more than 36 drinks per week) had a significantly higher risk (hazard ratio, 1.39) from cancer death compared to light drinkers. Therefore, dietary and alcohol consumption habits play an important role in determining cancer risk for individuals. In general, it is advisable to eat foods rich in antioxidants (generous amounts of fruits and vegetables each day) and high in fiber. In addition, avoidance of fatty meats and high-cholesterol foods also seems beneficial for the prevention of cancer. Moderate consumption of alcohol is beneficial for reducing the risk of cancer, but heavy drinking significantly increases the risk of cancer.

8.18 EXERCISE, CANCER, AND OXIDATIVE STRESS

Regular physical activities may modulate the risk of cancer. Epidemiological and laboratory studies suggest that there is an inverse relationship between regular exercise and malignancies such as intestinal, colon, pancreatic, breast, lung, skin, endometrial, and prostate cancer. Although physical activities are known to generate reactive oxygen species, mild oxidative stress generated as a result of moderate exercise can activate cellular stress response signaling, thus potentiating cellular antioxidant defense capacity, resulting in a reduction in cancer risk. However, strenuous exercise may result in the generation of large amounts of reactive oxygen species that can either directly damage DNA, causing mutation, or promote tumorigenesis by activating proinflammatory signaling [85]. Approximately 170 observational epidemiological studies have been conducted on the relationship between regular exercise and the prevention of cancer. The evidence of decreased risk of breast and colon cancer with physical activity has been established. Physical activities may also reduce the risk of prostate, lung, and endometrial cancer. Usually, 30 min of moderate to vigorous exercise at least 5 days a week is beneficial for reducing cancer risk [86]. Physical activity can improve quality of life in breast cancer survivors. Based on a large cohort study involving 2279 women with nonmetastatic breast cancer, Mandelblatt *et al.* [87] observed that regular physical activity was associated with a small but clinically significant increase in quality of life during active breast cancer therapy for Caucasian women, but no such relationship was observed for minority women. Physical activities may also delay the onset of breast cancer in women who are genetically susceptible to breast cancer. Several hypotheses have been proposed to explain the beneficial effects of exercise on breast cancer prevention. These include the mTOR network hypothesis (physical activity inhibits carcinogenesis by suppressing the mTOR signaling network in mammary carcinoma), the metabolic reprogramming hypothesis (physical exercise limits the amount of glucose and glutamine available to mammary carcinoma cells, thereby inducing apoptosis), and the hormesis hypothesis [88]. Exercise has also been shown to alter the metabolic profile of estrogen. Certain metabolites of estrogen have been noted to have reduced hormonal activity in comparison to the parent molecule, resulting in a relative suppression of the breast epithelial cell growth [89]. Habitual physical activity also reduces the risk of ovarian cancer [90].

Because the effect of physical exercise on cancer prevention varies, it is advisable to consult a medical professional before engaging in an exercise program for cancer prevention. Frequency, exercise load, and the type of exercise all seem to impact the influence of physical exercise on cancer prevention.

8.19 CONCLUSION

Oxidative stress has been implicated in the pathophysiology of many diseases, including carcinogenesis. Oxidative stress is a factor in the malignant transformation of cells. Efforts to minimize the cellular impact of reactive oxygen species have been evaluated, and diet seems to play a major role. Naturally increasing the body's antioxidant defense system through diet and exercise has been shown to have therapeutic and preventive benefit. However, antioxidant supplements may not be useful in preventing cancer risk. Lin *et al.* [91] noted that supplementation with vitamin C, vitamin E, or β-carotene offers no overall benefits in the primary prevention of total cancer incidence or mortality from cancer. Therefore, vitamin or antioxidant supplements must be taken only under medical advice if vitamin deficiency occurs in individuals.

REFERENCES

[1] Harman D. Aging: a theory based on free radical and radiation chemistry. J Gerontol 1956;11:298–300.

[2] Benz CC, Yau C. Ageing, oxidative stress and cancer: paradigms in parallax. Nat Rev Cancer 2008;8:875–9.

[3] Eheman C, Henley SJ, Ballard-Barbash R, Jacobs EJ, et al. Annual report to the nation on the status of cancer, 1975–2008, featuring cancers associated with excess weight and lack of sufficient physical activity. Cancer 2012;118:2338–66.

[4] Siegel R, Ward E, Brawley O, Jemal A. Cancer statistics 2011: the impact of eliminating socioeconomic racial disparities on premature cancer deaths. CA Cancer J Clin 2011;61:212–36.

[5] Perse M. Oxidative stress in the pathogenesis of colorectal cancer: cause or consequence? Biomed Res Int 2013;2013:725710.

[6] Cejas P, Casado E, Belda-Iniesta C, De Castro J, et al. Implications of oxidative stress and cell membrane lipid peroxidation in human cancer. Cancer Causes Control 2004;15:707–19.

[7] Finaud J, Lac G, Filaire E. Oxidative stress: relationship with exercise and training. Sports Med 2006;36:327–58.

[8] Federico A, Morgillo F, Tuccillo C, Ciardiello F, et al. Chronic inflammation and oxidative stress in human carcinogenesis. Int J Cancer 2007;121:2381–6.

[9] Grune T, Merker K, Sandig G, Davies KJ. Selective degradation of oxidatively modified protein substrates by the proteasome. Biochem Biophys Res Commun 2003;305:709–18.

[10] Brunk UT, Terman A. The mitochondrial–lysosomal axis theory of aging: accumulation of damaged mitochondria as result of imperfect autophagocytosis. Eur J Biochem 2002;269:1996–2002.

[11] Reuter S, Gupta SC, Chaturvedi MM, Agarwal B. Oxidative stress, inflammation and cancer: how are they are linked? Free Rad Biol Med 2010;49:1603–16.

[12] Bjelland S, Seeberg E. Mutagenicity, toxicity and repair of DNA base damage induced by oxidation. Mutat Res 2003;531:37–80.

[13] Scott TL, Rangaswamy S, Wicker CA, Izumi T. Repair of oxidative DNA damage and cancer: recent progress in DNA base excision pair. Antioxidant Redox Signal 2013. [Epub ahead of print].

[14] Valko M, Leibfritz D, Moncol J, Cronin M, et al. Free radicals and antioxidants in normal physiological functions and human disease. Int J Biochem Cell Biol 2007;39:44–84.

[15] Korde Choudhari S, Chaudhary M, Bagde S, Gadbail A, et al. Nitric oxide and cancer: a review. World J Surg Oncol 2013;11:118.

[16] Ozbenz T. Oxidative stress and apoptosis: impact on cancer therapy. J Pharm Sci 2007;96:2181–96.

[17] Perse M, Injac R, Erman A. Oxidative status and lipofuscin accumulation in urothelial cells of bladder in aging mice. PLoS ONE 2013;8:E59638.

[18] Gecit I, Aslan M, Gunes M, Pirincci N, et al. Serum prolidase activity, oxidative stress and nitric oxide levels in patients with bladder cancer. J Cancer Res Clin Oncol 2012;138:739–43.

[19] Willett WC, MacMahon B. Diet and cancer: an overview. N Engl J Med 1984;310:697–703.

[20] Lin CW, Yang LY, Shen SC, Chen YC. IGF-I plus E2 induces proliferation via activation of ROS-dependent ERKs and JNKs in human breast carcinoma cell. J Cell Physiol 2007;212:666–74.

[21] Seitz H, Pelucchi C, Bagnardi V, La Vecchia C. Epidemiology and pathophysiology of alcohol and breast cancer: updated 2012. Alc Alc 2012;47:204–12.

[22] Kedzierska M, Olas B, Wachowicz B, Jeziorski A, et al. The lipid peroxidation in breast cancer patients. Gen Physiol Biophys 2010;29:208–10.

[23] Ferlini C, Scambia G, Marone M, Distefano M, et al. Tamoxifen induces oxidative stress and apoptosis in oestrogen receptor negative human cancer cell lines. Br J Cancer 1999;79:257–63.

[24] Cao XH, Wang AH, Wang CL, Mao DZ, et al. Surfactin induces apoptosis in human breast cancer MCF-7 cells through a ROS/JNK-mediated mitochondrial/caspase pathway. Chem Biol Interact 2010;183:357–62.

[25] Deng YT, Huang HC, Lin JK. Rotenone induces apoptosis in MCF-7 human breast cancer cell-mediated ROS through JNK and p38 signaling. Mol Carcinog 2010;49:141–51.

[26] Souglakos J. Genetic alterations in sporadic and hereditary colorectal cancer: implementations for screening and follow-up. Dig Dis 2007;25:9–19.

[27] Johnson IT, Lund EK. Review article: nutrition, obesity, and colorectal cancer. Aliment Pharmacol Ther 2007;26:161–81.

[28] Foksinski M, Rozalski R, Guz J, Ruszkowska B, et al. Urinary excretion of DNA repair products correlates with metabolic rates as well as with maximum life spans of different mammalian species. Free Radic Biol Med 2004;37:1449–54.

[29] Oberreuther-Moschner DL, Rechkemmer G, Pool-Zobal BK. Basal colon crypt cells are more sensitive than surface cells toward hydrogen peroxide, a factor of oxidative stress. Toxicol Lett 2005;159:212–18.

[30] Haklar G, Sayin-Ozveri E, Yuksel M, Aktan AO, et al. Different kinds of reactive oxygen and nitrogen species were detected in colon and breast tumors. Cancer Lett 2001;26:219–24.

[31] Skrzydlewska E, Stankiewica Z, Sulkowska M, Sulkowski S, et al. Antioxidant status and lipid peroxidation in colorectal cancer. J Toxicol Environ Health A 2001;64:213–22.

[32] Guz J, Foklsinski M, Siomek A, Gackowski D, et al. The relationship between 8-oxo-7,8-dihydro-2′-deoxyguanosine level and extent of cytosine methylation in leukocytes DNA of healthy subjects and in patients with colon adenomas and carcinoma. Mutat Res 2008;640:170–3.

[33] Gopcevic KR, Rovcanin BR, Tatic SB, Krivokapic ZV, et al. Activity of superoxide dismutase, catalase, glutathione peroxidase and glutathione reductase in different stages of colorectal carcinoma. Dig Dis Sci 2013;58:2646–52.

[34] Ganesamoni R, Bhattacharya S, Kumar S, Chauhan A, et al. Status of oxidative stress in patients with renal cell carcinoma. J Urol 2012;187:1172–6.

[35] Jamal A, Bray F, Center MM, Ferlay J, et al. Global cancer statistics. CA Cancer J Clin 2011;61:69–90.

[36] El-Serag HB. Hepatocellular carcinoma: an epidemiologic view. J Clin Gastroenterol 2002;35:S72–8.

[37] Sasaki Y. Does oxidative stress participate in the development of hepatocellular carcinoma? J Gastroenterol 2006;41:1135–48.

[38] Finkel T, Holbrook NJ. Oxidants, oxidative stress and the biology of ageing. Nature 2000;408:239–47.

[39] Block TM, Mehta AS, Fimmel CJ, Jordan R. Molecular viral oncology of hepatocellular carcinoma. Oncogene 2003;22:5093–107.

[40] Li T, Zhao XP, Wang LY, Gao S, et al. Glutathione S transferase P1 correlated with oxidative stress in hepatocellular carcinoma. Int J Med Sci 2013;10:683–90.

[41] Abou-Alfa GK, Schwartz L, Ricci S, Amadori D, et al. Phase II study of sorafenib in patients with advanced hepatocellular carcinoma. J Clin Oncol 2006;24:4293–300.

[42] Marra M, Sordelli IM, Lamberti M, Tarantinol L, et al. Molecular targets and oxidative stress biomarkers in hepatocellular carcinoma: an overview. J Transl Med 2011;9:171–93.

[43] Imbesi M, Muscolino C, Allegra A, Saija A, et al. Oxidative stress in oncohematologic disease: an update. Expert Rev Hematol 2013;6:317–25.

[44] Petrola MJ, de Castro AJ, Pitombeira MH, Barbosa MC, et al. Serum concentrations of nitrite and malondialdehyde as markers of oxidative stress in chronic myeloid leukemia patients treated with tyrosine kinase inhibitors. Rev Bras Hematol Hemoter 2012;34:352–5.

[45] Ghalaut VS, Ghalaut PS, Singh S. Lipid peroxidation in leukemia. J Assoc Physicians India 1999;47:403–5.

[46] Ahmad R, Tripathi AK, Tripathi P, Singh R, et al. Studies on lipid peroxidation and non-enzymatic antioxidant status as indices of oxidative stress in patients with chronic myeloid leukemia. Singapore Med J 2010;51:110–15.

[47] Akinlolu A, Akingbola T, Salau B. Lipid peroxidation in Nigerians affected with hematological malignancies. Afr J Med Med Sci 2012;41(Suppl):145–8.

[48] Pryor WA, Hales BJ, Premovic PI, Church DF. The radicals in cigarette tar: their nature and suggested physiological implications. Science 1983;220:425–7.

[49] Pastor MD, Nogal A, Molina-Pinelo S, Melendez R, et al. Identification of oxidative stress related proteins as biomarkers for lung cancer and chronic obstructive pulmonary disease in bronchoalveolar lavage. Int J Mol Sci 2013;14:3340–55.

[50] Cobanoglu U, Demir H, Cebi A, Sayir F, et al. Lipid peroxidation, DNA damage and coenzyme Q10 in lung cancer patients: markers of risk assessment? Asian Pac J Cancer Prev 2011;12:1399–403.

[51] Jha R, Gaur P, Sharma SC, Das SN. Single nucleotide polymorphism in hMLH1 promoter and risk of tobacco-related oral carcinoma in high risk Asian Indians. Gene 2013;526:223–7.

[52] Dequanter D, Van de Velde M, Nuyens V, Nagy N, et al. Assessment of oxidative stress in tumors and histologically normal mucosa from patients with head and neck squamous cell carcinoma: a preliminary study. Eur J Cancer Prev 2013;22:558–60.

[53] Gupta A, Bhatt ML, Misra MK. Lipid peroxidation and antioxidant status in head and neck squamous cell carcinoma patients. Oxid Med Cell Longev 2009;2:68–72.

[54] Ratajczak-Wrona W, Jablonska E, Antonowicz B, Dziemianczyk D, et al. Levels of biological markers of nitric oxide in serum of patients with squamous cell carcinoma of the oral cavity. Int J Oral Sci 2013;5:141–5.

[55] Lin X, Zheng W, Liu J, Zhang Y, et al. Oxidative stress in malignant melanoma enhances tumor necrosis factor-α secretion of tumor associated macrophages that promote cancer cell invasion. Antioxid Redox Signal 2013;19:1337–55.

[56] Rass K, Reichrath J. UV damage and DNA repair in malignant melanoma and nonmelanoma skin cancer. Adv Exp Med Biol 2008;624:162–78.

[57] Brar SS, Kennedy TP, Sturrock AB, Huecksteadt TP, et al. An NAD(P)H oxidase regulates growth and transcription in melanoma cells. Am J Physiol Cell Physiol 2002;282:1212–24.

[58] Sander CS, Hamm F, Elsner P, Thiele JJ. Oxidative stress in malignant melanoma and non-melanoma skin cancer. Br J Dermatol 2003;148:913–22.

[59] Batista L, Gruosso T, Mechta-Grigoriou F. Ovarian cancer emerging subtypes: role of oxidative stress and fibrosis in tumor development and response to treatment. Int J Biochem Cell Biol 2013;45:1092–8.

[60] Xia C, Meng Q, Liu LZ, Rojanasakul Y, et al. Reactive oxygen species regulate angiogenesis and tumor growth through vascular endothelial growth factor. Cancer Res 2007;67:10823–30.

[61] Chan DW, Liu UW, Tsao GS, Yao KM, et al. Loss of MK3 mediated by oxidative stress enhances tumorigenicity and chemiluminescences of ovarian cancer cells. Carcinogenesis 2008;29:1742–50.

[62] Mateescu B, Batista L, Cardon M, Gruosso T, et al. miR-141 and miR-200a act on ovarian tumorigenesis by controlling oxidative stress response. Nat Med 2011;17:1262–72.

[63] Senthil K, Aranganathan S, Nalini N. Evidence of oxidative stress in the circulation of ovarian cancer patients. Clin Chim Acta 2004;339:27–32.

[64] Sanchez M, Torres JV, Tormos C, Iradi A, et al. Impairment of antioxidant enzymes, lipid peroxidation and 8-oxo-2′-deoxyguanosine in advanced epithelial ovarian carcinoma of a Spanish community. Cancer Lett 2006;233:28–35.

[65] Solopova NV, Samoylova AA, Titova NM, Savchenko AA, et al. Lipid peroxidation in erythrocyte membrane of women with benign and malignant neoplasm of the endometrium. Bull Exp Biol Med 2010;149:437–9.

[66] Leung PS, Chan XC. Role of oxidative stress in pancreatic inflammation. Antioxid Redox Signal 2009;11:135–65.

[67] Vaquero EC, Edderkaoui M, Pandol SJ, Gukovsky I, et al. Reactive oxygen species produced by NAD(P)H oxidase inhibit apoptosis in pancreatic cancer cells. J Biol Chem 2004;279:34643–54.

[68] Mochizuki N, Furuta S, Mitsushita J, Shang WH, et al. Inhibition of NADPH oxidase 4 activates apoptosis via the AKT/apoptosis signal-regulating kinase 1 pathway in pancreatic cancer PANC-1 cells. Oncogene 2006;25:3699–707.

[69] Kodydkova J, Vavrova L, Stankova B, Macasek J, et al. Antioxidant status and oxidative stress in pancreatic cancer and chronic pancreatitis. Pancreas 2013;42:614–21.

[70] Brar SS, Corbin Z, Kennedy TP, Hemendinger R, et al. NOX5 NAD(P)H oxidase regulates growth and apoptosis in DU 145 prostate cancer cells. Am J Physiol Cell Physiol 2003;285: C353–69.

[71] Kumar B, Koul S, Khandrika L, Meacham RB, et al. Oxidative stress is inherent in prostate cancer cells and is required for aggressive phenotype. Cancer Res 2008;68:1777–85.

[72] Battisti V, Maders LD, Bagatini MD, Reetz LG, et al. Oxidative stress and antioxidant status in prostate cancer patients: relation to Gleason score, treatment and bone metastasis. Biomed Pharmacother 2011;65:516–24.

[73] Akinci M, Kosova F, Cetin B, Sepici A, et al. Oxidant/antioxidant balance in patients with thyroid cancer. Acta Cir Bras 2008;23:551–4.

[74] Abdull Razis AF, Noor NM. Cruciferous vegetables: dietary phytochemicals for cancer prevention. Asian Pac J Cancer Prev 2013;14:1565–70.

[75] Williamson G, Faulkner K, Plumb GW. Glucosinolates and phenolics as antioxidants from plant foods. Eur J Cancer Prev 1998;7:17–21.

[76] Giacosa A, Barale R, Bavaresco L, Gatenby P, et al. Cancer prevention in Europe: the Mediterranean diet as a protective choice. Eur J Cancer Prev 2013;22:90–5.

[77] Darvesh AS, Bishayee A. Chemopreventive and therapeutic potential of tea polyphenols in hepatocellular cancer. Nutr Cancer 2013;65:329–44.

[78] Lai G, Weinstein SJ, Albanes D, Taylor PR, et al. The association of coffee intake with liver cancer incidence and chronic liver disease mortality in male smokers. Br J Cancer 2013;109:1344–51.

[79] Alexander B, Fishman AI, Eshghi M, Choudhury M, et al. Induction of cell death in renal cell carcinoma with combination of d-fraction and vitamin C. Integr Cancer Ther 2013;12:445–8.

[80] Azcarate-Peril MA, Sikes M, Bruno-Barcena JM. The intestinal microbiota gastrointestinal environment and colorectal cancer: a putative role for probiotics in prevention of colorectal cancer? Am J Physiol Gastrointest Liver Physiol 2011;301:G401–24.

[81] Cappellani A, Zanghi A, Di Vita M, Cavallaro A, et al. Strong correlation between diet and development of colorectal cancer. Front Biosci (Landmark Ed) 2013;18:190–8.

[82] Vieira FG, Di Pietro PF, Boaventura BC, Ambrosi C, et al. Factors associated with oxidative stress in women with breast cancer. Nutr Hosp 2011;26:528–36.

[83] Seitz HK, Pelucchi C, Bagnardi V, La Vecchia C. Epidemiology and pathophysiology of alcohol and breast cancer: update 2012. Alc Alc 2012;47:204–12.

[84] Jung EJ, Shin A, Park SK, Ma SH, et al. Alcohol consumption and mortality in the Korean Multi-Center Cancer Cohort Study. J Prev Med Public Health 2012;45:301–8.

[85] Na HK, Oliynyk S. Effects of physical activity on cancer prevention. Ann NY Acad Sci 2011;1229:176–83.

[86] Friedenreich CM, Orenstein MR. Physical activity and cancer prevention: etiologic evidence and biological mechanisms. J Nutr 2002;132(11Suppl):3456S–64S.

[87] Mandelblatt JS, Luta G, Kwan ML, Makgoeng SB, et al. Associations of physical activity with quality of life and functional ability in breast cancer patients during active adjuvant treatment: the Pathways Study. Breast Cancer Res Treat 2011;129:521–9.

[88] Thompson HJ, Jiang W, Zhu Z. Candidate mechanism accounting for effects of physical activity on breast carcinogenesis. IUBMB Life 2009;61:895–901.

[89] De Cree C, Ball P, Seidlitz B, Van Kranenburg G, et al. Responses of catecholestrogen metabolism to acute graded exercise in normal menstruating women before and after training. J Clin Endocrinol Metab 1997;82:3342–8.

[90] Lee AH, Su D, Pasalich M, Wong YL. Habitual physical activity reduces risk of ovarian cancer: a case controlled study in southern China. Prev Med 2013;57(Suppl):S31–3.

[91] Lin J, Cook NR, Albert C, Zaharris E, et al. Vitamin C and E and beta-carotene supplementation and cancer risk: a randomized clinical trial. J Natl Cancer Inst 2009;101:14–23.

CHAPTER 9

Diabetes and Oxidative Stress

CONTENTS

9.1 INTRODUCTION

Increasingly prevalent, diabetes mellitus represents a major public health concern and by 2030, an estimated 366 million people will be affected by diabetes worldwide [1]. Diabetes mellitus is characterized by relative or absolute deficiency of insulin secretion and/or insulin resistance that results in chronic hyperglycemia along with impaired metabolism of carbohydrates, lipids, and proteins. Oxidative stress, the imbalance between pro-oxidants and antioxidants, has been implicated in the pathogenesis of diabetes mellitus [2]. Studies suggest that oxidative stress plays an important role in systemic inflammation, impaired insulin secretion by pancreatic β cells, impaired glucose utilization by peripheral tissues, and endothelial dysfunction observed in patients with diabetes mellitus. Therefore, neutralization of oxidative stress can potentially reduce various complications of diabetes mellitus, including inhibiting the development of endothelial dysfunction, cardiomyopathy, retinopathy, nephropathy, and neuropathy in patients suffering from this disease [3]. Studies have shown increased concentrations of markers of protein, lipid, and DNA oxidation in serum, urine, or other body fluids in diabetic patients. Specifically, these markers are significantly more elevated in diabetic patients with macrovascular complications compared to other patients with diabetes [4]. Thus, oxidative stress has been identified as a causative factor for increased morbidity and mortality in patients with diabetes, and reducing oxidative stress in such patients may be useful for preventing some of the complications of diabetes.

9.2 VARIOUS TYPES OF DIABETES

Diabetes can be broadly classified into two categories—diabetes mellitus and diabetes insipidus. Diabetes mellitus is a syndrome, characterized by hyperglycemia due to relative insulin deficiency or insulin resistance. It is

A. Dasgupta and K. Klein: Antioxidants in Food, Vitamins and Supplements. DOI: http://dx.doi.org/10.1016/B978-0-12-405872-9.00009-4

important to note that diabetes insipidus, which is also characterized by polyuria, is different from diabetes mellitus because diabetes insipidus is not related to insulin secretion or insulin resistance but is an uncommon condition that occurs when the kidneys are unable to concentrate urine properly. The cause of diabetes insipidus is lack of secretion of antidiuretic hormone (ADH), as observed in patients with cranial diabetes insipidus (also known as central diabetes insipidus), or the inability of ADH to work at the collecting duct of the kidneys (nephrogenic diabetes insipidus). Diabetes mellitus can be primary or secondary in nature. Primary diabetes mellitus can be monogenic or polygenic. Monogenic diabetes mellitus covers a heterogeneous group of diabetes caused by a single gene mutation and characterized by impaired insulin secretion by β cells of the pancreas. Maturity-onset diabetes of the young (MODY), mitochondrial diabetes, and neonatal diabetes are examples of monogenic diabetes mellitus. The diagnosis of monogenic diabetes mellitus and differentiating this type of diabetes from type 1 and type 2 diabetes mellitus are essential. Monogenic diabetes accounts for 2–5% of all diabetes and is less common than type 1 and type 2 diabetes.

Type 1 diabetes, formerly known as insulin-dependent diabetes or juvenile-onset diabetes, is encountered in 5–10% of all patients with diabetes mellitus and is characterized by polyuria, polydipsia, and rapid weight loss. Type 1 diabetes is due to autoimmune destruction of pancreatic β cells by T lymphocytes. Both genetic susceptibility and environmental factors play important roles in the pathogenesis. Genetic susceptibility is polygenic, with the greatest contribution coming from the HLA region, but no single gene responsible for type 1 diabetes has been characterized. Type 2 diabetes mellitus is the most common form of diabetes mellitus, accounting for more than 90% of all cases. It was formerly referred to as non-insulin-dependent diabetes mellitus. Type 2 diabetes mellitus has adult onset and is characterized by insulin resistance and also may be accompanied by β cell dysfunction causing insulin deficiency. Many patients with type 2 diabetes mellitus are obese because obesity alone can cause insulin resistance. However, these patients may not need insulin initially after diagnosis or possibly throughout their lives. Timely diagnosis of type 2 diabetes mellitus is important because early intervention can prevent many complications, including neuropathy, nephropathy, and retinopathy. However, such diagnosis is often difficult because hyperglycemia develops gradually and during the early stage a patient may not notice any classic symptoms of diabetes. Ketoacidosis seldom occurs in type 2 diabetes, and, when seen, it is usually associated with a stress factor such as infection. Characteristics of types 1 and 2 diabetes mellitus and MODY are summarized in Table 9.1.

The American Diabetes Association recommends screening for diabetes mellitus for any individual older than age 45 years. Fasting blood glucose and

Table 9.1 Major Features of Type 1 and Type 2 Diabetes Mellitus and MODY

Clinical Feature	Type 1 DM	Type 2 DM	MODY
Typical age of diagnosis	<25 years	>25 years	<25 years
Body weight	Usually not obese	Overweight to obese	No obesity
Autoantibodies	Present (90%)	Absent	Absent
Insulin dependence	Yes	No	No
Family history	Infrequent	Frequent	Yes, in multiple generations
Diabetic ketoacidosis	High risk	Low risk	Low risk

DM, diabetes mellitus; MODY, maturity-onset diabetes of the young.

glycated hemoglobin A1c (HbA1c) are the best criteria for diagnosis of diabetes mellitus. Guidelines of the Expert Committee on Diagnosis and Classification of Diabetes Mellitus indicate that normal fasting glucose level should be 70–99 mg/dL (3.9–5.5 mmol/L). Individuals with fasting glucose levels between 100 mg/dL (5.6 mmol/L) and 125 mg/dL (6.9 mmol/L) are classified as having impaired fasting glucose. Impaired glucose tolerance and impaired fasting glucose are both considered to be prediabetic conditions because individuals are at higher risk of developing diabetes. Impaired glucose tolerance is present in an individual when the 2-h glucose value in the glucose tolerance test is in the range of 140 mg/dL (7.8 mmol/L) to 199 mg/dL (11.0 mmol/L). These conditions are regarded as prediabetic. HbA1c values between 5.7 and 6.4% may also be considered as a prediabetic condition in adults. The criteria for diagnosis of diabetes mellitus are fasting plasma glucose 126 mg/dL (7.0 mmol/L) or higher on more than one occasion with no calorie intake in the past 8 h. However, in a patient with classic symptoms of hyperglycemia, a random plasma glucose of 200 mg/dL (11.1 mmol/L) or greater is an indication of diabetes mellitus. In addition, in the glucose tolerance test, after an oral dose of 75 g of glucose, 2-h plasma glucose of 200 mg/dL (11.1 mmol/L) or greater indicates diabetes mellitus. Performance of the oral glucose tolerance test to establish the diagnosis of diabetes mellitus is only recommended for pregnant women. An international expert committee recommended use of the hemoglobin A1c (glycated hemoglobin) test for diagnosis of diabetes mellitus with a cutoff value of 6.5% [5]. Criteria for diagnosis of diabetes mellitus are the same for both adults and children. For diagnosis of gestational diabetes, typically the glucose tolerance test is performed using 75 g of oral anhydrous glucose during weeks 24–28 of gestation in pregnant women. The glucose tolerance test is typically performed in the morning after at least 8 h of overnight fasting. Various laboratory-based criteria for the diagnosis of diabetes mellitus and gestational diabetes are summarized in Table 9.2. If a woman has gestational

Table 9.2 Various Laboratory-Based Criteria for Diagnosis of Diabetes Mellitus and Gestational Diabetes Mellitus

Laboratory Test	Value	Comment
Diabetes mellitus		
Fasting blood glucose[a]	70–99 mg/dL	Normal value
	100–125 mg/dL	Impaired fasting glucose
	≥126 mg/dL	Determined on at least two occasions indicative of diabetes
Random blood glucose	≥200 mg/dL	Indicative of diabetes in a patient with suspected diabetes mellitus
Glucose tolerance test (2 h)[b]	<140 mg/dL	Normal
	140–199 mg/dL	Impaired glucose tolerance
	≥200 mg/dL	Indicative of diabetes
Hemoglobin A1c[c]	5.7–6.4%	Increased risk of diabetes
	≥6.5%	Indicative of diabetes
Gestational diabetes		
Glucose tolerance test	≥92 mg/dL (fasting)	Any of the criteria (fasting glucose or 1- or 2-h glucose) in glucose tolerance test is equal to or exceeding the limit
	≥180 mg/dL (1 h)	
	≥153 mg/dL (2 h)	

[a]*Fasting glucose means no calorie intake for at least 8 h.*
[b]*Glucose tolerance test is typically performed using 75 g of glucose given orally. The test is preferably performed in the morning in ambulatory patients after overnight fasting.*
[c]*The International Federation of Clinical Chemistry (IFCC) recommends expressing hemoglobin A1c value in millimoles of HbA1c/moles of hemoglobin units.*

diabetes mellitus, the risk of developing diabetes later is higher for both mother and child compared to the risk for a woman with no gestational diabetes.

Many investigators have demonstrated a link between oxidative stress and diabetes mellitus as well as increased levels of markers of oxidative stress in serum, urine, or other body fluids of patients with diabetes mellitus. However, there is no evidence of a link between oxidative stress and diabetes insipidus.

9.3 ELEVATED MARKERS OF OXIDATIVE STRESS IN PATIENTS WITH DIABETES MELLITUS

Because type 2 diabetes mellitus is more common than type 1 diabetes mellitus, more investigations have focused on demonstrating elevated levels of markers of oxidative stress in patients suffering from type 2 diabetes mellitus. A common end point of chronic hyperglycemia is cellular changes that lead to the generation of reactive oxygen free radicals, thus increasing oxidative stress. Formation of advanced glycation end products due to nonenzymatic glycation of proteins lipids and nucleic acids represents another source of generation of free radicals under hyperglycemic conditions. Advanced glycation products can also impair various organ functions, causing diabetic complications [6]. It has been speculated that four different pathways are responsible for complications of diabetes mellitus: polyol pathway, advanced glycation pathway, activation of protein kinase C (PKC) pathway, and hexosamine pathway. However, all four of these pathogenic mechanisms reflect a single hyperglycemia-induced process—overproduction of superoxide anion free radicals by the mitochondrial electron transport system. Studies indicate that hyperglycemia through continuous supply of proton donors from the tricarboxylic acid cycle reduces the threshold value for the proton gradient in the mitochondrial membrane, thus significantly increasing superoxide production. In addition to production of superoxide anion radicals, overproduction of other reactive oxygen and nitrogen species leads to increased oxidative stress and complications in diabetic patients, including diabetic peripheral neuropathy [7]. Moreover, overproduction of superoxide by the mitochondrial electron transport system and transition metal-catalyzed autoxidation of glucose can also generate free radicals such as superoxide anion radicals and hydrogen peroxide. Transition metal can also react with Amadori adducts of protein glycosylation, thus generating free radicals [8].

Kostolanska *et al.* [9] studied the relationship between glycation product, lipid peroxidation, and complications in 50 type 1 diabetes mellitus patients aged 7–19 years. Twelve healthy children were also studied as controls. The authors demonstrated higher concentrations of lipid peroxide, glycosylated hemoglobin (HbA1c), fructosamine, and advanced glycation end products in patients with type 1 diabetes mellitus compared to controls. Interestingly, these parameters were more elevated in patients who showed diabetic complication compared to patients with no complications, indicating that overproduction of advanced glycation products and lipid peroxidation products may be linked to complications of diabetes. Pitocco *et al.* [10] demonstrated high levels of isoprostane, a marker of lipid peroxidation product, in urine of both patients with type 1 diabetes and patients with type 2 diabetes.

Many investigators have demonstrated increased lipid peroxidation products and other markers of oxidative stress in blood of patients with type 2 diabetes mellitus. Likidlilid *et al.* [11] observed higher levels of malondialdehyde (end product of lipid peroxidation) in patients with type 2 diabetes mellitus compared to healthy controls, and the highest levels were observed in patients who also suffered from coronary heart disease in addition to type 2 diabetes mellitus. In another study, the authors found an excellent positive correlation between malondialdehyde (measured as thiobarbituric acid reactive substance) and glycosylated hemoglobin but a negative correlation between glycosylated hemoglobin and antioxidant enzymes such as glutathione peroxidase and glutathione reductase, in patients with type 2 diabetes, indicating increased oxidative stress in these patients [8]. Kalousova *et al.* [12] compared advanced glycation end products (AGEs) and advanced oxidation protein (AOPPs) and adducts in sera of 18 patients with type 1 and 34 patients with type 2 diabetes mellitus. They observed that AGEs were significantly elevated in patients with type 2 diabetes mellitus compared to controls but not in patients with type 1 diabetes. In contrast, AOPPs were significantly elevated in patients with both types of diabetes compared to controls, but levels were higher in type 1 patients compared to type 2 patients. AGEs correlated with fasting blood glucose and glycosylated hemoglobin only in patients with type 1 diabetes. The authors concluded that oxidative stress is increased in patients with both types of diabetes. Gradinaru *et al.* [13] also reported elevated levels of AOPPs, AGEs, oxidized low-density lipoprotein (LDL), and NOx (nitric oxide metabolic pathway products) in elderly patients with type 2 diabetes compared to controls.

9.4 HYPERGLYCEMIA, OXIDATIVE STRESS, AND IMPAIRED INSULIN SECRETION

Hyperglycemia is the major feature of diabetes mellitus caused by either insulin insufficiency or insulin resistance. Although numerous causative factors have been implicated in the pathophysiology of diabetes mellitus, oxidative stress seems to play a critical role [14]. Pancreatic β cells are particularly vulnerable to increased oxidative stress. The insulin gene is expressed almost exclusively in pancreatic β cells, and metabolic regulation of insulin gene expression is essential for proper secretion of insulin by pancreatic β cells. Although glucose is the major physiological regulator of insulin gene expression, chronic hyperglycemia (glucotoxicity) and elevated lipid concentrations (lipotoxicity) result in worsening of β cell functions in patients with type 2 diabetes, partly due to inhibition of insulin gene expression [15]. Fatty acid levels are also significantly elevated in patients with diabetes mellitus compared to healthy individuals. Prolonged exposure of pancreatic β cells to elevated levels of fatty acids impairs β cell function, causing impaired insulin secretion. Kelpe *et al.* [16] demonstrated that elevated levels of palmitate affect insulin gene expression via transcription mechanisms and ceramide synthesis. Hyperglycemia has also been noted to inhibit the functionality of the insulin promoter region because hyperglycemia can induce loss of two critical proteins (pancreatic and duodenal homeobox factor-1 and macrophage-activating factor A) that play important roles in the activation of the insulin promoter region [17,18]. Hyperglycemia can further compromise the promoter region of the insulin gene by upregulating the transcription factor C/EBP (also called CCAAT-enhancer binding protein; CCAAT stands for cytidine–cytidine–adenosine–adenosine–thymidine), which functions to repress insulin promoter activity [19]. Additional deleterious effects of hyperglycemia include induction of endoplasmic reticulum stress, mitochondrial dysfunction, and oxidative stress-induced insulin resistance [20]. Although each effect differs mechanistically, collectively they alter glucose homeostasis and lead to increasingly elevated physiological concentrations of glucose. Studies have shown an increased apoptotic rate in β islet cells cultured in varying concentrations of glucose [21].

9.5 DIABETIC COMPLICATIONS, MORBIDITY, AND OXIDATIVE STRESS

Just as oxidative stress has been linked to the destruction of pancreatic islet cells and the development of diabetes mellitus, its damaging effects have been shown to impact the functionality of other tissue types. Reactive oxygen and nitrogen species have been demonstrated to negatively impact the vascularity of many organ systems. The vascular complications of diabetes mellitus can be classified into two categories—microvascular and macrovascular complications. Examples of microvascular complications in diabetes are nephropathy, retinopathy, and neuropathy. Diabetic macrovascular complications include ischemic heart disease, cerebral vascular accidents, and peripheral vascular disease [22]. Although not all diabetic patients are affected, the increased morbidity and mortality associated with these complications highlights the importance of understanding the relationship between hyperglycemia, oxidative stress, and vascular injury.

Rema *et al.* [23] reported that levels of superoxide dismutase, vitamin E, and vitamin C were markedly reduced among patients with diabetes and those suffering from diabetic retinopathy compared to controls. This finding led the researchers to conclude that a weakened antioxidant defense system due to increased oxidative stress may have a causative role in the development of diabetes and diabetic retinopathy. Increased levels of 8-hydroxydeoxyguanosine (8-OHdG), a marker of DNA oxidation, have also been shown to be associated with an increase in the incidence of diabetic comorbidities; studies have shown increased levels of 8-OHdG in the urine and nonnuclear cells in patients with diabetic retinopathy and nephropathy [24–26]. Furthermore, tubulointerstitial lesions in diabetic nephropathy have been demonstrated to correlate with urinary 8-OHdG excretion [27,28]. Development of neuropathy has also been demonstrated to be associated with oxidative stress. Moreover, oxidative stress has been directly linked to apoptosis of hippocampal and dopaminergic neurons [29,30]. Although several pathways exist to explain the adverse effects of hyperglycemia on vascular tissues (glycation and advance glycation products, activation of the polyol pathway, activation of diacylglycerol and protein kinase C, redox imbalance, carbonyl stress, growth factor alternation, and altered gene expression), the underlying unifying theory remains the same that increased reactive oxygen species leading to significantly increased oxidative stress causes development of diabetic complications.

9.6 EXERCISE, OXIDATIVE STRESS, AND DIABETES

It has been estimated that approximately 60 million Americans may have prediabetic conditions (blood glucose levels above normal but below the criteria of diagnosis of diabetes mellitus), and approximately 24 million Americans are suffering from diabetes. Physical activity can reduce complications of diabetes and may also delay the onset of diabetes in prediabetic patients. *Physical activity* in a broad sense is defined as bodily movement produced by contraction of skeletal muscle that increases expenditure of energy, causing burning of calories. Both aerobic exercise and resistance training can improve insulin action and have beneficial effects on blood glucose and lipid levels. They can also reduce the risk of cardiovascular disease and mortality. However, exercise must be performed on a regular basis. According to the guidelines of the American College of Sports Medicine and American Diabetes Association, a weekly minimum of 150 min (2.5 h) of moderate to vigorous physical activity is recommended (for details see Box 9.1), but for weight loss, a minimum of 60 min of exercise per day may be needed [31]. Montesi *et al.* [32] commented that lifestyle intervention should be the first therapeutic strategy to prevent/treat metabolic syndrome (obesity, hypertension, dyslipidemia, diabetes, and nonalcoholic fatty liver disease). It has been demonstrated that both prediabetes and type 2 diabetes mellitus are associated with low cardiorespiratory fitness. Therefore, physical activity including walking is beneficial to these patients. Brisk walking is particularly appropriate because it can be practiced without any additional cost and has a low rate of injury. The benefits of physical activity for which there is strong clinical evidence are summarized in Box 9.1.

Various investigators have evaluated the efficacy of yoga and exercise in reducing oxidative stress in diabetic patients. Gordon *et al.* [33] evaluated the impact of Hatha yoga and conventional physical training exercise regimens on biochemical parameters, oxidative stress indicators, and oxidant status in patients with type 2 diabetes. To evaluate these perimeters, 77 type 2 diabetic patients in a Hatha yoga group were matched with a similar number of type 2 diabetic patients in a conventional exercise group and a control group. Various biochemical parameters were measured at baseline and at two consecutive 3-month intervals. Data analysis demonstrated that yoga and physical

Box 9.1 BENEFITS OF PHYSICAL ACTIVITY BASED ON STRONG CLINICAL EVIDENCE (GRADE A EVIDENCE)

- Physical activity causes increased glucose uptake as well as more reliance on carbohydrates to fuel muscle activity. This effect increases with increasing muscular activity.
- Muscular contraction during physical activity facilitates glucose uptake by a different mechanism (this mechanism is not affected in patients with type 2 diabetes, who have insulin resistance).
- Physical activity can acutely improve the action of insulin, and the effect may last 2–72 h.
- Resistance exercise increases muscle mass.
- People with type 2 diabetes should undertake at least 150 min/week of moderate to vigorous aerobic exercise spread over at least 3 days/week, with no more than 2 consecutive days in between aerobic activities.[a]

[a]*Recommendation of the American Diabetes Association.*

training exercises had both preventative and therapeutic effects. The concentrations of fasting glucose in the Hatha yoga and conventional exercise groups after 6 months were decreased by 29.58 and 27.43%, respectively, with significant reductions in serum triglyceride concentrations. In addition, the concentration of malondialdehyde, an indicator of oxidative stress, was reduced by 19.9% in yoga practitioners and 18.1% in those engaging in physical training exercises. Superoxide dismutase, an antioxidant enzyme, also showed improvement in these groups by 24.08 and 20.18%, respectively. The authors concluded that both Hatha yoga and exercise are effective in reducing oxidative stress in patients with type 2 diabetes. Hegde *et al.* [34] also evaluated the efficacy of yoga and exercise in reducing oxidative stress in type 2 diabetes patients. Of the 123 participants involved in this study, 60 were asked to attend yoga classes at least 3 days a week for a 3-month period. At the conclusion of the study, yoga participants had achieved significant reductions in body mass index and levels of fasting glucose, postprandial glucose, glycosylated hemoglobin, and malondialdehyde (a marker of lipid peroxidation) compared to the control group. In addition, yoga participants showed improvement in antioxidants such as glutathione and vitamin C (15 and 60%, respectively). The authors concluded that yoga is effective in reducing oxidative stress in patients with type 2 diabetes. In a follow-up study, Hegde *et al.* [35] further characterized the effects of exercise on patients with type 2 diabetes by examining the impact of diaphragmatic breathing exercises. Participants were asked to practice breathing exercises twice a day for a 3-month period. Indicators of oxidative stress and oxidant status, such as malondialdehyde (MDA), reduced glutathione (GSH), superoxide dismutase (SOD), and vitamins C and E, were assessed. The results demonstrated that malondialdehyde was reduced on average by 21%, whereas the antioxidants GSH and vitamin C were increased by 37 and 45%, respectively. Other parameters of general health also showed improvement. Participants had a decreased waist:hip ratio, body mass index, fasting blood glucose, and postprandial plasma glucose. In addition, reduction in glycated hemoglobin (HbA1c) concentrations was also noted.

Sigal *et al.* [36] evaluated the effects of aerobic and resistance training on glycemic control in patients with type 2 diabetes mellitus. A total of 251 adults with type 2 diabetes were enrolled in the Diabetes Aerobic and Resistance Exercise (DARE) clinical trial. Participants were asked to engage in aerobic, resistance, or both types of exercise three times a week for 22 weeks. At the conclusion of the study, the authors observed that aerobic and resistance training improved glycemic control. In addition, they found that individuals who engaged in both forms of exercise demonstrated the greatest improvement in the mean percentage reduction of HbA1c. Furthermore, exercise-induced improved glycemic control was found to correlate with baseline glycemic control, being greatest in individuals with higher baseline HbA1c levels.

9.7 DIET, OXIDATIVE STRESS, AND DIABETES

Like exercise, diet has been found to improve glycemic control and reduce oxidative stress in patients with diabetes. Vegetables and fruits are a good source of dietary fiber, and many fruits and some vegetables are a good source of antioxidants. In a study of 417 male participants age 65 years or older who also had type 2 diabetes, Takahashi *et al.* [37] demonstrated that consumption of 150 g of vegetables per day significantly reduced the levels of HbA1c. However, participants who consumed 200 g or more vegetables per day showed not only a significant reduction in HbA1c but also reduced serum triglyceride levels. In order to achieve the best glycemic control, the authors recommended a daily vegetable intake of 200 g or more and green vegetable intake of 70 g or more. In Hegde *et al.*'s [38] study of 123 patients with type 2 diabetes aged between 40 and 75 years. 60 patients consumed two low-calorie fruits (sweet lime, apple, or papaya) a day for a 3-month period, whereas 63 controls received standard diet and care. At the conclusion of the study, a drastic increase (64%) in plasma vitamin C was observed with significant reduction in malondialdehyde levels in patients who consumed fruits. In the control group, vitamin C concentration was reduced by 20% and malondialdehyde concentration was increased by 3%. However, other antioxidants, such as vitamin E and the antioxidant enzyme superoxide dismutase, did not show any significant change in the patients who consumed fruits. The authors also observed a reduction in HbA1c levels in diabetic patients who consumed fruits. The authors concluded that regular consumption of fruits can reduce oxidative stress in patients with type 2 diabetes. Moreover, vitamin C level can be used as a biomarker to evaluate fruit consumption by patients.

Vegetarian diet not only has a beneficial role in promoting health and preventing obesity but also can reduce the risk of developing diabetes mellitus in nondiabetic patients. Tonstad *et al.* [39] studied 15,200 men and 26,187 women who were diabetes free at the time of enrollment. After a 2-year follow-up period, the authors observed that vegetarian diet was associated with a substantial and independent reduction in the incidence of diabetes. Vegetarian diet may carry metabolic advantage in preventing type 2 diabetes, whereas intake of red meat and processed meat increases the risk of diabetes. Another study also reported that the prevalence of diabetes increased from 2.9% in vegans to 7.6% in non-vegetarians [40]. The association between oxidative stress, diabetes, and consumption of a vegetarian diet has also been assessed. Seventy-two patients with type 2 diabetes were randomized to a calorie-restricted vegetarian diet or a standard diabetic diet (both diets approximately 500 kcal/day). Patients who consumed the vegetarian diet during the 24-week study period demonstrated increased plasma vitamin C concentrations (16%), increased superoxide dismutase activity (49%), and decreased glutathione levels (42%). Decreased visceral fat was also noted, which probably explained the reduced oxidative stress experienced by patients who consumed the vegetarian diet. Additional health benefits cited in this study include an 8% reduction in LDL cholesterol and a 5% increase in high-density lipoprotein cholesterol in the vegetarian-diet patients [41].

The effect of dietary intake of cholesterol and saturated fats in patients with diabetes has also been evaluated. Studies have shown that decreasing dietary fat intake by 8% resulted in decreased HbA1c, increased plasma ascorbate levels, and a decreased concentration of malondialdehyde, a marker of oxidative stress [42]. Dierckx *et al.* [43] reported that dietary cholesterol and saturated fat intake positively correlated with peroxides in the blood, leading the authors to recommend consumption of more healthy unsaturated fats as a way to combat oxidative stress.

Collectively, these studies highlight the importance of dietary intake in patients with diabetes mellitus for improvement in glycated hemoglobin, decreased visceral fat, improved lipid profiles, and decreased oxidative stress, thus making proper nutrition an important auxiliary therapeutic intervention in diabetes.

9.8 ANTIOXIDANT SUPPLEMENTATION, OXIDATIVE STRESS, AND DIABETES

Antioxidant supplementation has also been postulated to be therapeutically beneficial in diabetes. As such, the role of micronutrient supplementation in patients with diabetes has been extensively investigated. In particular, vitamin E supplementation has been extensively examined because it is an excellent antioxidant. In a study of 30 patients with diabetes (4 patients with insulin-dependent [type 1] diabetes and 26 patients with non-insulin-dependent [type 2] diabetes), Sharma *et al.* [44] observed that vitamin E supplementation for 4 weeks significantly reduced oxidative stress, indicating that vitamin E supplementation may be useful for reduction of free radical-induced oxidative stress. Following vitamin E supplementation, patients with a controlled state of diabetes showed a threefold increase in vitamin E level and significantly decreased lipid peroxide levels measured as thiobarbituric acid reactive substances. Glutathione levels were also increased.

In addition to vitamin E, the supplementation effect of other micronutrients on patients with diabetes has been investigated. Several studies have postulated that calcium and vitamin D may play a role in glucose metabolism [45]. Pittas *et al.* [46] examined the effects of these micronutrients in relation to blood glucose and markers of inflammation. They observed that these dietary supplementations significantly decreased fasting glucose and insulin resistance in adults with impaired fasting blood glucose. Vitamin K (phylloquinone), an essential element in the synthesis of carboxylate clotting factors, also plays an important role in glucose metabolism and insulin sensitivity. Vitamin K (500 μm/day) supplementation has also been found to be beneficial in lowering fasting insulin concentrations and insulin resistance in older nondiabetic individuals [47]. In one study of 510 subjects, the authors observed that dietary phylloquinone uptake was associated with an improvement of cytokines and other markers related to insulin resistance and diabetes [48]. Hypomagnesemia has been reported in patients with type 2 diabetes. Hypomagnesemia may play a role in altered inulin function and endothelial dysfunction. Dasgupta *et al.* [49] noted that hypomagnesemia was associated with poor glycemic control, retinopathy, and nephropathy as well as foot ulcers. In another study, the authors reported that oral magnesium supplementation in type 2 diabetes patients with hypomagnesemia resulted in improved insulin sensitivity and metabolic control [50]. Epidemiological studies have also shown a strong inverse relationship between dietary magnesium intake and the risk of developing type 2 diabetes [51]. However, it is important to note that diabetic patients should consume antioxidant supplements, especially fat-soluble vitamins, under medical advice only. For a healthy individual, consuming fruits and vegetables as part of the daily diet is sufficient to achieve good antioxidant defense. Free radicals also have important biological activities because these molecules are involved in cell signaling pathways, and a proper balance between free radicals and the antioxidant system in the body must be maintained for optimal health. Some studies have indicated that intake of antioxidant supplements may even increase mortality [52].

9.9 CONCLUSION

Diabetes mellitus is a common metabolic disorder, and patients with diabetes often experience higher oxidative stress than healthy people. Complications of diabetes are not only related to the toxic effects of chronic hyperglycemia but also due to increased oxidative stress. Vegetarian diet, and consuming more fruits and vegetables, along with lifestyle change that incorporates regular physical activity are beneficial for diabetic patients because such practices not only reduce oxidative stress but also aid in improving glycemic control. American Diabetes Association guidelines recommend at least 150 min of moderate to vigorous physical activity per week (exercise must be performed at least three times each week; see Box 9.1) in patients with diabetes. Although documented and theorized benefits of antioxidant supplementation exist, it is important to note that antioxidant supplements, especially fat-soluble antioxidant vitamin supplements, must be consumed only under medical advice because use of antioxidant supplements has been associated with increased mortality in some studies.

REFERENCES

[1] Rathmann W, Giani G. Global prevalence of diabetes: estimates for the year 2000 and projections for 2030. Diabetes Care 2004;27:2568–9.

[2] Opara E. Role of oxidative stress in the etiology of type 2 diabetes and the effect of antioxidant supplementation on glycemic control. J Investig Med 2004;52:19–23.

[3] Zatalia SR, Sanusi H. The role of antioxidants in the pathophysiology, complications and management of diabetes mellitus. Acta Med Indones 2013;45:141–7.

[4] Maritim AC, Sanders RA, Watkins JB. Diabetes, oxidative stress and antioxidants: a review. J Biochem Mol Toxicol 2003;17:24–38.

[5] International Expert Committee. International Expert Committee report on the role of the A1C assay in the diagnosis of diabetes mellitus. Diabetes Care 2009;32:1327–34.

[6] Mohamed AK, Bierhaus A, Schiekofer S, Tritschler H, et al. The role of the oxidative stress and NF-kappa B activation in late diabetic complications. Biofactors 1999;10:157–67.

[7] Premkumar LS, Pabbidi RM. Diabetic peripheral neuropathy: role of reactive oxygen and reactive nitrogen species. Cell Biochem Biophys 2013;67:373–83.

[8] Kumawat M, Sharma TK, Singh I, Singh N, et al. Antioxidant enzymes and lipid peroxidation in type 2 diabetes mellitus patients with and without nephropathy. North Am J Med Sci 2013;5:213–19.

[9] Kostolanska J, Jakus V, Barak L. Glycation and lipid peroxidation in children and adolescents with type 1 diabetes mellitus with and without diabetic complications. J Pediatr Endocrinol Metab 2009;22:635–43.

[10] Pitocco D, Zaccardi F, Di Stasio E, Romitelli F, et al. Oxidative stress, nitric oxide and diabetes. Rev Diabet Stud 2010;7:15–25.

[11] Likidlilid A, Patchanans N, Peerapatdit T, Sriratana sathavorn C. Lipid peroxide and antioxidant enzyme activities in erythrocytes of type 2 diabetic patients. J Med Assoc Thai 2010;93:682–93.

[12] Kalousova M, Skrha J, Zima T. Advanced glycation end products and advanced oxidation protein products in patients with diabetes mellitus. Physiol Res 2002;51:597–604.

[13] Gradinaru D, Borsa C, Ionescu C, Margina D. Advanced oxidative and glycoxidative protein damage markers in the elderly with type 2 diabetes. J Proteomics 2013;92:313–22.

[14] Yang H, Jin X, Wai Kei Lam C, Yan S-K. Oxidative stress and diabetes mellitus. Clin Chem Lab Med 2011;49:1773–82.

[15] Poitout V, Hagman D, Stein R, Artner I, et al. Regulation of the insulin gene by glucose and fatty acids. J Nutr 2006;136:873–6.

[16] Kelpe CL, Moore PC, Parazzoli SD, Wicksteed B, et al. Palmitate inhibition of insulin gene expression is mediated at the transcriptional level via ceramide synthesis. J Biol Chem 2003;278:30015–21.

[17] Olson LK, Redmon JB, Towle HC, Robertson RP. Chronic exposure of HIT cells to high glucose concentrations paradoxically decreases insulin gene transcription and alters binding of insulin gene regulatory protein. J Clin Invest 1993;92:514–19.

[18] Matsuoka TA, Zhao L, Artner I, Jarrett HW, et al. Members of the large Maf transcription family regulate insulin gene transcription in islet beta cells. Mol Cell Biol 2003; 23:6049–62.

[19] Lu M, Seufert J, Habener JF. Pancreatic beta-cell-specific repression of insulin gene transcription by CCAAT/enhancer-binding protein beta: inhibitory interactions with basic helix–loop–helix transcription factor E47. J Biol Chem 1997;272:28349–59.

[20] Marshak S, Leibowitz G, Bertuzzi F, Socci C, et al. Impaired beta-cell functions induced by chronic exposure of cultured human pancreatic islets to high glucose. Diabetes 1999; 48:1230–6.

[21] Federici M, Hribal M, Perego L, Ranalli M, et al. High glucose causes apoptosis in cultured human pancreatic islets of Langerhans: a potential role for regulation of specific Bcl family genes toward an apoptotic cell death. Diabetes 2001;50:1290–301.

[22] Jakus V. The role of free radicals, oxidative stress and antioxidant systems in diabetic vascular disease. Bratisl Lek Listy 2000;101:541–51.

[23] Rema M, Mohan V, Bhaskar A, Shanmugasundaram K. Does oxidant stress play a role in diabetic retinopathy? Indian J Ophthalmol 1995;43:17–21.

[24] Dandona P, Thusu K, Cook S, et al. Oxidative damage to DNA in diabetes mellitus. Lancet 1996;347:444–5.

[25] Leinomen J, Lehtimaki T, Toyokuni S, et al. New biomarker evidence of oxidative DNA damage in patients with non-insulin-dependent diabetes mellitus. FEBS Lett 1997; 417:150–2.

[26] Pan H, Zhang L, Guo M, Sui H, et al. The oxidative stress status in diabetes mellitus and diabetic nephropathy. Acta Diabetol 2010;47:S71–6.

[27] Suzuki S, Hinokio Y, Komatu K, et al. Oxidative damage to mitochondrial DNA and its relationship to diabetic complications. Diabetes Res Clin Pract 1985;45:161–8.

[28] Kanauchi M, Nishioka H, Hasimoto T. Oxidative DNA damage and tubulointerstitial injury in diabetic nephropathy. Nephron 2002;91:327–9.

[29] Pugazhenthi S, Nesterova A, Jambal P, Audersirk G, et al. Oxidative stress-mediated down-regulation of bcl-2 promoter in hippocampal neurons. J Neurochem 2003;84: 982–96.

[30] Chong ZZ, Kang JQ, Maiese K. Apaf-1, Bcl-xl, cytochrome c, and caspase-9 form the critical elements for cerebral vascular protection by erythropoietin. J Cereb Blood Flow Metab 2003;23:320–30.

[31] Colberg SR, Sigal RJ, Fernhall B, Regensteiner JG, et al. Exercise and type 2 diabetes: the American College of Sports Medicine and the American Diabetes Association: joint position statement executive summary. Diabetes Care 2010;33:2692–6.

[32] Montesi L, Moscatiello S, Malavolti M, Mazocchi R. Physical activity for the prevention and treatment of metabolic disorders. Intern Emerg Med 2013;8(8):655–66.

[33] Gordon LA, Morrison EY, McGrowder DA, Young R, et al. Effect of exercise therapy on lipid profile and oxidative stress indicators in patients with type 2 diabetes. BMC Compl Alt Med 2008;8:1–10.

[34] Hegde SV, Adhikari P, Kotian S, Pinto V, et al. Effect of 3-month yoga on oxidative stress in type 2 diabetes with or without complications. Diabetes Care 2011;34:2208–10.

[35] Hegde SV, Adhikari P, Subbalakshmi NK, Nandini M, et al. Diaphragmatic breathing exercise as a therapeutic intervention for control of oxidative stress in type 2 diabetes mellitus. Compl Ther Clin Pract 2012;18:151–3.

[36] Sigal RJ, Kenny GP, Boule NG, Wells GA, et al. Effects of aerobic training, resistance training, or both on glycemic control in type 2 diabetes. Ann Intern Med 2007;147:357–69.

[37] Takahashi K, Kamada C, Yoshimura H, Okumura R, et al. Effects of total and green vegetable intakes on glycated hemoglobin A1C and triglycerides in elderly patients with type 2 diabetes mellitus: the Japanese Elderly Intervention Trial. Geriatr Gerontol Int 2012;12(Suppl. 1):50–8.

[38] Hegde SV, Adhikari PMN, D'Souza V. Effects of daily supplementation of fruits on oxidative stress indices and glycaemic status in type 2 diabetes mellitus. Compl Ther Clin Pract 2013;19:97–100.

[39] Tonstad S, Stewart K, Oda K, Batech M, et al. Vegetarian diets and incidence of diabetes in Adventist Health Study-2. Nutr Metab Cardiovasc Dis 2013;23:292–9.

[40] Tonstad S, Butler T, Yan R, Fraser GE. Type of vegetarian diet, body weight, and prevalence of type 2 diabetes. Diabetes Care 2009;32:791–6.

[41] Kahleova H, Matoulek M, Malinska H, Oliyarnik O, et al. Vegetarian diet improves insulin resistance and oxidative stress markers more than conventional diet in subjects with type 2 diabetes. Diabet Med 2011;28:549–59.

[42] Armstrong A, Chestnutt J, Young I. The effect of dietary treatment on lipid peroxidation and antioxidant status in newly diagnosed non-insulin dependent diabetes. Free Radic Biol Med 1996;21:719–26.

[43] Dierckx N, Horvath G, van Gils C, Vertommen J, et al. Oxidative stress status in patients with diabetes mellitus: relationship to diet. Eur J Clin Nutr 2003;57:999–1008.

[44] Sharma A, Kharb S, Chugh SN, Kakkar R, et al. Evaluation of oxidative stress before and after control of glycemia and after vitamin E supplementation in diabetic patients. Metabolism 2000;149:160–2.

[45] Chiu KC, Chu A, Go VL, Saad MF. Hypovitaminosis D is associated with insulin resistance and beta cell dysfunction. Am J Clin Nutr 2004;79:820–5.

[46] Pittas AG, Harris SS, Stark PC, Dawson-Hughes B. The effects of calcium and vitamin D supplementation on blood glucose and markers of inflammation in nondiabetic adults. Diabetes 2007;30:980–6.

[47] Yoshida M, Jacques PJ, Meigs JB, Saltzman E, et al. Effect of vitamin K supplementation on insulin resistance in older men and women. Diabetes Care 2008;31:2092–6.

[48] Juanola-Falgarona M, Salas-Salvado J, Estruch R, Portillo M, et al. Association between dietary phylloquinone intake and peripheral metabolic risk markers related to insulin resistance and diabetes in elderly subjects at high cardiovascular risk. Cardiovasc Diabetol 2013;12:7.

[49] Dasgupta A, Sarma D, Saikia UK. Hypomagnesemia in type 2 diabetes mellitus. Indian J Endocrinol Metab 2012;16:1000–3.

[50] Rodriguez-Moran M, Guerrero-Romero P. Oral magnesium supplementation improves insulin sensitivity and metabolic control in type 2 diabetic subjects: a randomized double-blind control trial. Diabetes Care 2003;26:1147–52.

[51] Rodriguez-Moran M, Mendia S, Galvan Z, Guerrero-Romero P. The role of magnesium in type 2 diabetes: a brief based clinical review. Magnes Res 2011;24:156–62.

[52] Poljsak B, Suput D, Milisav I. Achieving the balance between ROS and antioxidants: when to use the synthetic antioxidants. Oxid Med Cell Longev 2013;2013:956792.

CHAPTER 10

Role of Oxidative Stress in Neurodegenerative Diseases and Other Diseases Related to Aging

CONTENTS

10.1 INTRODUCTION

Oxidative stress has been implicated in numerous disease processes. Because the burden of disease in individuals increases with age, many investigators have postulated that oxidative stress may be involved in the process of aging as well as certain diseases related to advanced age, such as Alzheimer's disease. Numerous studies have directly linked aging to oxidative stress. This chapter discusses the role of oxidative stress in aging and the link between oxidative stress and various neurodegenerative diseases that are more common in advanced age. As discussed throughout the book, the best defense against oxidative stress is a balanced diet with generous servings of fruits and vegetables each day, regular exercise, and a healthy lifestyle. Antioxidant vitamins and supplements do not have any added benefit for healthy individuals. In fact, such supplements may cause more harm than good. Antioxidant vitamins and supplements must be taken only under medical advice. Various age-related diseases linked to oxidative stress are listed in Box 10.1.

10.2 ROLE OF OXIDATIVE STRESS IN AGING

In 1957, Harman proposed the free radical theory of aging [1]. Since then, there have been many publications in support of Harman's theory. The free radical theory of aging hypothesizes that free radicals are responsible for the age-related damage at the cellular and tissue levels. In general, equilibrium exists between free radicals and the body's antioxidant system, but when excess free radicals are generated, this equilibrium is destabilized, contributing to cellular damage. Fusco *et al.* [2] noted that just 1% of reactive oxygen species generated daily evade the body's antioxidant defenses. Mitochondria from postmitotic cells use oxygen at a higher rate, thus releasing oxygen radicals that exceed the cellular antioxidant defense. In the mitochondrial theory

A. Dasgupta and K. Klein: Antioxidants in Food, Vitamins and Supplements. DOI: http://dx.doi.org/10.1016/B978-0-12-405872-9.00010-0

Box 10.1 VARIOUS AGE-RELATED DISEASES LINKED TO OXIDATIVE STRESS

Age-Related Neurodegenerative Diseases

- Alzheimer's disease
- Amyotrophic lateral sclerosis
- Huntington disease
- Parkinson's disease

Age-Related Ophthalmological Disorders

- Glaucoma
- Cataract formation
- Macular degeneration

Age-Related Hearing Disorder

- Presbycusis

Age-Related Musculoskeletal System Diseases

- Osteoarthritis
- Sarcopenia

Age-Related Skin Disorders[a]

- Common skin lesions (dry skin, freckling, senile purpura, lentigines, atrophicus, etc.)
- Benign tumor
- Premalignant tumor
- Malignant tumor (basal cell carcinoma, squamous cell carcinoma, malignant melanoma, etc.)
- Infectious disease (e.g., dermatophytosis)
- Autoimmune disease (e.g., contact dermatitis)
- Pressure ulcers, lower extremity ulcer
- Pruritus

[a]*Only common age-related skin disorders are listed.*

of aging, the key role of mitochondria in cell aging has been outlined by the degeneration induced in cells microinjected with mitochondria isolated from fibroblast of old rats, as well as by the inverse relationship between the rate of mitochondrial production of hydroperoxide and maximum life span of the species. The concurrent enhancement of lipid peroxidation and oxidative modification of proteins in mitochondria results in oxidative damage and contributes to mutation of mitochondrial DNA (mtDNA) during aging. The respiratory enzymes containing the defective mtDNA coded protein subunits may in turn increase the production of free radicals by mitochondria, thus further aggravating the oxidative stress. In addition, superoxide generated during mitochondrial respiration may react with nitric oxide inside mitochondria, producing highly reactive peroxynitrite [3]. Furthermore, oxidative stress has been shown to alter cellular metabolism through cell signaling, and the cumulative effect of these reactive oxygen and nitrogen species results in progressive impairment of cellular functions, thus directly linking oxidative stress to various age-related disorders [4].

10.3 OXIDATIVE STRESS AND AGE-RELATED NEURODEGENERATIVE DISEASES

Oxidative stress has been directly linked to neurodegenerative diseases, which occur mostly in the older population. The nervous system consists of numerous components (brain, spinal cord, nerves, and sensory receptors) that operate collectively to allow for movement and perception. The neuroinflammatory process induced in part by proinflammatory cytokines yields enhanced reactive oxygen species and reactive nitrogen species that have neurotoxic properties [5]. Moreover, damage of neuronal structures and their function due to oxidative stress progresses with aging, thus directly contributing to neurodegenerative diseases of old age. The central nervous system is vulnerable to damage by oxygen free radicals because of the high levels of polyunsaturated lipids in the neuronal cell membranes and poor antioxidant defense system present in such cells. In addition, neurons are postmitotic cells and gradually accumulate oxidative damage over time, which explains why neurodegenerative diseases more commonly occur in older individuals.

10.4 PARKINSON'S DISEASE AND OXIDATIVE STRESS

Parkinson's disease was first described in 1817 by James Parkinson. It is a neurodegenerative disorder clinically characterized by resting tumor, rigidity, akinesia, and bradykinesia, and it is the second most common form of dementia. In Parkinson's disease, protein aggregates of α-synuclein, a protein that plays an important role in mitochondrial function and vesicle formation, are found in Lewy bodies located primarily in the putamen and substantia nigra. These brain regions are mainly involved in learning and motor movement. In addition, the neurotransmitter dopamine is dramatically lost in neurons, causing substantial neuronal death in Parkinson's disease [6]. In Parkinson's disease, α-synuclein released from the cell membrane is modified by acrolein (lipid peroxidation generates both acrolein and 4-hydroxy-2-nonenal) and accumulates in the cytoplasm, forming Lewy bodies. Therefore, lipid peroxidation plays a role in the pathophysiology of Parkinson's disease [7]. The end products of lipid peroxidation, such as 4-hydroxy-2-nonenal, are responsible for cytotoxicity in conjunction with oxidative stress acting as a second toxic messenger of free radicals. In addition, 4-hydroxy-2-nonenal is also present in Lewy bodies [8], which may partly explain another paramount finding in this disease—the loss of pigmented neurons. Experiments using the rat model have shown that 4-hydroxyl-2-nonenal can alter dopamine transport, and such alteration of dopamine can cause neuronal death (dopamine is the main neurotransmitter involved in motor function). Therefore, it can be postulated that this oxidative damage may partly contribute to the pathophysiology of this disease [9].

Elevated markers of lipid peroxidation and oxidative damage of DNA in patients with Parkinson's disease have also been investigated. In a study of 211 patients with Parkinson's disease and 135 healthy controls, Chen *et al.* [10] observed significantly elevated levels of malondialdehyde (measured by high-performance liquid chromatography) in plasma of patients with Parkinson's disease compared to controls. Leukocyte levels of 8-hydroxy-2-deoxyguanosisne, a marker of oxidative damage to DNA, were also elevated in patients with Parkinson's disease compared to healthy individuals. Leukocyte 8-hydroxy-2-deoxyguanosisne level continuously increased with advanced Parkinson's disease, whereas plasma malondialdehyde level peaked at an early stage of the disease. In addition, erythrocyte glutathione peroxidase levels and plasma vitamin E levels were reduced in patients with Parkinson's disease compared to controls. The authors concluded that increased oxidative stress and decreased antioxidant capacity in peripheral blood of patients with Parkinson's disease and significant correlation between leukocyte 8-hydroxy-2-deoxyguanosisne and various stages of the disease indicate that oxidative stress plays an important role in the pathogenesis of Parkinson's disease. In a study of 80 patients with Parkinson's disease and 80 controls, Sanyal *et al.* [11] demonstrated elevated plasma levels of malondialdehyde in patients with Parkinson's disease (mean, 7.48 nmol/mL) compared to controls (mean, 5.1 nmol/mL).

10.5 HUNTINGTON DISEASE AND OXIDATIVE STRESS

Huntington disease is a fatal autosomal dominant neurodegenerative disease (life expectancy approximately 20 years after visible symptoms are seen, and it may take 10 years for full manifestation of the disease) of midlife onset that is caused by an expanded DNA segment containing a polymorphic trinucleotide CAG (cytosine–adenine–guanine) that encodes huntingtin protein. Therefore, this mutation increases the size of the Huntingtin gene. The normal gene has 35 or fewer CAG repeats, but people with Huntington disease have 36 or more repeats, and a case with more than 120 repeats has been reported. People with 36–39 repeats may not develop the signs and symptoms of Huntington disease, but people with 40 or more repeats almost always develop the disorder. The expanded CAG repeat in the Huntingtin gene results in the production of an abnormally long version of the huntingtin protein containing the polyglutamine domain (the normal N-terminus of the huntingtin protein contains 10–35 glutamines, but defective protein may have 40 or more glutamine repeats). The elongated protein is cut into smaller toxic fragments that bind together and accumulate in neurons, particularly those in the striatum and cerebral cortex region of the brain, disrupting their functions. Clinical manifestations include involuntary jerky limb movements, increased eye blinking, and fidgeting. Although genetic inheritance has been strongly implicated in the pathogenesis of the disease, oxidative stress also seems to play a role in the pathogenesis of Huntington disease, as evidenced by the observation of 4-hydroxy-2-nonenal adducts in the caudate nucleus and putamen of brains and striatum of male transgenic Huntington disease mice [12].

Postmortem specimens obtained from brains of patients with Huntington disease and from age- as well as sex-matched controls demonstrated elevated levels of protein carbonyls in the brains of Huntington disease patients compared to controls [13]. Elevated levels of malondialdehyde, 8-hydroxy-2-deoxyguanosisne, 3-nitrotyrosine, and heme oxygenase in areas of degeneration in the brains of Huntington disease patients as well as increased free radical production in animal models indicate the involvement of oxidative stress either as a causative event or as a secondary constituent of cell death cascade in Huntington disease [14]. In a study of 16 patients with Huntington disease and 36 age- and gender-matched controls, Chen *et al.* [15] observed elevated plasma malondialdehyde levels as well as elevated leukocyte 8-hydroxy-2-deoxyguanosine in patients with Huntington disease compared to controls. Moreover, activities of antioxidant enzymes copper/zinc superoxide dismutase and glutathione peroxidase were significantly lower in erythrocytes of patients with Huntington disease compared to controls. Plasma malondialdehyde level also correlated with the severity of Huntington disease in these patients.

10.6 AMYOTROPHIC LATERAL SCLEROSIS AND OXIDATIVE STRESS

Amyotrophic lateral sclerosis (ALS), also known as Lou Gehrig's disease, was discovered in 1869 by Jean-Martin Charcot. Clinically, it manifests as a chronic progressive loss of all voluntary movement resulting in muscle weakness, atrophy, and spasticity due to killing of motor neurons, thus causing progressive paralysis in mid-life. Generally, it is fatal within 1–5 years of onset, and the risk of the disease increases significantly after 60 years of age. In 90–95% of patients with ALS, there is no apparent genetic link (sporadic ALS), whereas in 5–10% of cases, the disease is inherited in a dominant manner (familial ALS). The typical onset of both forms of disease is at 50–60 years of age [16]. Oxidative damage may play a role in ALS because mutation in cytoplasmic copper–zinc superoxide dismutase (SOD1, which neutralizes superoxide radical) has been documented in approximately 20% of patients with ALS. Mutation of other genes may also be related to ALS [17]. Studies have shown that alteration of this antioxidant enzyme results in increased lipid peroxidation in ALS. In addition, increased modification of proteins by 4-hydroxy-2-nonenal, a product of membrane lipid peroxidation, was observed in the lumbar spinal cord of sporadic ALS patients compared to controls. Immunoprecipitation analysis revealed that one such modified protein is the astrocytic glutamate transporter, the modification of which may result in severely compromised function of this protein. The authors commented that oxidative stress leads to the production of 4-hydroxy-2-nonenal, which may impair the function of the glutamate transporter and excitotoxic motor neuron degeneration in ALS [18]. In addition, by using transgenic mice that overexpress mutated superoxide dismutase, investigators have shown that oxidative stress can induce modification of numerous proteins that results in alteration of energy metabolism, mitochondrial dysfunction, proteasomal overload, and alteration of cytoskeletal integrity [19]. Baillet *et al.* [20] studied 31 patients with ALS, 24 patients with Parkinson's disease, and 30 healthy subjects. They observed elevated plasma lipid peroxidation product (measured as malondialdehyde) in patients with ALS (mean, 1.78 μmol/L) and Parkinson's disease (mean, 1.74 μmol/L) compared to controls (mean, 1.41 μmol/L). A trend toward increased oxidized glutathione was observed in both ALS and Parkinson's disease patients compared to controls. These results indicated a role of oxidative stress in the pathogenesis of both ALS and Parkinson's disease.

10.7 ALZHEIMER'S DISEASE AND OXIDATIVE STRESS

Alzheimer's disease is an age-related neurodegenerative disease that was first described by Alois Alzheimer in 1907. It is the most common form of dementia, affecting 30 million people worldwide. Clinically, it is characterized by irreversible progressive memory loss, and histopathological features include the presence of neurofibrillary tangles, senile plaques, and brain atrophy. Alzheimer's disease rarely has an early onset (ages 30–60 years), with only 5% of all cases being early onset. Some cases of early onset have no known cause, but most cases are inherited—a type known as familial Alzheimer's disease caused by any one of three defective genes: amyloid precursor protein gene (APP; chromosome 21), presenilin 1 (PSEN1; chromosome 14), and presenilin 2 (PSNE 2; chromosome 1). The apolipoprotein E gene (ApoE, especially the ApoE ε4 allele) has been recognized as a risk factor for sporadic late onset of Alzheimer's disease. Other genetic mutations may also be linked to Alzheimer's disease [21]. However, the exact mechanism of pathogenesis of Alzheimer's disease has not been elucidated. In addition to genetic susceptibility, hypotheses such as amyloid cascade, excitotoxicity, oxidative stress, and inflammation have been proposed to explain the pathogenesis of Alzheimer's disease.

In Alzheimer's disease, amyloid β peptide, a major component of senile plaque, is generated by the proteolytic cleavage of amyloid precursor protein by the action of β- and γ-secretase. The amyloid β peptide has been shown to induce oxidative stress (methionine at residue 35 is particularly important for inducing oxidative stress) in both *in vitro* and *in vivo* studies. The amyloid β peptide exists in various aggregated stages, such as monomer, oligomer, protofibrils, and fibrils, but the oligomeric form is highly toxic. Evidence that the brains of patients with Alzheimer's disease are under increased oxidative stress is supported by the fact that levels of lipid peroxidation products, such as 4-hydroxy-2-nonenal, acrolein, and isoprostanes, are elevated in these brains. Moreover, 4-hydroxy-2-nonenal can covalently modify the histidine side chains of amyloid β peptide, leading to increased aggregation of this peptide. In addition, 4-hydroxy-2-nonenal can bind to many proteins in the Alzheimer's brain, causing progression of the disease [22]. Furthermore, increased oxidative stress in the Alzheimer's brain is evidenced by increases in iron, lipid peroxidation, protein oxidation, and DNA oxidation; a decline in membrane polyunsaturated fatty acid levels; and the presence of markers of oxidative stress in neurofibrillary tangles and senile (neuritic) plaques. In a study of postmortem brain specimens of patients with Alzheimer's disease, Markesbery and Lovell [23] observed significantly increased free 4-hydroxy-2-nonenal in multiple brain regions of these patients. Using a transgenic mouse model, Pratico *et al.* [24] demonstrated that lipid peroxidation precedes amyloid plaque formation, suggesting that brain oxidative damage contributes to the pathogenesis of Alzheimer's disease before amyloid β accumulation in the Alzheimer's brain.

Elevated markers of lipid peroxidation have also been documented in the plasma of patients with Alzheimer's disease. In a study of 46 patients with Alzheimer's type dementia and 46 controls, Puertas *et al.* [25] observed increased levels of thiobarbituric acid reactive substances and protein carbonyl content in sera of patients with Alzheimer's type dementia compared to controls. Plasma levels of glutathione were also decreased in patients with Alzheimer's type dementia compared to controls. The authors suggested that there is a defect in the antioxidant defense system that is incapable of responding to increased free radical production, which may lead to oxidative damage and the development of pathological alteration that results in the neurodegeneration process in patients with Alzheimer's type dementia. In another study, the authors also observed increased levels of malondialdehyde in plasma of patients with Alzheimer's disease and patients with vascular dementia compared to controls. Interestingly, malondialdehyde levels were increased more in patients with vascular dementia (2.8-fold higher) compared to patients with Alzheimer's disease [26].

The role of nutrition in preventing or delaying the onset of Alzheimer's disease has been explored. In general, epidemiological studies indicate that intake of fish, fruits, and vegetables and moderate consumption of wine may reduce the risk of cognitive decline and dementia. In addition, adherence to a Mediterranean diet may be associated with slower cognitive decline and reduced risk of progression from mild cognitive impairment to Alzheimer's disease [27]. Cognitive engagement and physical activities have been associated with a decreased risk of Alzheimer's disease. However, diabetes, ε4 allele of the ApoE gene, smoking, and depression can increase the risk of Alzheimer's disease. Whereas consumption of fish, fruits, vegetables, and coffee and light to moderate alcohol consumption may reduce the risk of Alzheimer's disease, eating foods rich in saturated fatty acids and high in calories and excess alcohol consumption can increase the risk of Alzheimer's disease [28]. Several epidemiological studies indicate that moderate consumption of red wine is associated with a lower incidence of dementia and Alzheimer's disease because it is enriched with antioxidant polyphenolic compounds, which have a neuroprotective effect. Resveratrol, a polyphenolic antioxidant abundant in red wine, is especially beneficial for reducing the risk of dementia and Alzheimer's disease [29]. Curcumin, a polyphenolic antioxidant in *Curcuma longa* (Indian spice turmeric), has anti-amyloidogenic, antioxidant, anti-inflammatory, and metal chelating properties and may have a protective effect against Alzheimer's disease [30].

10.8 OXIDATIVE STRESS AND AGE-RELATED EYE DISEASE

Glaucoma and cataract formation are usually observed in the older population. Glaucoma is a neurodegenerative disorder characterized by increased intraocular pressure and selective loss of retinal ganglion cells, whose cell bodies lie in the inner surface of the retina and project axons along the optic nerve to the lateral geniculate nucleus. The visual loss in glaucoma indicates that the primary site of injury is within the axons at the optic nerve head. Evidence indicates that oxidative stress occurs in glaucoma, which contributes to selective loss of retinal ganglion cells. Mitochondrial dysfunction in glaucoma patients may be responsible for oxidative stress, and free radicals may be directly toxic to retinal ganglion cells or may contribute to secondary degeneration by affecting other ocular cells and structures, such as glia, immune cells, or the trabecular meshwork [31]. Izzotti *et al.* [32] reported significantly increased oxidative DNA damage in the ocular epithelium of patients with glaucoma compared to controls. The pathogenic role of reactive oxygen species in glaucoma is also supported by multiple observations, including resistance to aqueous humor outflow caused by hydrogen peroxide-induced degeneration of the trabecular meshwork, altered antioxidant defense capacity of the trabecular meshwork in glaucoma patients, and increased ocular pressure as well as severity of visual field defect in glaucoma parallel oxidative DNA damage affecting the trabecular meshwork.

Cataract is the opacification of the lens of the eye. There are three types of cataract (Table 10.1). Exposure to sunlight may increase the risk of cataract

Table 10.1 Types of Cataract

Cataract Type	Comment
Nuclear	This type is located at the center of the lens and is usually an age-related disorder.
Cortical	This type extends from outside the lens to the center and is usually associated with patients who have diabetes.
Subcapsular	This type develops at the back of the lens and may be associated with individuals working with microwave radiation, patients with diabetes, or patients taking steroids.

formation, and ultraviolet (UV) filter in human eyes decreases with age. Therefore, aging eyes may experience higher oxidative stress from sunlight exposure. Glutathione is the primary lenticular antioxidant, and loss of glutathione is probably a crucial feature that may precede cataract formation. In addition, more than 90% of protein sulfhydryl groups are lost in the most advanced cataracts, and almost half of all the methionine residues in nuclear proteins become oxidized to methionine sulfoxide in the most advanced cataracts [33]. In a study of 50 patients with cataracts and 50 controls, Kaur *et al.* [34] observed significantly increased serum lipid peroxidation product in the form of malondialdehyde (measured as thiobarbituric acid reactive substances) in cataract patients (range, 2.3–7.5 nmol/mL) compared to controls (range, 1.5–2.8 nmol/mL). Significantly decreased blood superoxide dismutase and glutathione peroxidase levels were also observed in these patients compared to controls. The authors concluded that oxidative stress plays an important role in the onset and progression of cataracts.

Age-related macular degeneration is a leading cause of irreversible visual impairment and blindness in the older population (age 60 years or older). Based on clinical and pathological features, macular degeneration can be subdivided into atrophic (dry) and neovascular (wet) types. Various risk factors for macular degeneration include cigarette smoking, nutritional factors, obesity, and insufficient antioxidants in the diet. In a study of 25 patients with age-related macular degeneration and 25 controls without age-related macular degeneration, Yildirim *et al.* [35] observed significantly higher levels of malondialdehyde (measured as thiobarbituric acid reactive substances) and advanced oxidation protein products (measured spectrophotometrically) in sera of patients with age-related macular degeneration compared to controls. The activity of superoxide dismutase was significantly lower in sera of these patients compared to controls. The authors concluded that the decreased effectiveness of the antioxidant defense system and the increased oxidative stress may play a role in the pathogenesis of age-related macular degeneration.

10.9 OXIDATIVE STRESS AND AGE-RELATED HEARING LOSS

Age-related hearing loss or presbycusis—one of the most prevalent age-related chronic conditions, affecting an estimated 10 million people worldwide—is a complex degenerative disease caused by impaired function of the cochlea, the auditory portion of the inner ear [36]. Characteristic features include bilateral hearing loss, particularly in relation to high-frequency sounds, and microscopic degeneration of hair cells, stria vascularis, and afferent spiral ganglion neurons. Risk factors include atherosclerosis, smoking, hereditary factors, and exposure to loud noises [37]. Oxidative stress has been linked to each of these risk factors, and studies have shown that exposure to loud noises, atherosclerotic reduction of cochlear blood supply, and inherited hypofunctioning antioxidant defenses result in oxidative damage and even death of hair cells [38]. Kidd and Bao [39] noted that numerous studies have focused on the hypothesis of age-related mitochondrial dysfunction as the cause of presbycusis. This is supported by the observation that many genetic conditions that are linked to hearing loss also cause impaired mitochondrial function, and maternally inherited mutations to the mitochondrial genome may result in deafness. The antioxidant network of cells becomes less efficient with advancing age, and oxidative stress may play a role in the pathogenesis of age-related hearing loss. Animal experiments indicate that the cochlea is hypersensitive to reactive oxygen species induced by mitochondrial damage. Mice missing the gene coding of SOD1 show premature presbycusis. Similarly, mice deficient in antioxidant enzyme glutathione peroxidase suffer accelerated age-related hearing loss. Glutathione S-transferase and *N*-acetyltransferase are two classes of antioxidant enzymes active in cochlea, and polymorphism of genes encoding these two enzymes may increase the risk of presbycusis. Bared *et al.* [40] observed that the risk of presbycusis was increased among white subjects carrying the GSTM1 and GSTT1 null genotype and the NAT*6 mutant allele. Individuals with GSTM1 null genotype were almost three times more likely to develop presbycusis than those with the wild type. GSTM1 null genotype was more prevalent in white Hispanics than in white non-Hispanics, but GSTT1 and NAT*6 were equally represented in both groups.

10.10 OXIDATIVE STRESS AND AGE-RELATED MUSCULOSKELETAL SYSTEM

The musculoskeletal system is composed of bone, cartilage, muscle, ligaments, and tendons. It primarily functions to provide support, allow motion, and provide protection for vital organs. With aging, the integrity of this system declines. Osteoarthritis is a degenerative disease of joints that causes joint stiffness and tenderness. This disease is closely associated with aging, and it is estimated that nearly half of the world's population aged 65 years or older is affected. The mechanism of osteoarthritis is not completely understood, but several hypothesis have been proposed, including the normal wear and tear theory, senescence-related secretory phenotype, low reactivity of chondrocytes to growth factors, abnormal accumulation of advanced glycation product, and oxidative stress due to mitochondrial dysfunction. Mitochondria are a major source of reactive oxygen species in the cell, and mitochondrial dysfunction has been speculated to play an important role in age-related diseases including osteoarthritis. Mitochondrial DNA damage in osteoarthritis is promoted by inflammatory cytokines such as interleukin-1β and tumor necrosis factor-α (TNF-α), and contributes to chondrocyte death. Mechanical injury to cartilage, such as articular cartilage crushing and shearing force injury, may also cause increased production of reactive oxygen species. Moreover, reactive oxygen species may play a key role in age-related chondrocyte changes in several signaling pathways. A high level of reactive oxygen species can inhibit activation of the insulin receptor substrate-1 (IRS-1)–phosphatidylinositol-3-kinase (PI-3 kinase)–Akt (also known as protein kinase B) signaling pathway, which normally promotes matrix synthesis. In addition, reactive oxygen species activate the extracellular signal-regulated kinase (ERK)–also known as mitogen-activated protein (MAP) kinase–signaling pathway, which has the opposite effect of the PI-3 kinase–Akt pathway and suppresses production of aggrecan, a type II collagen [41]. It has been shown in animal models that lectin-like oxidized low-density lipoprotein receptor-1 (LOX-1) is expressed in chondrocytes in osteoarthritis cartilage, and binding of oxidized low-density lipoprotein to LOX-1 increases intracellular oxidative stress [42].

In a study of articular cartilage from patients undergoing joint replacement surgery, Gavriilidis *et al.* [43] observed higher levels of lipid peroxidation products (malondialdehyde measured as thiobarbituric acid reactive substances) in osteoarthritis cartilage compared to control cartilage. In addition, lipid peroxidation products were similarly elevated on SOD2-depleted chondrocytes. Depletion of SOD2 also resulted in significant mtDNA strand breaks in chondrocytes. Osteoarthritic chondrocytes showed less spare respiratory capacity and higher protein leak compared to control chondrocytes. The authors concluded that SOD2 depletion in osteoarthritis chondrocytes led to oxidative damage and mitochondrial dysfunction, suggesting that SOD2 downregulation is a potential contributor to the pathogenesis of osteoarthritis. Surapaneni and Venkataramana [44] reported significantly elevated levels of erythrocytes malondialdehyde (measured as thiobarbituric acid reactive substances) and significant declines in erythrocyte glutathione as well as ascorbic acid levels in patients with osteoarthritis compared to controls. Plasma vitamin E levels and activities of antioxidant enzyme catalase were also significantly reduced in patients with osteoarthritis compared to controls, indicating that these patients experienced elevated oxidative stress compared to controls.

Sarcopenia is an age-related decline in muscle mass and strength, leading to significant impairment in the ability to carry out normal daily functions. Individuals with sarcopenia are also at a higher risk of falls and fractures, which may eventually lead to loss of independence. It has been estimated that approximately one-fourth to one-half of the population older than 65 years of age may suffer from sarcopenia. Aging has been shown to predispose skeletal muscle to increased oxidative stress both at rest and during disuse atrophy, indicating that oxidative stress may play a role in disease-induced and sarcopenia-associated muscle loss. Increased oxidative stress and inflammation in skeletal muscle may also lead to mitochondrial dysfunction, decreased protein synthesis, increased protein degradation, and apoptosis by activation or inhibition of some major signaling pathways. One such pathway involved in age-related muscle dystrophy is the insulin-like growth factor 1/phosphatidylinositol-3/protein kinase B signaling pathway, which regulates protein synthesis. Forkhead transcription factors encompass a large family of proteins characterized by a DNA binding domain termed "forkhead box" (FOXO). Forkhead transcription factors are redox regulated, and FOXO proteins may play a role in the loss of muscle mass or muscle nuclei with aging [45]. Sriram *et al.* [46] noted that abnormal levels of reactive oxygen species and inflammatory cytokines are present in skeletal muscle during muscle wasting including sarcopenia, and myostatin, a pro-oxidant, signals the generation of reactive oxygen species in muscle cells. Myostatin induces oxidative stress through TNF-α via nuclear factor-κB (NF-κB) and NADPH oxidase. Likewise, modulation of mitogen-activated protein kinases by oxidative stress has also been implicated in the pathogenesis of sarcopenia. A balanced diet rich in antioxidants may be helpful in reducing the risk of developing sarcopenia. Pillard *et al.* [47] noted that physical activity can be a valuable countermeasure to sarcopenia in its treatment and prevention.

10.11 OXIDATIVE STRESS, THE INTEGUMENTARY SYSTEM, AND SKIN APPENDAGES

The integumentary system is a complex organ system composed of numerous components (skin, hair, nails, and glands). It functions primary to protect the body from the external environment, excrete waste, and regulate temperature. Moreover, skin produces vitamin D and a variety of hormones, such as growth factors and sex steroids. Skin mirrors the first sign of aging. With aging, dysfunction of this system may occur, resulting in increased dryness of the skin, thinning, age spots, wrinkles, and decreased skin elasticity. With time, the epidermis develops an abnormality in permeability barrier homeostasis, which is further accentuated in photo-aged skin. Commonly, wound healing disorders due to comorbidities are also encountered in the elderly. The sebaceous glands responsible for sebum production and lubrication of the skin lose their morphological as well as functional characteristics with aging. Elderly people are also characterized by a significant reduction in sweat output, and higher core and skin temperatures, as well as a reduction in sensory thermal sensitivity. With advanced age, the capacity of the skin to synthesize vitamin D from the sunlight also declines [48]. Skin aging can be classified into two broad categories: intrinsic aging and extrinsic aging. Hereditary genetic influences probably contribute to no more than 3% of aging, making epigenetic and post-translational mechanisms the most important pathway of aging. Human skin is constantly exposed to air, solar radiation, environmental pollutants, and mechanical and chemical insult, which are capable of inducing oxidative stress. Extrinsic skin damage develops due to several factors, including ionizing radiation, exposure to UV radiation, severe physical and psychological stress, alcohol intake, poor nutrition, and exposure to environmental pollutions. UV radiation results in photochemical generation of reactive oxygen species, mainly superoxide anion, hydrogen peroxide, hydroxyl radical, and singlet oxygen, all of which can cause skin damage and skin aging [49]. Air pollutants such as car exhaust can induce oxidative stress in human skin. Skin may also be exposed to oxidative stress due to physical damage to the skin, such as burns and wounds. The endogenous sources of reactive oxygen species are xanthine oxidase and nitric oxide synthase (which can produce nitric oxide directly in the skin), which can cause further skin damage. Therefore, oxidative stress plays a role in aging of the skin [50].

10.12 ROLE OF GENETIC VARIABILITY AND LIFESTYLE IN RESPONSE TO OXIDATIVE STRESS/AGING

Oxidative stress is linked to the aging process and longevity. In general, long-lived species tend to show reduced oxidative damage, reduced mitochondrial free radical production, increased resistance to oxidative stress, and increased antioxidant defense. Heritability studies have indicated that genetic factors may cause an approximately 25% variation in human life span. Numerous genetic variants of antioxidant defense mechanisms may impact longevity, but a major role in longevity seems to be attributed to genes that encode the activities of two important antioxidant enzymes—SOD2 and GPX1 (glutathione peroxidase 1). Alterations of these enzymes seem to be prevalent in people of

Danish/European descent, in whom they positively impact longevity [51]. Soerensen *et al.* [52] genotyped 1650 individuals from the Danish 1905 Cohort Study and observed decreased mortality in individuals with either the manganese–superoxide dismutase single nucleotide polymorphism rs4880 C or the glutathione peroxidase 1 single nucleotide polymorphism rs1050450 T alleles. The authors concluded that genetic variation in SOD2 and GPX1 may be associated with aging and longevity. In general, females live longer than males, and it has been postulated that higher levels of estrogen in females protect them from aging by upregulating the expression of antioxidant, longevity-related genes such as those expressing manganese–superoxide dismutase and glutathione peroxidase. Moreover, mitochondrial oxidative stress is lower in females than in males [53]. In addition to these genes that encode antioxidant enzymes SOD2 and GPX1, many other genes associated with longevity have been reported. Longevity genes can be separated into the following categories: inflammatory and immune related, stress response elements, and mediators of glucose and lipid metabolism, DNA repair components, and cellular proliferation, as well as DNA haplotypes [54].

Telomere length in humans is emerging as a biomarker of aging because shortening of telomere is associated with age-related diseases as well as early mortality. In a cohort study involving Ashkenazi Jewish centenarians and their offspring as well as offspring-matched controls, Atzmon *et al.* [55] observed that centenarians and their offspring maintain longer telomeres compared to controls with advancing age and that longer telomeres are also associated with protection from age-related diseases, better cognitive function, and lipid profiles indicating healthy aging. Sequence analysis of two major genes associated with telomerase enzyme activities—hTERT (human telomerase reverse transcriptase) and hTERC (human telomerase RNA component)—indicated overexpression of synonymous and intronic mutations among centenarians compared to controls. In addition, the authors identified an hTERT haplotype that was associated with both exceptional longevity and longer telomere length. Due to its high replication rate, lack of histone-like proteins, and scarcity of DNA repair enzymes, mtDNA is much more exposed to mutagenic events than is nuclear DNA. Moreover, high amounts of reactive oxygen species are generated in mitochondria due to oxidative phosphorylation. mtDNA is a maternally inherited 16,568-base pair haploid genome that encodes 13 subunits of the respiratory chain complex. The mutation rate of mtDNA is 10 times faster than that of nuclear DNA. With aging, mutation of mtDNA increases, thus causing impaired function of mitochondria and overproduction of reactive oxygen species. However, polymorphism in mtDNA may cause subtle differences in the encoded proteins, thus altering the generation of free radicals. An association between certain mitochondrial haplogroups and longevity has been reported, and haplogroup J is significantly higher among male centenarians in northern Italy. In addition, haplogroups D, U, and K are also associated with longevity [56]. Moreover, C150T somatic transition is overrepresented in centenarians [57].

Modifiable lifestyle factors also impact longevity. These factors include physical activity, dietary intake of antioxidants, and healthy lifestyle. Individuals who consume large quantities of fruits and vegetables may live longer and have a lower incidence of cancer, stroke, and cardiovascular disease. Vasto *et al.* [58] noted that adherence to the Mediterranean nutrition profile with low glycemic index food consumption may increase longevity. To age in a healthy manner, the authors recommend following a diet with a low quantity of saturated fat and high amounts of fruits and vegetables rich in phytochemicals. Increased physical activity has been consistently linked with longevity; studies have shown that exercise attenuates the risk of several age-related diseases. SIRTs are nicotinamide adenine dinucleotide (NAD)-dependent histone deacetylases that are activated by cellular stress and play an important role in reactivating cellular defense and repair. The founding member of the SIRT protein family was first characterized in yeast [Silent Information Regulator 2 (Sir2)]. Increasing evidence suggests a role of SIRTs in mediating the adaptive response to physical exercise. Moreover, SIRT3 has been shown to increase after endurance exercise in order to prevent age-related decline in mitochondrial oxidative capacity in human skeletal muscle. Modulation of SIRTs by physical activity can explain many beneficial effects of exercise in preventing cardiovascular diseases, metabolic diseases, tumors, and aging [58,59]. See Chapters 12 and 17 for more in-depth discussion of the benefits of diet and healthy lifestyle in combating oxidative stress.

10.13 CONCLUSION

The free radical theory of aging hypothesizes that free radicals are responsible for age-related damage at the cellular and tissue levels. In general, equilibrium exists between free radicals and the body's antioxidant system, but when excess free radicals are generated, this equilibrium is destabilized, contributing to cellular damage. Oxidative stress has been documented to play an important role in the pathogenesis of various age-related disorders, including neurodegenerative disorders, ophthalmological disorders, age-related hearing loss, musculoskeletal system disease, and skin diseases. Although various genetic factors may favorably affect longevity, only approximately 25% of longevity can be attributed to genetic variability. Therefore, incorporating a balanced diet, physical activity, and healthy lifestyle is the best way to age successfully.

REFERENCES

[1] Harman D. Prolongation of the normal life span by radiation protection chemicals. J Gerontol 1957;12:257–63.

[2] Fusco D, Colloca G, Lo Monaco MR, Cesari M. Effects of antioxidant supplementation on the aging process. Clin Interv Aging 2007;2:377–87.

[3] Sastre J, Pallardo FV, Vina J. Mitochondrial oxidative stress plays a key role in aging and apoptosis. IUBMB Life 2000;49:427–35.

[4] Fried LP, Tangen CM, Walston J, Newman AB, et al. Fragility in older adults: evidence for a phenotype. J Gerontol A Biol Sci Med Sci 2001;56:M146–56.

[5] Floyd RA. Antioxidants, oxidative stress, and degenerative neurological disorders. Exp Biol Med 1999;222:236–45.

[6] Reed TT. Lipid peroxidation and neurodegenerative disease. Free Radic Biol Med 2011; 51:1302–19.

[7] Shamoto-Nagi M, Maruyama W, Hashizume Y, Yoshida M, et al. In parkinsonian substantia nigra, alpha-synuclein is modified by acrolein, a lipid-peroxidation product, and accumulates in the dopamine neurons with inhibition of proteasome activity. J Neural Transm 2007;114:1559–67.

[8] Zarkovic K. 4-Hydroxynonenal and neurodegenerative diseases. Mol Aspects Med 2003; 24:293–303.

[9] Morel P, Tallineau C, Pontcharraud R, Piriou A, et al. Effects of 4-hydroxynonenal, a lipid peroxidation product, on dopamine transport and NA^{+}/ K^{+} ATPase in rat striatal synaptosomes. Neurochem Int 1998;33:531–40.

[10] Chen CM, Liu JL, Wu YR, Chen YC, et al. Increased oxidative damage in peripheral blood correlates with severity of Parkinson's disease. Neurobiol Dis 2009;33:429–35.

[11] Sanyal J, Bandyopadhyay SK, Banerjee TK, Mukherjee SC, et al. Plasma levels of lipid peroxidation in patients with Parkinson's disease. Eur Rev Med Pharmacol Sci 2009;12:129–32.

[12] Lee J, Kosaras B, Del Signore SJ, Cormier K, et al. Modulation of lipid peroxidation and mitochondrial function improves neuropathology in Huntington's disease mice. Acta Neuropathol 2011;121:487–98.

[13] Sorolla MA, Reverter-Branchat G, Tamarit J, Ferrer I, et al. Proteomic and oxidative stress analysis in human brain samples of Huntington disease. Free Radic Biol Med 2008;45:667–8.

[14] Brown SE, Ferrante RJ, Beal MF. Oxidative stress in Huntington's disease. Brain Pathol 1999;9:147–63.

[15] Chen CM, Wu YR, Cheng ML, Liu JL, et al. Increased oxidative damage and mitochondrial abnormalities in the peripheral blood of Huntington's disease patients. Biochem Biophys Res Commun 2007;359:335–40.

[16] Cleveland DW, Rothstein JD. From Charcot to Lou Gehrig: deciphering selective motor neuron death in ALS. Nat Rev Neurosci 2001;2:806–19.

[17] Nirmalananthan N, Greensmith L. Amyotrophic lateral sclerosis: recent advances and future therapies. Curr Opin Neurol 2005;18:712–19.

[18] Pedersen WA, Fu W, Keller JN, Markesbery WR, et al. Protein modification by the lipid peroxidation product 4-hydroxynonenal in the spinal cords of amyotrophic lateral sclerosis patients. Ann Neurol 1998;44:819–24.

[19] Jain MR, Ge WW, Elkabes S, Li H. Amyotrophic lateral sclerosis: protein chaperone dysfunction revealed by proteomic studies of animals models. Proteomics Clin Appl 2008;2:670–84.

[20] Baillet A, Chanteperdrix V, Trocme C, Casez P, et al. The role of oxidative stress in amyotrophic lateral sclerosis and Parkinson's disease. Neurochem Res 2010;35:1530–7.

[21] Rocchi A, Pellegrini S, Sicilano G, Murri L. Causative and susceptibility genes for Alzheimer's disease: a review. Brain Res Bull 2003;61:1–24.

[22] Butterfield DA, Bader Lange ML, Sultana R. Involvement of lipid peroxidation product, HNE, in the pathogenesis and progression of Alzheimer's disease. Biochim Biophys Acta 2010;1801:924–9.

[23] Markesbery WR, Lovell MA. Four-hydroxynonenal, a product of lipid peroxidation, is increased in the brain in Alzheimer's disease. Neurobiol Aging 1998;19:33–6.

[24] Pratico D, Uryu K, Leight S, Trojanoswki JQ, et al. Increased lipid peroxidation precedes amyloid plaque formation in an animal model of Alzheimer amyloidosis. J Neurosci 2001;21:4183–7.

[25] Puertas MC, Martinez-Martos JM, Cobo MP, Carrera MP, et al. Plasma oxidative stress parameters in men and women with early stage of Alzheimer type dementia. Exp Gerontol 2012;47:625–30.

[26] Gustaw-Rothenberg K, Lowalczuk K, Stryjecka-Zimmer M. Lipids' peroxidation markers in Alzheimer's disease and vascular dementia. Geriatr Gerontol Int 2010;10:161–6.

[27] Solfrizzi V, Panza F, Frisardi V, Seripa D, et al. Diet and Alzheimer's disease risk factors for prevention: the current evidence. Expert Rev Neurother 2011;11:677–708.

[28] Hu N, Yu JT, Tan L, Wang YL, et al. Nutrition and the risk of Alzheimer's disease. Biomed Res Int 2013;2013:524820.

[29] Vingtdeux V, Dreses-Werringloer U, Zhao H, Davies P, et al. Therapeutic potential of resveratrol in Alzheimer's disease. BMC Neurosci 2008;9(Suppl. 2):S6.

[30] Chin D, Huebbe P, Pallauf K, Rimbach G. Neuroprotective properties of curcumin in Alzheimer's disease: merits and limitations. Curr Med Chem 2013;20:3955–85.

[31] Chrysostomou V, Rezania F, Trounce IA, Crowston JG. Oxidative stress and mitochondrial dysfunction in glaucoma. Curr Opin Pharmacol 2013;13:12–15.

[32] Izzotti A, Bagnis A, Sacca SC. The role of oxidative stress in glaucoma. Mutat Res 2006;612:105–14.

[33] Vinson JA. Review of oxidative stress in cataracts. Pathophysiology 2006;13:151–62.

[34] Kaur J, Kukreja S, Kaur A, Malhotra N, et al. The oxidative stress in cataract patients. J Clin Diagn Res 2012;6:1629–32.

[35] Yildirim Z, Ucgun NI, Tildirim F. The role of oxidative stress and antioxidants in the pathogenesis of age-related macular degeneration. Clinics (Sao Paulo) 2011;m66:743–6.

[36] Yamasoba T, Lin FR, Someya S, Kashio A, et al. Current concepts in age-related hearing loss: epidemiology and mechanistic pathways. Hear Res 2013;303:30–8.

[37] Cruickshanks KJ, Klein R, Klein BE, Wiley TL, et al. Cigarette smoking and hearing loss: the epidemiology of hearing loss study. JAMA 1998;279:1715–19.

[38] Baker K, Staecker H. Low dose oxidative stress induces mitochondrial damage in hair cells. Anat Rec 2012;295:1868–76.

[39] Kidd AR, Bao J. Recent advances in the study of age related hearing loss: a mini review. Gerontology 2012;58:490–6.

[40] Bared A, Ouyang X, Angeli S, Du LL, et al. Antioxidant enzymes, presbycusis, and ethnic variability. Otolaryngol Head Neck Surg 2010;143:263–8.

[41] Li Y, We X, Zhou J, Wei L. The age related changes in cartilage and osteoarthritis. Biomed Res Int 2013;2013:916530.

[42] Kishimoto H, Akagi M, Zushi A, Teramura T, et al. Induction of hypertrophic chondrocyte-like phenotypes by oxidized LDL in cultured bovine articular chondrocytes through increase in oxidative stress. Osteoarthritis Cartilage 2010;18:1284–90.

[43] Gavriilidis C, Miwa S, von Zglinicki T, Taylor RW, et al. Mitochondrial dysfunction in osteoarthritis is associated with downregulation of superoxide dismutase 2. Arthritis Rheum 2013;65:378–87.

[44] Surapaneni KM, Venkataramana G. Status of lipid peroxidation, glutathione, ascorbic acid, vitamin E and antioxidant enzymes in patients with osteoarthritis. Indian J Med Sci 2007;61:9–14.

[45] Meng SJ, Yu L-J. Oxidative stress, molecular inflammation and sarcopenia. Int J Mol Sci 2010;11:1509–26.

[46] Sriram S, Subramanian S, Sathiakumar D, Venkatesh R, et al. Modulation of reactive oxygen species in skeletal muscle by myostatin is mediated through NF-κB. Aging Cell 2011; 10:931–48.

[47] Pillard F, Laoudj-Chenivesse D, Carnac G, Mercier J, et al. Physical activity and sarcopenia. Clin Geriatr Med 2011;27:449–70.

[48] Zouboulis CC, Makrantonaki E. Clinical aspects and molecular diagnostics of skin aging. Clin Dermatol 2011;29:3–14.

[49] Poljsak B, Dahmane R. Free radicals and extrinsic skin aging. Dermatol Res Pract 2012;2012:135206.

[50] Kohen R, Gati I. Skin low molecular weight antioxidants and their role in aging and in oxidative stress. Toxicology 2000;148:149–57.

[51] Dato S, Crocco P, D'Aquila P, de Rango F, et al. Exploring the role of genetic variability and lifestyle in oxidative stress response for healthy aging and longevity. Int J Mol Sci 2013;14:16443–72.

[52] Soerensen M, Christensen K, Stevnsner T, Christiansen L. The Mn superoxide dismutase single nucleotide polymorphism rs4880 and the glutathione peroxidase 1 single nucleotide polymorphism rs1050450 are associated with aging and longevity in the oldest old. Mech Ageing Dev 2009;130:308–14.

[53] Vina J, Gambini J, Lopez-Grueso R, Abdelaziz KM, et al. Females live longer than males: role of oxidative stress. Curr Pharm Des 2011;17:3959–65.

[54] Vargas-Alarcon G, Flores-Dominguez C. Detecting polymorphisms in human longevity studies: HLA typing and SNP genotyping by Amplicon sequencing. Methods Mol Biol 2013;1048:215–28.

[55] Atzmon G, Cho M, Cawthon RM, Budagov T, et al. Evolution in health and medicine Sackler colloquium: genetic variation in human telomerase is associated with telomere length in Ashkenazi centenarians. Proc Natl Acad Sci USA 2010;107:1710–17.

[56] Neiemi AK, Hervonen A, Hurme M, Karhunen PJ, et al. Mitochondrial DNA polymorphisms associated with longevity in a Finnish population. Hum Genet 2003;112:29–33.

[57] Rose G, Passarino G, Scornaienchi V, Romeo G, et al. The mitochondrial DNA control region shows genetically correlated levels of heteroplasmy in leukocytes to centenarians and their offspring. BMC Genomics 2007;8:293.

[58] Vasto S, Rizzo C, Caruso C. Centenarians and diet: what they eat in the Western part of Sicily. Immun Ageing 2012;9(10):1–106.

[59] Pucci B, Villanova L, Sansone L, Pellegrini L, et al. Sirtuins: the molecular basis of beneficial basis of physical activity. Intern Emerg Med 2013;8(Suppl. 1):S23–5.

CHAPTER 11

Oxidative Stress Related to Other Diseases

CONTENTS

11.1 INTRODUCTION

In addition to cardiovascular diseases, cancer, diabetes, neurodegenerative diseases, and psychiatric illness, for which the role of oxidative stress has been well documented, many other diseases may also be linked to increased oxidative stress (Box 11.1). The role of oxidative stress is firmly established for diseases such as alcohol and drug abuse, asthma, chronic obstructive pulmonary disease (COPD), various seizures, hepatitis and liver diseases, rheumatoid arthritis, kidney diseases, and various eye disorders. It is also established for transplant recipients. However, for other disorders, the link is indirect or more studies are needed to firmly establish a relationship between such diseases and oxidative stress. This chapter discusses diseases for which the role of oxidative stress is firmly established.

11.2 ALCOHOL ABUSE AND OXIDATIVE STRESS

As discussed in Chapter 14, when consumed in moderation, alcohol has many health benefits. Moreover, red wine in particular, beer, and other alcohol beverages are rich in antioxidants. However, when alcohol is consumed in excess, it acts as a pro-oxidant. Most alcohol is metabolized through the alcohol dehydrogenase pathway, yielding acetaldehyde, a toxic metabolite. Acetaldehyde is then converted into acetic acid by aldehyde dehydrogenase. However, in alcoholics, an alternative pathway of alcohol metabolism through CYP2E1 is known to generate free radicals. In addition, catalase may also metabolize alcohol in the presence of hydrogen peroxide and nicotinamide adenine dinucleotide phosphate (NADPH) oxidase or xanthine oxidase, generating free radicals [1]. Studies have shown that the superoxide anion radical, and hydroxyl, peroxyl, and α-hydroxyl radicals, as well as hydrogen peroxide are implicated in ethanol-induced oxidative tissue damage. Although ethanol-induced lipid peroxidation is considered a major

A. Dasgupta and K. Klein: Antioxidants in Food, Vitamins and Supplements. DOI: http://dx.doi.org/10.1016/B978-0-12-405872-9.00011-2

Box 11.1 COMMON DISEASES LINKED TO OXIDATIVE STRESS

- Alcohol and drug abuse disorders[a]
- Asthma
- β-Thalassemia
- Chronic obstructive pulmonary disease (COPD)
- Chronic obstructive sleep apnea
- Chronic fatigue syndrome
- Crohn's disease
- Convulsion (seizure)
- Cystic fibrosis
- Developmental brain disorders
- Eye diseases (cataract formation, macular degeneration)
- Fibromyalgia
- Hepatitis and liver diseases
- Inflammations and rheumatoid arthritis
- Kidney disease
- Menopause
- Neurodegenerative diseases
- Obesity and metabolic syndrome
- Pancreatitis
- Phenylketonuria
- Psychiatric disorders
- Skin disease
- Stroke
- Sterility and infertility
- Transplant recipients

[a] *Alcohol and drug abuse are considered psychiatric illnesses.*

pathogenic mechanism in ethanol-induced hepatotoxicity, it also occurs in extrahepatic tissues. It has been documented that ethanol causes oxidative hemolysis, and ethanol-induced erythrocyte instability is probably due to a combination of the direct effect of ethanol on cells and oxidation of ethanol by catalase present in erythrocytes.

Ucar *et al.* [2] studied lipid peroxidation products and antioxidant enzyme activities in two types of alcoholics. Type I alcoholics have a later onset of abuse, less psychiatric dysfunction, and a more benign alcohol-related problem, whereas type II alcoholics have onset at an early age and have serious alcohol-related problems, including socially disruptive behavior while drinking. The authors observed that lipid peroxidation products as determined by malondialdehyde (MDA) level in erythrocyte membranes were strikingly elevated in type II alcoholics (mean, 0.410 nmol MDA/mg of protein) compared to controls (mean, 0.197 nmol MDA/mg of protein), although lipid peroxidation levels were also significantly elevated in type I alcoholics (mean, 0.225 nmol MDA/mg of protein). Erythrocyte lysate lipid peroxidation products were also significantly elevated in both groups of alcoholics compared to controls. In addition, MDA levels in erythrocytes were strongly and positively correlated with age of onset of alcoholism and duration of alcohol intake in both type I and type II alcoholics, indicating that many alcohol-related complications in alcoholics may be associated with increased oxidative stress. In addition, antioxidant glutathione levels in erythrocyte membranes were significantly lower in both types of alcoholics compared to controls, but, as expected, the levels were reduced more in type II alcoholics. Moreover, antioxidant enzymes levels (superoxide dismutase and catalase) measured in erythrocyte membranes were reduced in both types of alcoholics, with more significant reductions observed in type II alcoholics.

Chronic alcohol ingestion increases the risk of acute respiratory distress syndrome characterized by alveolar epithelial and endothelial barrier disruption and severe inflammation. Alcohol abuse also increases the incidence of sepsis or pneumonia, which results in higher hospital admission rates and higher mortality compared to non-alcoholics. These complications of alcohol abuse are related to increased oxidative stress due to the generation of reactive oxygen species, depletion of critical antioxidants such as glutathione, and oxidation of thiol/disulfide redox potential in the alveolar epithelial lining fluid. Alcohol-induced depletion of glutathione plays an important role in the mechanism of many respiratory dysfunctions associated with alcohol abuse [3]. In addition to oxidative stress and depletion of reduced glutathione due to oxidation, alcohol abuse suppresses innate and adaptive immunity in lungs, causing alcoholics to be more susceptible to infection, pneumonia, and COPD [4]. Alcoholic liver diseases (fatty liver, alcoholic hepatitis, and liver cirrhosis) are major complications of alcohol abuse. The oxidative metabolism of alcohol (both alcohol dehydrogenase and CYP2E1 pathways) generates excess NADH, resulting in an increased ratio of NADH to nicotinamide adenine dinucleotide (NAD^+) in hepatocytes that leads to inhibition of fatty acid oxidation and promotion of lipogenesis. In addition, alcohol alters transcriptional control of lipid metabolism, and, as a combination of both effects, fats accumulate in hepatocytes. Alcohol also causes augmented formation of reactive oxygen and reactive nitrogen species from multiple sources, including hepatocytes, inflammatory cells, and other types of liver tissues along with depletion of antioxidants. All these mechanisms contribute to alcoholic liver diseases. Human studies have confirmed that excessive consumption of alcohol leads to increases in markers of oxidative stress in plasma and urine, including lipid peroxidation products, F_2-isoprostane, and 4-hydroxy-2-nonenal. The oxidative stress markers are also closely associated with severity of liver injury. Alcohol generates 1-hydroxyethyl radicals, and protein adducts of these radicals have been detected in patients with alcoholic cirrhosis. In addition, acute alcohol consumption in healthy individuals caused increased levels of biomarkers of oxidative stress, indicating that oxidative stress may precede alcoholic liver diseases. Moreover, levels of antioxidants such as glutathione, vitamin C, and vitamin E are also depleted in patients with alcoholic liver diseases, indicating that oxidative stress plays an important role in the pathophysiology of alcoholic liver diseases [5].

11.3 DRUG ABUSE AND OXIDATIVE STRESS

The 2010 National Survey on Drug Use and Health conducted by the Substance Abuse and Mental Health Services Administration estimated that 22.6 million Americans (8.9% of the US population) aged 12 years or older were current illicit drug users. Illicit drugs abused by Americans include marijuana/hashish, cocaine (including crack cocaine), heroin, hallucinogens, inhalants, and prescription-type psychoactive drugs obtained illegally without a valid prescription. Marijuana is the most commonly used illicit drug. In 2010, there were 17.4 million past-month users of marijuana [6]. Although marijuana is the most commonly abused drug in the United States, abuse of prescription medications for nonmedical purposes is also a serious public health issue. Cocaine abuse continues to be widespread among the general population. The drugs that are commonly abused in the United States are classified into different categories (Table 11.1). Recently, abuse of emerging designer drugs such as bath salts, spices, and K2 Blonde has become epidemic. Bath salts are β-keto (bk) derivatives of amphetamine and can also be considered synthetic cathinones, active ingredients of the khat plant that is abused (leaf chewing) in some Middle Eastern and north African countries. These synthetic cathinones include methylenedioxypyrovalerone (MDPV), 4-methylmethcathinone, and other related synthetic drugs with intense amphetamine-like effects. However, MDPV and mephedrone are two major compounds sold as bath salts. Like amphetamine and cocaine, these compounds have stimulant-producing effects. Euphoria after bath salt use lasts for 2–4 h. Users have reported that euphoria was better compared to that produced by the abuse of 3,4-methylenedioxymethamphetamine (MDMA) [7]. Since 2008, synthetic marijuana compounds sold as Spice, K2 Blonde, or herbal high have

Table 11.1 Commonly Abused Drugs in the United States

Category	Drugs
Stimulants	Amphetamine, methamphetamine, MDMA, MDA, bath salts, cocaine
Narcotic analgesics	Heroin, codeine, morphine, oxycodone, hydrocodone, hydromorphone, buprenorphine, meperidine, methadone, fentanyl
Sedative–hypnotics	Barbiturates (pentobarbital, secobarbital), benzodiazepines (e.g., diazepam, alprazolam, clonazepam, lorazepam, temazepam, zolpidem, flunitrazepam), γ-hydroxybutyric acid
Hallucinogens	LSD, marijuana, synthetic marijuana (spices, K2 Blonde)
Anesthetics	Ketamine, phencyclidine

LSD, lysergic acid diethylamide; MDA, 3,4-methylenedioxyamphetamine; MDMA, 3,4-methylenedioxymethamphetamine.

gained popularity among drug abusers. Although less toxic than bath salts, these compounds are very addictive and may cause severe overdose and even death. The first synthetic compound in this category, JWH-018, was synthesized by Dr. John W. Huffman at Clemson University to study its effect on cannabinoid receptors. It has been speculated that someone saw his paper and copied his method to produce JWH-018 illegally for abuse. Synthetic cannabinoids are called herbal high or legal high, and usually plant materials are sprayed with these compounds so that they are present at the surface of the product for maximum effect. However, pure compounds, which are white powders, are also available for purchase form the black market. Currently, more than 100 compounds are available that belong to this class of abuse drugs, but the most common are JWH-018, JWH-073, JWH-250, JWH-015, JWH-081, HU-210, HU-211 (synthesized at Hebrew University), and CP-47,497 (synthesized at Pfizer). When a synthetic cannabinoid compound is smoked, it produces an effect similar to or greater than that of smoking marijuana. By 2010, there were 1057 cases of synthetic marijuana-related toxicity in 18 states and the District of Columbia. Synthetic marijuana compounds are abused due to their agonist activity at cannabinoid receptors 1 and 2 (CB1 and CB2, respectively). Therefore, the mechanism that causes euphoria and hallucination is similar to that produced by natural marijuana. However, users also report adverse effects, such as rapid heartbeat, irritability, and paranoia. In addition, tremor, seizure, slurred speech, dilated pupils, hypokalemia, tachycardia, hypertension, and chest pain may also be present in patients who have overdosed with Spice. Long-term adverse effects are not known [8].

There is a link between oxidative stress and certain drugs of abuse, such as amphetamine, cocaine, and marijuana. Evidence indicates that cardiac oxidative stress is a prominent early event of cocaine abuse that severely compromises the cardiac antioxidant cellular system, causing cellular injury. Moreover, other oxidative damage, such as peroxidation of membrane phospholipids and depletion of antioxidants such as glutathione, has been observed in chronic cocaine-treated animals and patients who abuse cocaine [9]. In addition to cardiac damage, involvement of oxidative stress in cocaine-induced damage of liver, kidney, and the central nervous system has been reported in human and animal models. Metabolism of cocaine through cytochrome P450 yields the pro-oxidant cocaine metabolite norcocaine and other oxidative metabolites. The role of stable radical norcocaine nitroxide in causing cocaine-induced hepatotoxicity has been established. Depletion of cellular antioxidants and impairment of mitochondrial respiration by cocaine also lead to excessive generation of reactive oxygen species [10]. Experiments with rat hepatocytes indicated that a 2-h exposure to amphetamine caused increased lipid peroxidation (measured as thiobarbituric acid reactive substances) and depleted levels of antioxidant glutathione. In addition, amphetamine showed potential cytotoxicity against rat hepatocytes [11]. Govitrapong *et al.* [12] reported that plasma lipid peroxidation products as determined by thiobarbituric acid reactive substances were increased by 21% in amphetamine abusers compared to controls. In addition, the activities of erythrocyte antioxidant enzymes glutathione peroxidase (32% decreased), catalase (14% decrease), and superoxide dismutase (31% decreased) were significantly decreased in amphetamine users compared to controls. Marijuana administration to rats resulted in increased thiobarbituric acid reactive substances and decreased glutathione levels in brain after exposure to marijuana [13]. Elevated plasma isoprostane levels in heroin abusers compared to healthy individuals have also been reported [14].

11.4 ROLE OF OXIDATIVE STRESS IN LIVER AND KIDNEY DISEASE

The role of oxidative stress in the pathophysiology of various liver diseases, including hepatitis infection, has been well-established. The liver is an important organ that plays a vital role in metabolic homeostasis because it is responsible for synthesis of many proteins and storage and redistribution of nutrients, carbohydrates, fats, and vitamins. Moreover, the liver is the major organ than detoxifies drugs and other chemicals. During normal function of various liver enzymes, such as cytochrome P450, diamine oxidase, liver dehydrogenases, and tryptophan dual oxidase, free radicals are generated. However, reactive oxygen species are generated more commonly than reactive nitrogen species. Inflammation of the liver produces excessive oxygen free radicals to attack host cells, which results in cell damage. Activities of antioxidant enzymes such as superoxide dismutase and glutathione peroxidase are significantly reduced in patients with chronic liver cirrhosis and hepatitis compared to healthy controls. Oxidative stress also plays an important role in non-alcoholic liver diseases. Increased malondialdehyde level in patients with chronic hepatitis B infection has also been reported [15]. Paracha *et al.* [16] noted that during liver diseases such as hepatocellular or cholestatic dysfunction, reactive oxygen species are produced that are involved in transcriptional activation of a large number of cytokines and growth factors, and production of reactive nitrogen species contributes further to the damage. Hepatitis virus infection can also induce oxidative stress and depletion of antioxidants such as cellular glutathione, thus lowering antioxidant defense. As a result, important components of cells, such as proteins, DNA, and lipids, are oxidatively damaged. Increased nitroxidative stress causes mitochondrial dysfunction due to oxidative damage to mitochondrial DNA, lipids, and proteins that contributes to a more severe disease state in both alcoholic and non-alcoholic liver diseases. Many mitochondrial proteins, including enzymes involved in fat oxidation and energy supply, may be oxidatively modified under increased nitroxidative stress and may be inactivated, leading to increased fat accumulation in the liver and adenosine triphosphate ATP depletion [17].

Many end-stage renal disease patients and chronic hemodialysis patients are in a state of chronic inflammation as evidenced by elevated markers of inflammation, such as elevated serum C-reactive protein levels. The chronic inflammation in these patients with proinflammatory cytokines as an important mediator has been implicated in long-term complications such as atherosclerosis, amyloidosis, and resistance to erythropoietin. Chronic inflammation is known to produce oxidative stress, especially through priming of leukocytes. Overproduction of free radicals in patients with end-stage renal disease has been linked to inflammation, metabolic alterations associated with uremia, and bioincompatibility of the hemodialysis process. In addition, impairment of antioxidant defense mechanism due to depletion of superoxide dismutase and glutathione peroxidase in erythrocytes further contributes to increased oxidative stress in these patients. The basal level of superoxide anion is higher in patients receiving hemodialysis compared to healthy subjects. Antioxidants are also lost during hemodialysis [18]. Oxidative stress and inflammation associated with chronic kidney disease are linked to cardiovascular and other complications. Increased oxidative stress in patients with chronic kidney disease is due to the increased production of reactive oxygen species and diminished antioxidant defense probably caused by impaired activation of Nrf2 (nuclear factor-erythroid-2-related factor 2), the transcription factor that regulates genes that code antioxidant and detoxifying molecules [19].

In patients with declining kidney function, increased levels of lipid and protein oxidation can be observed as soon as the glomerular filtration rate declines below 20 mL/min. In a study of 49 nondiabetic stable chronic kidney dialysis patients (16 on hemodialysis, 17 on continuous ambulatory peritoneal dialysis, and 16 nondialyzed patients) and 13 healthy subjects, Capusa *et al.* [20] observed the highest level of lipid peroxidation products (measured as thiobarbituric acid reactive substances) in the sera of hemodialysis patients (mean, 78.3 nmol/g protein) compared to controls (mean, 42.0 nmol/g protein). Patients with chronic kidney disease who were receiving continuous ambulatory peritoneal dialysis also showed elevated lipid peroxidation products (mean, 58.3 nmol/g protein) compared to controls. Patients with chronic kidney disease but not receiving hemodialysis showed a lower level of lipid peroxidation products (mean, 53.1 nmol/g protein) compared to patients receiving dialysis. Plasma total free thiol was decreased in all patients with chronic kidney disease compared to controls. The authors concluded that oxidative stress occurs in all patients with chronic kidney disease and worsens as renal function declines. Pavlova *et al.* [21] observed 4-fold higher levels of conjugate dienes in blood of patients with glomerulonephritis and 2-fold higher levels in renal failure patients compared to controls. Concentrations of conjugated dienes were more strikingly elevated in urine compared to blood. In glomerulonephritis patients, conjugated dienes were 12-fold higher and in renal failure patients 4-fold higher compared to controls. Concentrations of thiobarbituric acid reactive substances in urine showed a similar trend, with 7-fold higher values in glomerulonephritis patients, 2-fold higher in pyelonephritis patients, and 1.5-fold higher in renal failure patients compared to controls. Kidney diseases also caused a 2.5-fold inhibition of antioxidant enzyme catalase activities in blood. These data clearly indicate that children with kidney disease experience significantly elevated oxidative stress compared to healthy children [21].

11.5 ELEVATED OXIDATIVE STRESS IN PATIENTS WITH DISEASES LINKED TO CHRONIC INFLAMMATION

Chronic inflammation is encountered in many disease states, and in general it leads to increased oxidative stress. The sources of inflammation are varied and include microbial infection, viral infection, allergen exposure, and exposure to toxic chemicals, as well as autoimmune diseases, obesity, and various chronic diseases. Acute inflammation is an initial stage that persists for a short time. It is beneficial to the host because it may kill invading pathogens, whereas chronic inflammation is detrimental to health because it may predispose an individual to various chronic illnesses. Chronic inflammation is linked to various diseases, including arthritis, appendicitis, gastritis, laryngitis, pancreatitis, dermatitis, meningitis, and other diseases related to a dysfunctional immune system, such as lupus and various other autoimmune diseases. In addition, inflammation is linked to asthma, pneumonia, heart disease, diabetes, cancer, and even Alzheimer's disease.

During inflammation, mast cells and leukocytes are recruited to the site of damage, leading to a "respiratory burst" due to increased uptake of oxygen, which releases a large amount of reactive oxygen species at the site of damage. In addition, inflammatory cells produce mediators such as cytokines and chemokines through activation of nuclear factor (NF)-κB, and both cytokines and chemokines act by recruiting more inflammatory cells at the site of damage, producing more reactive species. Oxidative stress also promotes recruitment and activation of leukocytes and resident cells, thus eliciting inflammation. In addition, oxidative stress produces proinflammatory oxidized lipids and advanced protein oxidation and glycation end products, thus promoting inflammation. This sustained inflammatory and oxidative system can damage healthy neighboring cells, and if inflammation is chronic for a long period of time, this may lead to carcinogenesis [22].

Oxidative stress contributes to chronic inflammation of tissues and plays a central role in atherosclerosis. Moreover, oxidative stress also causes immunomodulations that result in various autoimmune diseases, such as rheumatoid arthritis and lupus erythematosus. Rheumatoid arthritis is characterized by chronic joint inflammation and a variable degree of bone and cartilage erosion. Currently, several lines of evidence suggest that reactive oxygen species play important roles in the pathophysiology of rheumatoid arthritis. Epidemiological studies have shown an inverse relationship between dietary intake of antioxidants and the incidence of rheumatoid arthritis. If the ferrous ion that catalyzes hydroxyl radical production is found in synovial fluid, it indicates a poor prognosis for the disease. Studies of synovial fluid in patients with rheumatoid arthritis showed oxidative damage to hyaluronic acid, and oxidized low-density lipoprotein, along with elevated lipid peroxidation products. Oxidative stress-induced damage of cartilage as well as extracellular collagen and intracellular DNA has also been observed. Oxidative stress has been demonstrated to induce T cell hyporesponsiveness in rheumatoid arthritis through effects on proteins and proteasomal degradation [23].

Overproduction of tumor necrosis factor-α (TNF-α) is thought to be associated with increased release of reactive oxygen species in patients with rheumatoid arthritis [24]. Miesel *et al.* [25] observed a fivefold increase in mitochondrial reactive oxygen species production measured as chemiluminescence of lucigenin, which is accumulated in mitochondria within cells and reacts with superoxide to form a chemiluminescent product. Using specific inhibitors, the authors could distinguish between the production of reactive oxygen species by mitochondria and that by NADPH oxidase. In addition, enhanced mitochondrial radical production in patients with rheumatoid arthritis correlated with increased levels of TNF-α in plasma, confirming the role of TNF-α in inducing free radical production by mitochondria [25]. In a study of 30 female patients with rheumatoid arthritis and 30 patients with systemic lupus erythematosus and appropriately age-matched controls, Hassan *et al.* [26] observed significantly higher levels of serum malondialdehyde (measured as thiobarbituric acid reactive substances) in patients with rheumatoid arthritis (mean, 4.09 nmol/mL) and lupus (mean, 2.78 nmol/mL) compared to controls (mean, 1.57 nmol/mL), indicating that both groups of patients experienced increased oxidative stress compared to controls. Serum glutathione levels were reduced in both patients with rheumatoid arthritis (mean, 23.37 mg/dL) and patients with lupus (mean, 27.17 mg/dL) compared to controls (mean, 41.02 mg/dL). Activities of antioxidant enzymes glutathione peroxide were also reduced in both groups of patients compared to controls. The markers of increased oxidative stress and impaired antioxidant activity significantly correlated with the severity of disease in both groups of patients. In addition, oxidative stress was increased more in patients with rheumatoid arthritis than in patients with systemic lupus erythematosus. In patients with lupus, alopecia and nephritis were significantly associated with increased oxidative stress.

Inflammatory bowel diseases are characterized by chronic intestinal inflammation and may be transmural, involving any portion of the gastrointestinal tract (Crohn's disease) or limited to the colonic mucosa (ulcerative colitis) or both (intermediate). Although the exact etiology of these diseases is uncertain, dysfunctional immunoregulation of the gut is believed to be the main cause of such conditions. Among the immunoregulatory factors, reactive oxygen species are produced in abnormally high levels in patients with inflammatory bowel diseases [27]. Lih-Brody *et al.* [28] analyzed mucosal biopsy of individuals undergoing colonoscopic biopsies (36 normal individuals undergoing routine colonoscopy, 38 patients with Crohn's disease, and 29 patients with ulcerative colitis) for protein carbonyl content, DNA oxidation product (8-hydroxy-2′-deoxyguanosine), reactive oxygen species (measured as chemiluminescence of luminol due to reaction with reactive oxygen species), trace metals (copper, zinc, and iron), and superoxide dismutase. The authors observed significantly increased levels of protein carbonyls, 8-hydroxy-2′-deoxyguanosine, reactive oxygen species, and iron with decreased copper and superoxide dismutase in biopsies of patients with Crohn's disease compared to controls. For patients with ulcerative colitis, protein carbonyls, reactive oxygen species, and iron were elevated in inflamed tissues with decreased zinc and copper compared to controls. These results indicated that patients with inflammatory bowel diseases experienced increased oxidative stress compared to controls.

Obesity is also characterized by increased oxidative stress and exacerbated inflammatory outcome due to inflammation of immune cells to adipocytes. Oxidative stress and inflammatory signals are interconnected in obesity and result in insulin resistance with metabolic syndrome and increased risk of cardiovascular diseases. Disorders of the mitochondrial electron transport chain and overproduction of reactive oxygen species and lipoperoxide, as well as alterations in antioxidant defense have been reported in obesity [29]. Another group of investigators commented that obesity is a low-grade inflammation of white adipose tissue, resulting in chronic activation of the innate immune system that can subsequently lead to insulin resistance, impaired glucose tolerance, and even diabetes [30]. Metabolic syndrome is characterized by different combinations of three or more of the following features: hyperglycemia, hypertriglyceridemia, a low level of high-density lipoprotein, hypertension, and abdominal obesity. Patients with metabolic syndrome exhibit activation of biochemical pathways that leads to overproduction of reactive oxygen species, altered antioxidant defense, and increased lipid peroxidation. Yubero-Serrano *et al.* [31] divided a total of 91 patients with metabolic syndrome into four groups based on the number of components of metabolic syndrome. The authors observed higher amounts of hydrogen peroxide and lipid peroxidation products in plasma of patients with five components of metabolic syndrome compared to patients with two and three components. However, plasma superoxide dismutase activities were lower in patients with two components of metabolic syndrome compared to patients with three, four, or five components. Plasma glutathione peroxide activities were lower in patients with two or three components of metabolic syndrome compared to patients with four or five components. The authors concluded that patients with more metabolic syndrome components experienced higher oxidative stress.

Sleep apnea is characterized by sympathetic activation, inflammation, and oxidative stress. Animal data clearly indicate that increased oxidative stress plays an important role in the development of cardiovascular complications. Presumably, oxidative stress is due to activation of NADPH and other reactive oxygen species producing enzymes within the cardiovascular system as evidenced by data obtained from knockout mice and pharmacological interventions [32].

11.6 INCREASED OXIDATIVE STRESS IN PATIENTS WITH ASTHMA, COPD, AND RELATED DISEASES

Asthma is a chronic inflammatory disorder of the respiratory airways induced by cellular mechanisms that cause increased production of reactive oxygen species. According to the Allergy Foundation of America, every day 44,000 people in the United States have an asthma attack and 9 die from such attacks. Reactive oxygen species play an important role not only in the pathogenesis of asthma but also in various other airway disorders, including cystic fibrosis, COPD, idiopathic pulmonary fibrosis, and adult respiratory distress syndrome. Pollen, animal dander, dust mites, and fungi are among many airborne allergens that can trigger airway inflammation associated with asthma mediated by immunoglobulin E (IgE).

Pulmonary reactive oxygen formation plays an important role in the molecular mechanism of asthma because reactive oxygen species may act as signaling modifiers of transcription factors such as NF-κB and activator protein-1 in epithelial cells. Expression of genes for many proinflammatory cytokines, enzymes, and adhesion molecules may be potentially activated due to signaling modification of these transcription factors by reactive oxygen species, thus inducing more inflammatory response in asthma patients. Therefore, there is a dynamic mechanism between excessive formation of reactive oxygen species and the subsequent molecular mechanism that

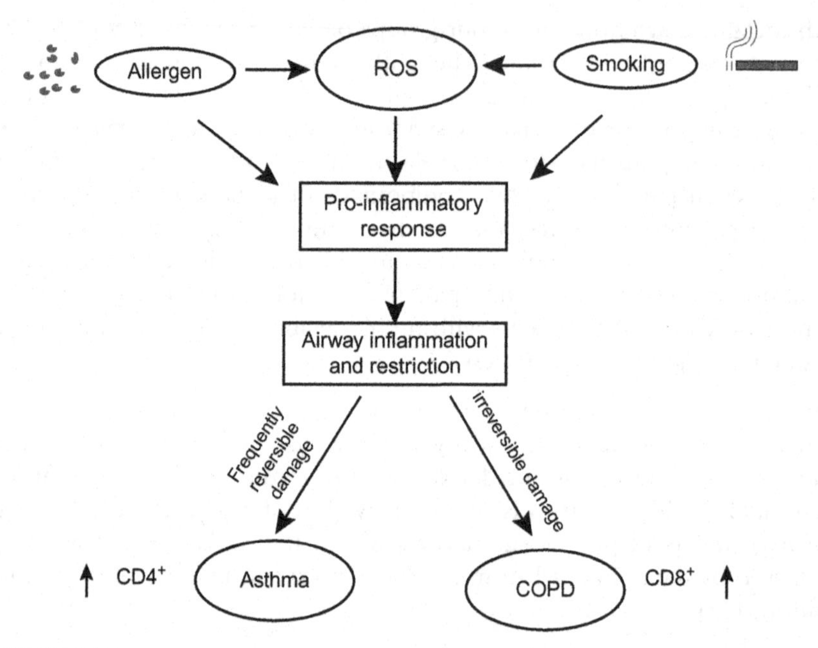

FIGURE 11.1

Schematic diagram showing major differences in and a proposed redox activation mechanism of asthma and COPD. Source: *Zuo L, Otenbaker NP, Rose BA, Salisbury KS. Molecular mechanisms of reactive oxygen species-related pulmonary inflammation and asthma. Mol Immunol 2013;56:57–63. © 2013 Elsevier.*

produces symptoms of both asthma and COPD (Figure 11.1). Although asthma and COPD share many of the same symptoms induced by reactive oxygen species via proinflammatory cytokine pathways, there are also differences between these two diseases. In COPD, which is associated with long-term smoking, there is persistent limited expiratory airflow that can worsen progressively with age, whereas in asthma airflow limitation is momentary and frequently can be reversed. In addition, in COPD, the level of $CD8^+$ lymphocytes increases, whereas in asthma the $CD4^+$ lymphocyte level is characteristically increased. Mast cells play a critical role in asthma. NADPH oxidase, a key enzyme complex in membranes of phagocytes and nonphagocytic cells, is involved in the production of superoxide required for respiratory burst and also involved in cell signaling regulation. This enzyme complex is thought to be the key factor in producing reactive oxygen species causing various respiratory inflammatory diseases, including asthma. In addition, nitric oxide is linked to patients with severe asthma because it has a significant physiological role in proinflammatory activities associated with asthma. Exhaled nitric oxide in asthma patients has been proposed as a useful marker for treatment [33].

In a study of 15 mild asthmatic patients and 15 age- and sex-matched controls, Wood *et al.* [34] showed that isoprostane (8-iso-PGF2α) levels were elevated in plasma of patients with asthma (mean, 213 pg/mL) compared to controls (139 pg/mL), and isoprostane levels also positively correlated with severity of asthma, indicating that oxidative stress plays an important role in the pathophysiology of asthma. The same group also reported that plasma 8-iso-PGF2α levels were elevated in patients with cystic fibrosis compared to controls (mean. 214 pg/mL in patients with cystic fibrosis vs. 135 pg/mL in controls). In addition, neutrophil, monocyte, and total white cell counts were elevated in patients with cystic fibrosis compared to controls, and these elevated values correlated with plasma isoprostane levels. Isoprostane levels also correlated negatively with serum antioxidant vitamins (vitamins E and C and β-carotene). The authors concluded that oxidative stress was increased in these cystic fibrosis patients despite normal dietary antioxidant intake. The immune response appears to play an important role in causing increased oxidative stress in these patients [35]. Ahmad *et al.* [36] reported higher levels of malondialdehyde and total protein carbonyls in plasma of patients with COPD compared to controls. In addition, erythrocyte superoxide dismutase and catalase activities were also reduced in patients with COPD compared to controls. These results support the hypothesis that patients with COPD experience higher oxidative stress due to oxidant–antioxidant imbalance. Bartoli *et al.* [37] reported higher levels of malondialdehyde exhaled breath condensate in patients with asthma, bronchiectasis, and COPD compared to controls. In addition, malondialdehyde levels were significantly higher in patients with COPD compared to patients with asthma and bronchiectasis. Asthmatic patients who received corticosteroids showed lower levels of malondialdehyde.

Kanoh *et al.* [38] measured exhaled ethane, a marker for lipid peroxidation, in 34 patients with interstitial lung diseases, including idiopathic pulmonary fibrosis, cryptogenic organizing pneumonia, collagen vascular disease-associated interstitial pneumonia, and pulmonary sarcoidosis. They observed elevated ethane levels in all patients (mean, 8.5 pmol/dL, $n = 34$) compared to healthy volunteers (mean, 2.9 pmol/dL, $n = 16$). Serial measurements of ethane in breath specimens correlated with the clinical picture, and 4 patients with persistently high ethane levels died. These data provide evidence of elevated oxidative stress in patients with interstitial lung diseases. In a study of 34 infants with acute bronchiolitis and 25 age-matched healthy controls, Gurkan *et al.* [39] reported elevated malondialdehyde levels in sera of patients with acute bronchiolitis at the post-bronchiolitis stage (mean, 4.2 nmol/L) compared to controls (mean, 0.7 nmol/L), indicating increased oxidative stress in these patients. Non-asthmatic patients with seasonal allergic rhinitis also demonstrated elevated oxidative stress as evidenced by elevated levels of isoprostane, leukotriene B4, and nitric oxide in exhaled air in these patients compared to controls [40].

11.7 OXIDATIVE STRESS IN PATIENTS WITH SEIZURE

Epilepsy is one of the most common and serious brain disorders and can be classified as idiopathic, symptomatic, or provoked. Reactive oxygen and nitrogen species are involved in the pathogenesis of central nervous system neurodegenerative disorders such as Parkinson's disease, stroke, dementia, and epilepsy. Studies indicate an increase in mitochondrial production of reactive oxygen and nitrogen species after persistent seizure. Experiments based on the genetic rat model and seizure induced in the animal model using kainic acid, pilocarpine, pentylenetetrazole, and trimethyltin indicate that oxidative stress is an etiological factor in epileptic seizure related to mitochondrial dysfunction, involvement of redox-active metal such as iron, arachidonic acid pathway activation, and impairment of the antioxidant system. Aging may also play a role [41]. In a study of 40 children (24 males and 16 females) with refractory generalized epileptic seizure and 40 age- and sex-matched healthy children, Saad *et al.* [42] demonstrated significantly elevated levels of plasma malondialdehyde (measured as thiobarbituric acid reactive substance) in children with seizure (mean, 6.32 nmol/L) compared to healthy children (mean, 2.54 nmol/L). Plasma zinc and selenium as well as erythrocyte glutathione peroxidase enzyme levels were lower in patients than in controls. However, erythrocyte superoxide dismutase values did not differ significantly between patients and controls. Menon *et al.* [43] evaluated oxidative parameters in 100 patients with epilepsy by comparing values with those of an equal number of age- and sex-matched healthy subjects. They observed that malondialdehyde and protein carbonylation products were higher in epilepsy patients compared to healthy subjects. Interestingly, no significant difference was observed in nitric oxide levels between patients and healthy subjects. As expected, 25 untreated patients showed higher levels of malondialdehyde and protein carbonylation products compared to patients treated with anticonvulsants. However, the number of antiepileptic drugs taken did not cause any differences because patients receiving one antiepileptic drug had similar markers of oxidative stress as patients taking more than one antiepileptic drug. The authors concluded that oxidative stress experienced by these patients was induced by seizure.

11.8 OXIDATIVE STRESS IN TRANSPLANT PATIENTS

It has been well-established that patients with chronic renal failure experience significantly more oxidative stress than do healthy individuals. Kidney transplantation is one of the best treatments for chronic renal disease that not only leads to improved renal functions but also can partially reduce the oxidative stress that was experienced by these patients prior to kidney transplantation. Vostalova *et al.* [44] measured various oxidative stress parameters in 70 patients (46 males and 24 females) before and 1 year after kidney transplantation. They observed improved kidney function as well as significant reductions in various parameters of oxidative stress in these patients 1 year after transplant. The authors concluded that improved renal function was the cause of improved parameters of oxidative stress in these patients after transplantation, which was independent of immunosuppressive therapy that they received. However, cardiac operations either involving heart transplantation or without heart transplantation are associated with increased oxidative stress and enhanced production of both proinflammatory and anti-inflammatory cytokines. Lipid peroxidation products are also increased in both groups of patients [45]. In a study of 12 patients undergoing liver transplantation and 12 healthy controls, Thorat *et al.* [46] observed that before reperfusion, antioxidant levels were significantly decreased along with significantly elevated lipid peroxidation products compared to healthy controls. Upon reperfusion of liver graft, further highly elevated lipid peroxidation products were observed compared to pre-perfusion specimens. The authors concluded that patients undergoing liver transplantation have a lower antioxidant defense and evidence of free radical damage, which results in additional insult of reperfusion injury. In a study of 14 pediatric liver transplant patients (ages 5–16 years receiving tacrolimus) and 18 healthy subjects, Al-Qabandi *et al.* [47] observed significantly elevated malondialdehyde (mean, 226.9 μmol/mg of protein) in peripheral blood lymphocytes of liver transplant patients compared to controls (mean, 162.8 μmol/mg of protein), indicating that these patients were experiencing elevated cellular oxidative stress. In addition, levels of the antioxidant enzymes catalase and glutathione peroxidase were significantly lower in liver transplant recipients compared to controls.

11.9 OXIDATIVE STRESS IN PSYCHIATRIC DISORDERS

Elevated oxidative stress has been documented in many psychiatric and neurodegenerative diseases. Alcohol- and drug-related disorders are also considered psychiatric illnesses, and elevated levels of oxidative stress in alcoholics and drug abusers have been discussed earlier in this chapter. Tsaluchidu *et al.* [48] noted that elevated oxidative stress is associated with the majority of psychiatric disorders, and peroxidation of lipid membranes may play a role in the pathophysiological process of these psychiatric illnesses. The largest and most multifaceted body of literature supports a strong association between schizophrenia and oxidative stress, followed by bipolar

Box 11.2 ASSOCIATION OF OXIDATIVE STRESS WITH COMMON PSYCHIATRIC DISORDERS

- Alcohol and drug abuse
- Attention deficit hyperactivity disorder
- Autistic disorder
- Anxiety disorder
- Bipolar disorder
- Dementia and Alzheimer's disease
- Depression and mood disorders
- History of suicide attempt
- Psychosis
- Schizophrenia
- Sleep disorders

disorder and depression [49]. The various psychiatric disorders associated with oxidative stress are listed in Box 11.2. Neurodegenerative diseases such as age-related dementia, Parkinson's disease, and Alzheimer's disease were discussed in Chapter 10.

Schizophrenia is a debilitating neurodevelopmental disorder that strikes at a young age, and pathophysiology may be related to deregulation of synaptic plasticity, with downstream alteration of cytokine-regulated inflammatory immune processes that results in impaired antioxidant defense. Consequently, these patients experience elevated oxidative stress as evidenced by increased lipid peroxidation. In a study of 22 patients in the early stage of disease (within the first 10 years of psychotic episodes), 39 patients in the late stage (minimum 10 years after diagnosis), and their respective age- and sex-matched controls, Pedrini *et al.* [50] observed significantly elevated blood levels of thiobarbituric acid reactive substances, interleukin-6, and protein carbonyl content in patients in both early and late stages of schizophrenia compared to controls, indicating that these patients were experiencing significantly elevated oxidative stress compared to controls. In a study of 12 euthymic patients with bipolar disorder type I, 18 patients with stable chronic schizophrenia, and 30 matched healthy volunteers, Gubert *et al.* [51] observed significantly increased lipid peroxidation products (measured as thiobarbituric acid reactive substances) in patients with schizophrenia compared to controls, but no significant difference was observed in lipid peroxidation products between patients with bipolar disorder and controls. In addition, activities of complexes I–III in mitochondria isolated from peripheral blood mononuclear cells were decreased in patients with schizophrenia compared to healthy controls, indicating that mitochondrial complex I dysfunction and oxidative stress may play important roles in the pathophysiology of schizophrenia. However, Andreazza *et al.* [52] reported increased levels of oxidative stress markers, including thiobarbituric acid reactive substances and nitric oxide, in patients with bipolar disorders, indicating that oxidative stress may play a role in the pathophysiology of bipolar disorders.

Oxidative stress is increased in adult patients with attention deficit hyperactivity disorder. In a study of 35 adult patients and 29 healthy volunteers, Bulut *et al.* [53] demonstrated significantly elevated serum malondialdehyde levels in patients with attention deficit hyperactivity disorders compared to healthy controls, whereas the serum levels of the antioxidant enzymes paraoxonase and arylesterase were significantly lower in patients compared to healthy controls. These observations are consistent with the hypothesis that patients with attention deficit hyperactivity disorder experience more oxidative stress than healthy individuals. Autistic children are subjected to increased oxidative stress. In a study of 30 autistic children (aged 3–15 years) and 30 healthy children as a control group, the authors observed a significantly elevated level of lipid peroxidation products in the plasma of autistic children compared to controls. On the other hand, plasma vitamin E and glutathione levels were remarkably lower in autistic children, although vitamin C levels did not differ between autistic children and controls [54]. Rett syndrome occurs with a frequency up to 1/10,000 live female births, probably due to mutation in the X-linked methyl-CpG binding protein 2 gene (MECP2). These infants show clinical features following 6–18 months of normal development, and these patients also lose their acquired cognitive, social, and motor skills and develop autistic behavior accompanied by stereotype hand movement. These patients may experience increased oxidative stress, and elevated plasma malondialdehyde level has been reported. Elevated F_2-isoprostane and 4-hydroxy-2-nonenal levels may also be observed in these patients due to increased oxidative stress [55].

Patients with generalized anxiety disorder also experience elevated oxidative stress as evidenced by elevated lipid hydroperoxide level in blood and reduced activity of the antioxidant enzyme paraoxonase compared to healthy subjects. In addition, generalized anxiety disorder could be predicted when lipid hydroperoxide level exceeded 7.74 μmol/L in blood [56]. Elevated F_2-isoprostane levels were observed in depressed older men but not depressed older women [57]. Reduced antioxidant defense has also been reported in early onset of first episode of psychosis. Glutathione deficit seems to be implicated in psychosis [58]. In a study of 30 patients with insomnia and 30 healthy volunteers, Gulec *et al.* [59] observed significantly higher malondialdehyde levels and lower antioxidant enzyme glutathione peroxidase activities in patients with insomnia compared to healthy volunteers, indicating an important role of sleep in attenuating oxidative stress. Individuals with a history of suicide attempts showed significantly higher levels of nitric oxide metabolite and lipid hydroperoxide and lower total antioxidant potential of blood compared to individuals without suicide attempts. Logistic regression demonstrated that both unipolar and bipolar disorder, female gender, smoking behavior, and lipid hydroperoxides were significantly associated with suicide attempts [60].

11.10 OTHER DISEASES LINKED TO ELEVATED OXIDATIVE STRESS

Association between oxidative stress and many diseases has been reported in the literature. In a study of 32 subjects with β-thalassemia minor and 28 healthy subjects, Selek *et al.* [61] reported lower antioxidant capacity of serum as well as lower activities of the antioxidant enzymes paraoxonase and arylesterase in serum of patients with β-thalassemia minor compared to healthy subjects. In addition, levels of lipid hydroperoxide were higher in these patients compared to the healthy subjects. Phenylketonuria is an autosomal recessive disease caused by phenylalanine 4-hydroxylase deficiency that leads to accumulation of phenylalanine in tissues and plasma of patients. Sirtori *et al.* [62] reported that patients with phenylketonuria showed higher amounts of thiobarbituric acid reactive substances in plasma compared to healthy controls. Decreased erythrocyte glutathione peroxidase levels were also observed in these patients, indicating that oxidative stress is involved in the pathophysiology of tissue damage found in patients with phenylketonuria.

Postmenopausal women have increased oxidative stress and decreased antioxidant status. Estrogen acts as an antioxidant, and hormone replacement therapy can significantly reduce oxidative stress in postmenopausal women, as evidenced by significantly lower 8-hydroxy-2-deoxyguanosine (marker of DNA damage) in women receiving hormone replacement therapy (combined estrogen/progestin therapy) compared to women not receiving any hormone replacement therapy. In addition, lipid peroxidation products (measured as thiobarbituric acid reactive substances) were lower in women receiving estrogen therapy compared to women not receiving any therapy [63]. Naher *et al.* [64] reported significantly higher malondialdehyde levels in erythrocytes of infertile males compared to healthy fertile males. Glutathione levels were also decreased in infertile males.

Skin is chronically exposed to both endogenous and exogenous pro-oxidant agents. Skin antioxidant defense protects cells from oxidative stress-related injuries and prevents production of oxidation products such as malondialdehyde and 4-hydroxy-2-nonenal that may induce protein damage, apoptosis, or release of proinflammatory cytokines. When such defense is compromised, excessive oxidative stress modifies cellular redox balance, leading to alteration of cell homeostasis and initiation of the degeneration process. Several skin diseases, such as atopic dermatitis, contact dermatitis, irritant contact dermatitis, acne, and psoriasis, are linked to elevated oxidative stress [65].

Oxidative stress also plays a role in acute pancreatitis because elevated lipid peroxide and superoxide radicals have been demonstrated in blood of patients with acute pancreatitis. Elevated oxidative stress in these patients may last longer than the clinical manifestation [66]. Patients with fibromyalgia experience elevated oxidative stress as evidenced by higher levels of malondialdehyde and lower activity of antioxidant enzyme superoxide dismutase compared to controls [67].

Oxidative stress is not only one of the predisposing factors in adult neurological disorders but also plays an important role in the pathophysiology of child-onset neurodegenerative disorders such as Cockayne syndrome, spinal muscular atrophy, subacute sclerosing panencephalitis, myoclonic epilepsy, and Lafora disease [68]. Free radical production is increased in ischemic and hemorrhagic stroke, resulting in excessive oxidative stress that contributes to brain damage. Cherubini *et al.* [69] noted that malondialdehyde concentration correlated with the size of ischemic stroke and clinical outcome. Another study found that plasma 4-hydroxynonenal levels in patients with ischemic stroke were higher than those of control subjects, and the authors suggested that plasma 4-hydroxynonenal is a potential biomarker for ischemic stroke [70]. Ritzenthaler *et al.* [71] noted that oxidative stress is a leading cause of neuronal damage in ischemic stroke, and melatonin may play a role in the antioxidant response. Increased intake of fruits and vegetables may be associated with decreased risk of stroke [72].

11.11 CONCLUSION

Oxidative stress plays an important role in many disease states. In this chapter, diseases for which the role of oxidative stress is firmly established have been discussed. Certainly, this list may grow in the future. In addition, there are also many diseases for which an indirect association with oxidative stress has been suggested, but further research is needed to establish a firm correlation between these diseases and the role of oxidative stress in their pathophysiology. These diseases have not been addressed in this chapter. Currently, healthy lifestyle and intake of generous amounts of antioxidant-rich fruits and vegetables are strongly recommended to combat oxidative stress. Antioxidant supplements may be helpful in certain diseases in which oxidative stress plays an important role, but antioxidant supplements must be taken only if recommended by a physician.

REFERENCES

[1] Comporti M, Signorini C, Leoncini S, Gardi C, et al. Ethanol-induced oxidative stress: basic knowledge. Genes Nutr 2010;5:101–9.

[2] Ucar G, Demir B, Ulug B. Lipid peroxidation and antioxidant enzyme activities in erythrocytes of type I and II alcoholics. Cell Biol Funct 2005;23:29–37.

[3] Liang Y, Yeligar SM, Brown LA. Chronic alcohol abuse-induced oxidative stress in the development of acute respiratory distress syndrome. Sci World J 2012;2012:740308.

[4] Kaphali L, Calhoun WJ. Alcoholic lung injury: metabolic, biochemical and immunological aspects. Toxicol Lett 2013;222:171–9.

[5] Zhu H, Jia Z, Misra H, Li YR. Oxidative stress and redox signalling mechanisms of alcoholic liver diseases: updated experimental and clinical evidence. J Dig Dis 2012;13:133–42.

[6] Results from the 2010 National Survey on Drug Use and Health: Summary of National Findings. Rockville, MD: US Department of Health and Human Services Substance Abuse and Mental Health Services Administration (SAMHSA), Office of Applied Studies.

[7] Fass JA, Fass AD, Garcia A. Synthetic cathinones (bath salts): legal status and patterns of abuse. Ann Pharmacother 2012;46:436–41.

[8] Wells D, Ott CA. The "new" marijuana. Ann Pharmacother 2011;45:414–17.

[9] Cerretani D, Fineschi V, Bello S, Riezzo I, et al. Role of oxidative stress in cocaine-treated cardiotoxicity and cocaine-related death. Curr Med Chem 2012;19:5619–23.

[10] Valente MJ, Carvalho F, de Pinho PG, Carvalho M. Contribution of oxidative metabolism to cocaine-induced liver and kidney damage. Curr Med Chem 2012;19:5601–6.

[11] El-Tawil OS, Abou-Hadeed AH, El-Bab MF, Shalaby AA. D-Amphetamine induced cytotoxicity and oxidative stress in isolated rat hepatocytes. Pathophysiology 2011;18:279–85.

[12] Govitrapong P, Boontem P, Kooncumchoo P, Pinweha S, et al. Increased blood oxidative stress in amphetamine users. Addict Biol 2010;15:100–2.

[13] Calderon Guzman D, Labra Ruiz N, Hernandez Garcia E, Juarez Olguin H. Levels of 5-hydroxyindole acetic acid and lipid peroxidation in brain after administration of marijuana and nalbuphine in male and female rats. Proc West Pharmacol Soc 2010;53:20–5.

[14] Kovasti L, Njau S, Nikolaou K, Tsolakidou A, et al. Isoprostane as a marker of oxidative stress in chronic heroin users: correlates with duration of heroin use or concomitant hepatitis C infection. Am J Drug Alcohol Abuse 2010;36:13–17.

[15] Zhu R, Wang Y, Zhang L, Guo Q. Oxidative stress and liver disease. Hepatol Res 2012; 42:741–9.

[16] Paracha UZ, Fatima K, Akqahtani M, Chaudhary A, et al. Oxidative stress and hepatitis C. Virol J 2013;10:251.

[17] Song BJ, Abdelmegeed MA, Henderson LE, Yoo SH, et al. Increased nitroxidative stress promotes mitochondrial dysfunction in alcoholic and nonalcoholic fatty liver diseases. Oxid Med Cell Longev 2013;2013:781050.

[18] Morena M, Delbosc S, Dupuy AM, Canaud B. Overproduction of reactive oxygen species in end stage renal disease patients: a potential component of hemodialysis-associated inflammation. Hemodial Int 2005;9:37–46.

[19] Ruiz S, Pergola PE, Zager RA, Vaziri ND. Targeting the transcription factor Nrf2 to ameliorate oxidative stress and inflammation in chronic kidney disease. Kidney Int 2013;86:1029–41.

[20] Capusa C, Stoian I, Rus E, Lixandru D, et al. Does dialysis modality influence the oxidative stress of uremic patients? Kidney Blood Press Res 2012;35:220–5.

[21] Pavlova EL, Lilova MI, Savov VM. Oxidative stress in children with kidney disease. Pediatr Nephrol 2005;20:1599–604.

[22] Hitchon CA, El-Gabalawy HS. Oxidation in rheumatoid arthritis. Arthritis Res 2004; 6:265–78.

[23] Reuter S, Gupta SC, Chaturvedi MM, Agarwal BB. Oxidative stress, inflammation and cancer: how they are linked? Free Radic Biol Med 2010;49:1603–16.

[24] Mirshafiey A, Mohsenzadegan M. The role of reactive oxygen species in immunopathogenesis of rheumatoid arthritis. Iran J Allergy Asthma Immunol 2008;7:195–202.

[25] Miesel R, Murphy MP, Kroger H. Enhanced mitochondrial radical production in patients with rheumatoid arthritis correlates with elevated levels of tumor necrosis factor alpha in plasma. Free Radic Res 1996;25:161–9.

[26] Hassan SSZ, Gheita TA, Kenawy SA, Fahim AT, et al. Oxidative stress in systematic lupus erythematosus and rheumatoid arthritis: relationship to disease manifestation and activity. Int J Rheumatol 2011;14:325–31.

[27] Rezaie A, Parker RD, Abdollahi M. Oxidative stress and pathogenesis of inflammatory bowel diseases: an epiphenomenon or the cause? Dig Dis Sci 2007;52:2015–21.

[28] Lih-Brody L, Powell SR, Collier KP, Reddy GM, et al. Increased oxidative stress and decreased antioxidant defense in mucosa of inflammatory bowel disease. Dig Dis Sci 1996;41:2078–86.

[29] Bondia-Pons I, Ryan L, Martinez JA. Oxidative stress and inflammation interactions in human obesity. J Physiol Biochem 2012;68:701–11.

[30] Bastard JP, Maachi M, Lagathu C, Kim MJ, et al. Recent advances in the relationship between obesity, inflammation and insulin resistance. Eur Cytokine Netw 2006;17:4–12.

[31] Yubero-Serrano EM, Delgado-Lista J, Pena-Orihuela P, Perez-Martinez PP, et al. Oxidative stress is associated with the number of components of metabolic syndrome: LIPGENE study. Exp Mol Med 2013;45:e28.

[32] Dumitrascu R, Heitmann J, Seeger W, Weissmann N, et al. Obstructive sleep apnea, oxidative stress and cardiovascular disease: lessons from animal studies. Oxid Med Cell Longev 2013;2013:234631.

[33] Zuo L, Otenbaker NP, Rose BA, Salisbury KS. Molecular mechanism of reactive oxygen species-related pulmonary inflammation and asthma. Mol Immunol 2013;56:57–63.

[34] Wood LG, Fitzgerald DA, Gibson PG, Cooper DM, et al. Lipid peroxidation as determined by plasma isoprostane is related to disease severity in mild asthma. Lipids 2000; 35:967–74.

[35] Wood LG, Fitzgerald DA, Gibson PG, Cooper DM, et al. Oxidative stress in cystic fibrosis: dietary and metabolic factors. J Am Coll Nutr 2001;20(Suppl. 2):157–65.

[36] Ahmad A, Shameem M, Husain Q. Altered oxidant–antioxidant levels in the disease prognosis of chronic obstructive pulmonary disease. Int J Tuberc Lung Dis 2013;17:1104–9.

[37] Bartoli ML, Novelli F, Costa F, Malagrino L, et al. Malondialdehyde in exhaled breath condensate as a marker of oxidative stress in different pulmonary diseases. Mediat Inflamm 2011;2011:891752.

[38] Kanoh S, Kobayashi H, Motoyoshi K. Exhaled ethane: an *in vivo* biomarker of lipid peroxidation in interstitial lung diseases. Chest 2005;128:2387–92.

[39] Gurkan F, Atamer Y, Ece A, Kocyigit Y, et al. Relationship among serum selenium levels, lipid peroxidation and acute bronchiolitis in infancy. Biol Trace Elem Res 2004;100:97–104.

[40] Gratziou C, Rovina N, Makris M, Simoes DC, et al. Breath markers of oxidative stress and airway inflammation in seasonal allergic rhinitis. Int J Immunopathol Pharmacol 2008;21: 949–57.

[41] Shin EJ, Jeong JH, Chung YH, Kim WK, et al. Role of oxidative stress in epileptic seizure. Neurochem Int 2011;59:122–37.

[42] Saad K, Hammad E, Hassan AF, Badry R. Trace element, oxidant, and antioxidant enzyme values in blood of children with refractory epilepsy. Int J Neurosci 2013. [E-pub ahead of print]

[43] Menon B, Ramalingam K, Kumar RV. Oxidative stress in patients with epilepsy is independent of antiepileptic drugs. Seizure 2012;21:780–4.

[44] Vostalova J, Galandakova A, Svobodova AR, Kajabova M, et al. Stabilization of oxidative stress 1 year after kidney transplantation: effect of calcineurin immunosuppressive. Ren Fail 2012;34:952–9.

[45] Kubala L, Ciz M, Vondracek J, Cerny J, et al. Perioperative and postoperative course of cytokines and the metabolic activity of neutrophils in human cardiac operations and heart transplantation. J Thorac Cardiovasc Surg 2002;124:1122–9.

[46] Thorat VN, Suryakar AN, Naik P, Tiwale BM. Total antioxidant capacity and lipid peroxidation in liver transplantation. Indian J Clin Biochem 2009;24:102–4.

[47] Al-Qabandi W, Owayed AF, Dhaunsi GS. Cellular oxidative stress and peroxisomal enzyme activities in pediatric liver transplant patients. Med Princ Pract 2012;21:264–70.

[48] Tsaluchidu S, Cocchi M, Tonello L, Puri BK. Fatty acid oxidative stress in psychiatric disorders. BMC Psychiatry 2008;8(Suppl. 1):S5.

[49] Ng F, Berk M, Dean O, Bush AI. Oxidative stress in psychiatric disorders: evidence base and therapeutic implications. Int J Neuropsychopharmacol 2008;11:851–76.

[50] Pedrini M, Massuda R, Fries GR, de Bittencourt-Pasquali MA, et al. Similarities in serum oxidative stress markers and inflammatory cytokines in patients with overt schizophrenia at early and late stage. J Psychiatr Res 2012;46:819–24.

[51] Gubert C, Stertz L, Pfaffenseller B, Panizzutti BS, et al. Mitochondrial activity and oxidative stress markers in peripheral blood mononuclear cells of patients with bipolar disorder, schizophrenia and healthy subjects. J Psychiatr 2013;47:1396–402.

[52] Andreazza AC, Kauer-Santanna M, Fret BN, Bond DJ, et al. Oxidative stress markers in bipolar disorder: a meta-analysis. J Affect Disord 2008;111:135–44.

[53] Bulut M, Selek S, Bez Y, Cermal Kaya M, et al. Lipid peroxidation markers in adult attention deficit hyperactivity disorder: new findings for oxidative stress. Psychiatry Res 2013;209:638–42.

[54] Al-Gadani Y, El-Ansary A, Attas O, Al-Ayadhi L. Metabolic biomarkers related to oxidative stress and antioxidant status in Saudi autistic children. Clin Biochem 2009;42:1032–40.

[55] Signorini C, De Felice C, Durand T, Oger C, et al. Isoprostane and 4-hydroxy-2-nonenal: markers or mediators of disease? Focus on Rett syndrome as a model of autism spectrum disorder. Oxid Med Cell Longev 2013;2013:343824.

[56] Bulut M, Selek S, Bez Y, Karababa IF, et al. Reduced PON1 enzymatic activity and increased lipid hydroperoxide levels that point out oxidative stress in generalized anxiety disorder. J Affect Disord 2013;150:829–33.

[57] Milaneschi Y, Cesari M, Simonsick EM, Vogelzangs N, et al. Lipid peroxidation and depressed mood in community dwelling older men and women. PLoS ONE 2013;11: e65406.

[58] Mico JA, Rojas-Corrales MO, Gibert-Rahola J, Parellada M, et al. Reduced antioxidant defense in early onset first episode psychosis: a case controlled study. BMC Psychiatry 2011;11:26.

[59] Gulec M, Ozkol H, Selvi Y, Tuluce Y, et al. Oxidative stress in patients with primary insomnia. Prog Neuropsychopharmacol Biol Psychiatry 2012;37:247–51.

[60] Odebrecht Vargas H, Vargas Nunes SO, Pizzo de Castro M, Cristina Bortolasci C, et al. Oxidative stress and lowered total antioxidant status are associated with a history of suicide attempts. J Affect Disord 2013;150:923–30.

[61] Selek S, Aslan M, Horoz M, Gur M, et al. Oxidative stress and serum PON1 activity in beta-thalassemia minor. Clin Biochem 2007;40:287–91.

[62] Sirtori LR, Dutra-Filho CS, Fitarelli D, Sitta A, et al. Oxidative stress in patients with phenylketonuria. Biochim Biophys Acta 2005;1740:68–73.

[63] Escalanye Gomez C, Quesada Mora S. HRT decreases DNA and lipid oxidation in postmenopausal women. Climacteric 2013;16:104–10.

[64] Naher ZU, Ali M, Biswas SK, Mollah FH. Effect of oxidative stress in male infertility. Mymensingh Med J 2013;22:136–42.

[65] Briganti S, Picardo M. Antioxidant activity, lipid peroxidation and skin disease: what's new? J Eur Acad Dermatol Venerol 2003;17:663–9.

[66] Tsai K, Wang SS, Chen TS, Kong CW, et al. Oxidative stress: an important phenomenon with pathogenetic significance in the progression of acute pancreatitis. Gut 1998;42:850–5.

[67] Bagis S, Tamer L, Sahin G, Bilgin R, et al. Free radicals and antioxidants in primary fibromyalgia: an oxidative stress disorder? Rheumatol Int 2005;25:188–90.

[68] Hayashi M. Oxidative stress in developmental brain disorders. Neuropathology 2009;29:1–8.

[69] Cherubini A, Ruggiero C, Polidori MC, Mecocci P. Potential markers of oxidative stress in stroke. Free Radic Biol Med 2005;39:841–52.

[70] Lee WC, Wong HY, Chai YY, Shi CW, et al. Lipid peroxidation dysregulation in ischemic stroke: plasma 4-HNE as a potential biomarker? Biochem Biophys Res Commun 2012;425:842–7.

[71] Ritzenthaler T, Lhommeau I, Douillard S, Cho TH, et al. Dynamics of oxidative stress and urinary excretion of melatonin and its metabolites during acute ischemic stroke. Neurosci Lett 2013;544:1–4.

[72] Hariri M, Darvishi L, Maghsoudi Z, Khorvash F, et al. Intakes of vegetables and fruits are negatively correlated with risk of stroke in Iran. Int J Prev Med 2013;4(Suppl. 2):S300–5.

CHAPTER 12

Fruits, Vegetables, and Nuts

Good Sources of Antioxidants

CONTENTS

12.1 INTRODUCTION

The US Department of Agriculture (USDA) publishes dietary guidelines for Americans every 5 years. The most recent guidelines, published in 2010, recommend increased daily intake of fruits and vegetables. Various food groups recommended for consumption in the dietary guidelines are given in Table 12.1. The goal is to balance calories and also to eat a balanced diet with regular exercise at least 30 min a day. Suggested servings of each food group for people of various ages are given in Table 12.2, and suggested consumption of dairy products is summarized in Table 12.3. The average daily calorie requirement for an adult is 2000 calories, although people who are not active may require fewer calories and people who are involved in intensive exercise and physical activity may require more. Women require a slightly lower daily calorie intake than do men [1].

The benefits of consuming generous servings of fruits and vegetables on a daily basis have been clearly established. Esfahani *et al.* [2] reviewed 22 reports on the effect of diets rich in fruits and vegetables on reducing risks of chronic diseases. They concluded that fruits and vegetables significantly increase blood levels of antioxidant vitamins (β-carotene and vitamins C and E) and folate and reduce the level of homocysteine as well as reduce concentrations of various markers that are elevated during oxidative stress, such as malondialdehyde (a marker of lipid peroxidation) and 8-hydroxydeoxyguanosine (a marker of DNA oxidation). Several studies also reported that daily consumption of fruits and vegetables increased the oxygen radical absorption capacity (ORAC) of blood, indicating superior antioxidant defense of the body. It has been estimated that one-third of all cancer deaths in the United States could be prevented through appropriate dietary modification, especially by enriching diets with fruits and vegetables, because various dietary antioxidants have shown considerable promise as cancer-preventing agents. Curcumin, genistein, resveratrol, and lycopene are effective in preventing cancers of skin, prostate, breast,

A. Dasgupta and K. Klein: Antioxidants in Food, Vitamins and Supplements. DOI: http://dx.doi.org/10.1016/B978-0-12-405872-9.00012-4

Table 12.1 Food Groups and Subgroups of Food Recommended for Consumption

Food Group	Subgroups and Examples
Vegetables	*Dark green vegetables*: all fresh, frozen, and canned dark green leafy vegetables and broccoli, cooked or raw. Examples: broccoli and spinach *Red and orange vegetables*: all fresh, frozen, and canned red and orange vegetables, cooked or raw. Examples: tomatoes, red peppers, carrots, sweet potatoes, winter squash, and pumpkin *Beans and peas*: all cooked and canned beans and peas. Examples: kidney beans, lentils, chickpeas, and pinto beans *Starchy vegetables*: all fresh, frozen, and canned starchy vegetables: for example, white potatoes, corn, and green peas *Other vegetables*: all fresh, frozen, and canned other vegetables, cooked or raw. Examples: iceberg lettuce, green beans, and onions
Fruits	All fresh, frozen, canned, and dried fruits and fruit juices. Examples: oranges and orange juice, apples and apple juice, bananas, grapes, melons, berries, and raisins
Grains	*Whole grains*: all whole grain products and whole grains used as ingredients. Examples: whole wheat bread, whole grain cereals and crackers, oatmeal, and brown rice *Enriched grains*: all enriched refined grain products and enriched refined grains used as ingredients. Examples: white breads, enriched grain cereals and crackers, enriched pasta, and white rice
Dairy	All milks, including lactose-free and lactose-reduced products and fortified soy beverages; yogurts; frozen yogurts; dairy desserts; and cheeses. Most choices should be fat-free or low-fat. Cream, sour cream, and cream cheese are not included due to their low calcium content.
Protein foods	All meat, poultry, seafood, eggs, nuts, seeds, and processed soy products. Meat and poultry should be lean or low-fat. Beans and peas are considered part of this group, as well as the vegetable group, but should be counted in one group only.

Source: *US Department of Agriculture/US Department of Health and Human Services* [1].

lung, and liver [3]. Jansen *et al.* [4] reported that lower consumption of fruits, vegetables, whole grains, and fiber was associated with higher risk of pancreatic cancer. Therefore, the authors recommended a diet rich in fruits, vegetables, whole grains, and fibers. A study based on the Chinese population in Hong Kong indicated that consumption of more fruits, vegetables, and fish was associated with lower risk of developing type 2 diabetes, whereas consumption of more meat and milk products was associated with higher risk [5]. In another study, the authors observed that high intake of whole wheat bread and whole grain cereals and their products was associated with a 20–30% reduction in the risk of type 2 diabetes [6]. Higher consumption of nuts was

Table 12.2 USDA Recommended Daily Consumption of Various Foods[a]

Age (Years)	Fruits	Vegetables	Grain[b]	Protein[c]
Children				
2–3	1 cup	1 cup	3 oz equivalent	2 oz equivalent
4–8	1–1½ cups	1½ cups	5 oz equivalent	4 oz equivalent
Girls				
9–13	1½ cups	2 cups	5 oz equivalent	5 oz equivalent
14–18	1½ cups	2½ cups	6 oz equivalent	5 oz equivalent
Boys				
9–13	1½ cups	2½ cups	6 oz equivalent	5 oz equivalent
14–18	2 cups	3 cups	8 oz equivalent	6½ oz equivalent
Women				
19–30	2 cups	2½ cups	6 oz equivalent	5½ oz equivalent
31–50	1½ cups	2½ cups	6 oz equivalent	5 oz equivalent
51 +	1½ cups	2½ cups	5 oz equivalent	5 oz equivalent
Men				
19–30	2 cups	3 cups	8 oz equivalent	6½ oz equivalent
31–50	2 cups	3 cups	7 oz equivalent	6 oz equivalent
51 +	2 cups	2½ cups	6 oz equivalent	5½ oz equivalent

[a]*This guideline is for people who do daily activities plus 30 min of exercise. People with more activity and who perform more exercise can eat more to balance calories.*
[b]*In general, 1 slice of bread, 1 cup of ready-to-eat cereal, or ½ cup of cooked rice, cooked pasta, or cooked cereal can be considered as 1 oz equivalent from the grains group. It is recommended to consume at least half of all grains as whole grains, such as whole wheat flour, oatmeal, brown rice, and whole cornmeal. White flour, white rice, and white bread are refined grain products and may be less nutritious than whole grain products.*
[c]*One ounce equivalent is 1 oz of lean beef, chicken, pork, fish, seafood, etc. cooked or one egg, ½ oz of nuts, 1 tablespoon of peanut butter, 2 tablespoons of hummus, ¼ cup of cooked beans (black, kidney, pinto, or white), or ¼ cup of cooked peas (chickpeas, split peas, or lentil) or baked beans. One small steak is approximately 4 oz equivalent, one small chicken breast is 3 oz equivalent, and one salmon steak is 4–6 oz equivalent. Hamburger is 2 or 3 oz equivalent.*
Source: *US Department of Agriculture/US Department of Health and Human Services* [1].

associated with lower risk of cardiovascular disease in women with type 2 diabetes. Diets rich in nuts resulted in a lower concentration of total cholesterol and low-density lipoprotein (LDL) cholesterol [7]. Vegetarians often have lower blood pressure than non-vegetarians. Based on their study of 500 white subjects, Pettersen *et al.* [8] concluded that vegetarians as well as vegans do have lower systolic and diastolic blood pressure and a reduced risk of developing hypertension than omnivores. In an 18-year follow-up study of 93,600 women ages 25–42 years from the Nurses' Health Study II who were healthy at baseline, Cassidy *et al.* [9] observed an inverse relationship between higher intake of anthocyanins from red and blue fruits and the risk of myocardial infarction, and especially strawberries tended to be associated with decreased

Table 12.3 USDA Recommended Daily Consumption of Dairy Products[a]

Age (Years)	Amount
Children	
2–3	2 cups
4–8	2½ cups
Girls	
9–13	3 cups
14–18	3 cups
Boys	
9–13	3 cups
14–18	3 cups
Women	
19–30	3 cups
31–50	3 cups
51+	3 cups
Men	
19–30	3 cups
31–50	3 cups
51+	3 cups

[a]*In general, 1 cup of milk, yogurt, or soy milk (soy beverage), 1½ oz of natural cheese, or 2 oz of processed cheese can be considered as 1 cup from the dairy group.*
Source: *US Department of Agriculture/US Department of Health and Human Services* [1].

risk of myocardial infarction. The intake of a combination of two anthocyanin-rich fruits, strawberry and blueberry, three or more times a week further reduced risk.

Many clinical trials have been conducted to investigate whether taking antioxidant supplements in addition to a healthy diet is helpful in preventing cardiovascular disease and cancer. To date, results of these clinical trials have shown no additional benefits of taking antioxidant supplements in preventing disease. In contrast, higher intake of vitamin E and β-carotene may cause harm by increasing mortality. Accumulating evidence indicates that reactive oxygen species that cause oxidative stress also play an essential physiological function, for example, in cell-to-cell communication and other metabolic functions. Therefore, removing all reactive oxygen species by consuming excess antioxidant dietary supplement may actually increase the risk of chronic diseases, including cardiovascular diseases and cancer, rather that reducing the risk of such diseases [10].

12.2 VARIOUS ANTIOXIDANTS IN FRUITS AND VEGETABLES

More than 25,000 bioactive food constituents are present in diet, and many such compounds play some role in modifying the multitude of processes that are related to various diseases. Most bioactive compounds present in the typical diet are derived from plants, and these chemicals are collectively termed *phytochemicals*. The majority of phytochemicals are also antioxidants. It has been postulated that these antioxidants originating from various foods can also work *in vivo* as individual compounds, exerting their antioxidant property by neutralizing reactive oxygen or reactive nitrogen compounds and contributing to antioxidant defense of the body, thus inducing longevity, cell maintenance, and DNA repair [11].

Dietary phytochemicals can be broadly classified as phenolic compounds, nitrogen-containing compounds, organosulfur compounds, phytosterols, and carotenoids. Phenolic compounds consist of one or more aromatic rings with one or more hydroxyl groups in the chemical structure, and they are categorized as phenolic acids, flavonoids, stilbene, coumarin, and tannins. Phenolic acids can be subdivided into two groups: hydroxybenzoic acid and hydroxycinnamic acid derivatives. Among common fruits, wild blueberry and blackberry contain the highest amount of phenolic compounds, followed by pomegranate, cranberry, plum, raspberry, and strawberry. Flavonoids are one of the largest groups of phenolic compounds in plant food. This complex class of compounds usually has a generic structure containing two aromatic rings (rings A and B) linked by three carbons that are usually an oxygenated heterocyclic ring or C-ring. Based on differences in the structure of the heterocyclic C-ring, these compounds are classified as flavonols, flavones, anthocyanidins, and isoflavonoids. These compounds often exist naturally as conjugates in glycosylated or esterified form. More than 80 sugars have been associated with flavonoids. Red and blue colors in some fruits and vegetables are due to anthocyanidin compounds, whereas major flavonoids in apples are quercetin, epicatechin, and cyanidin [12]. Phenolic compounds are secondary metabolites of plants and are used as a defense against the sun's ultraviolet light as well as against some pathogens. Extensive research indicates that these compounds can protect against cancers, cardiovascular diseases, diabetes, hypertension, asthma, and even infection if consumed in abundance through fruits and vegetables. Fruits such as grapes, pears, cherries, and berries are rich in these compounds and may contain up to 200–300 mg polyphenolic compounds per 100 g of fresh fruits. Typically, a glass of red wine or a cup of tea may contain more than 100 mg of these compounds. Legumes and chocolate are also rich in antioxidants. More than 8000 polyphenolic compounds have been characterized. Caffeic acid is a phenolic acid that is an important antioxidant present in coffee. One of the most important compounds in the stilbene class is resveratrol, which is an important antioxidant in grapes and red wine. Flavonoids are also responsible for the various colors of fruits and vegetables because these compounds are also pigment [13].

There are more than 600 different carotenoids, of which β-carotene is the most common carotenoid antioxidant. Other major carotenoids are lycopene (abundant in tomato), lutein, β-cryptoxanthin, zeaxanthin, and astaxanthin. Carotenoids are the principal pigments responsible for colors of some fruits and vegetables. Lycopene is responsible for the red color of red tomatoes and other fruits, such as watermelon, papaya, pink grapefruit, apricot, and pink guava. β-Carotene is also found in many foods that are orange in color. Lutein, best known for its association with healthy eyes, being found in the retina, is abundant in green, leafy vegetables such as collard greens, spinach, and kale. Lutein is also present in egg yolks and animal fats. β-Cryptoxanthin is found in papaya, apples, egg yolks, and butter. Zeaxanthin is a red-colored pigment that is responsible for the red color of salmon. It is also found in trout, shrimp, krill, and crayfish [14]. Vitamin E is a complex mixture of eight compounds (four tocopherols and four tocotrienols). However, α-tocopherol is most common. Major antioxidants found in plants are summarized in Table 12.4.

12.3 ORGANIC FOOD VERSUS PROCESSED FOOD

For many years, the media and also herbalists have advocated that consuming organic foods is healthier than eating processed foods. These niche-market food products are rapidly gaining market share in today's society because consumers are willing to pay higher prices for foods perceived to be safer, healthier, and more nutritious than conventionally produced foods. In general, it can be assumed that foods grown organically should not contain any pesticides and other harmful chemicals and should be higher in nutrients than foods grown on farms under conventional farming conditions. The current state of research indicates that although food grown in organic conditions may be more nutritional, food quality also depends on farming conditions, weather, and the quality of seeds. Whereas some reports have indicated that organically grown foods are more nutritious than conventionally grown foods, other reports have found no nutritional advantage to organically grown foods.

In one report, the authors compared antioxidants present in sweet bell pepper grown under organic and conventional growing conditions and observed that organically grown sweet bell pepper had an increased level of antioxidants such as carotenoids and phenolic compounds, as well as vitamin C, compared

Table 12.4 Major Antioxidants in Plants

Antioxidant Class	Individual Compounds
Antioxidant vitamins	Vitamin C, β-carotene (pro-vitamin A), vitamin E
Phenolic compounds	
Flavonols	Quercetin, kaempferol, myricetin, isorhamnetin
Flavan-3-ols	Catechins, epicatechin, epicatechin-3-gallate, gallocatechin, epigallocatechin, theaflavin, theaflavin-3-gallate, thearubigin
Flavones	Luteolin, apigenin
Flavanones	Eriodictyol, hesperetin, naringenin
Isoflavones	Genistein, daidzein, glycitein, biochanin A, formononetin
Anthocyanins	Cyanidin, delphinidin, malvidin, pelargonidin, peonidin, petunidin
Stilbene	Resveratrol
Phenolic acids/esters	Hydroxybenzoic acid derivatives (*p*-hydroxybenzoic, gallic, vanillic, syringic acid, etc.), hydroxycinnamic acid derivatives (curcumin; chlorogenic, ferulic, sinapic, caffeic acid, etc.)

to conventionally grown sweet pepper [15]. Hallmann [16] investigated the influence of organic and conventional cultivation systems on the nutritional value and content of bioactive compounds in select types of tomatoes and observed that those grown organically had more vitamin C and antioxidants compared to those grown under conventional conditions. Hunter *et al.* [17] evaluated the micronutrient content of food produced under organic and conventional agricultural conditions and concluded that organically produced food often had higher micronutrient content compared to food grown using conventional agricultural methods. Micronutrient content was an average of 5.7% higher in organic foods. Raigon *et al.* [18] reported that organically grown eggplant had higher concentrations of potassium, calcium, magnesium, and antioxidants (total phenolic compounds) than eggplant grown under conventional conditions. The authors further commented that, overall, the results showed that organic management and fertilization had a positive effect on the accumulation of certain beneficial minerals and phenolic compounds in eggplant grown under these conditions. Hallmann *et al.* [19] observed that organic tomato juice contained significantly more β-carotene, chlorogenic acid, and rutin as well as more total phenolic acids, gallic acid, *p*-coumaric acid, total flavonoids, quercetin 3-O-glucoside, and quercetin compared to nonorganic tomato juice. However, another study found no difference in the content of bioactive compounds in tomatoes grown under organic conditions and those grown under conventional conditions [20]. Briviba *et al.* [21] compared the content of antioxidant polyphenolic compounds in organically grown and conventionally grown apples and found no difference in antioxidant content between organically grown and conventionally grown apples

(308 μm/g fresh weight in organically grown and 321 μm/g in conventionally grown apples). Consumption of organically grown versus conventionally grown apples by volunteers did not cause any significant change in antioxidant capacity of blood as measured by various parameters as well as protection against DNA damage.

Fox *et al.* [22] noted that organic and natural beef production is becoming increasingly popular due to high consumer demand. The main difference between organically grown beef and conventionally grown beef is the use of antibiotics and growth-promoting hormones in conventionally grown beef. The authors concluded that based on the literature, there is no clear evidence that organically produced products are safer and better than conventionally produced foods. However, in another study, the authors observed that if a foodborne pathogen is present in a food, usually organically (antibiotic-free) grown food is safer because bacteria are susceptible to antibiotics, whereas bacteria found in conventionally produced food may be resistant to certain antibiotics [23]. Rist *et al.* [24] studied the composition of breast milk of mothers who consumed organic dairy/meat products versus mothers who consumed conventionally produced dairy/meat products and observed that mothers who consumed organic food had higher amounts of beneficial fatty acids in their breast milk. Switching diet from conventional to organic also contributed to higher concentrations of these beneficial fatty acids, most commonly conjugated linoleic acid isomers and *trans*-vaccenic acid. Conjugated linoleic acids are beneficial because experimental evidence indicates that these fatty acids may have anticancer, antidiabetic, and antiatherosclerosis effects and may also have an immune modulatory function. It has been demonstrated that cows' milk produced under organic conditions has higher concentrations of conjugated linoleic acids than cows' milk produced under conventional conditions [24].

12.4 FRUITS AND VEGETABLES RICH IN ANTIOXIDANT VITAMINS

Many fruits and vegetables are rich in antioxidant vitamins. When possible, it is better to consume raw fruits and vegetables because cooking may destroy some vitamins, although this is not true for all vitamins and nutrients. Moreover, degradation of vitamin C and other vitamins may occur with prolong storage, and it is better to store whole fruit than cut fruit because vitamins are more stable in whole fruit [25]. Vegetables and fruits, especially citrus, are important sources of vitamin C. However, vitamin C content may vary depending on the type of plant, cultivating condition, and post-harvest condition. The vitamin C content of various fruits and vegetables is summarized in Table 12.5. The daily requirement of vitamin C is 90 mg for men and 75 mg for women; one glass of orange

Table 12.5 Vitamin C Content of Various Fruits and Vegetables as per Common Measure

Food	Weight (g)	Common Measure	Vitamin C (mg)
Orange juice, frozen	213	6-oz can	293.7
Orange juice, raw	248	1 cup	124.0
Orange, raw	131	1 orange	69.7
Grapefruit juice, frozen	207	6-oz can	248.0
Grapefruit, raw	118	½ grapefruit	39.3
Peaches, frozen, sweetened	250	1 cup	235.5
Pepper, sweet, cooked	136	1 cup	190.3
Papaya, raw	304	1 papaya	185.1
Strawberries, raw	166	1 cup	97.6
Mangos, raw	207	1 mango	75.3
Kiwifruit, green, raw	76	1 medium fruit	70.5
Melons, cantaloupe, raw	160	1 cup	58.7
Melons, honeydew, raw	170	1 cup	30.6
Raspberries, raw	123	1 cup	32.2
Blackberries, raw	144	1 cup	30.2
Pineapple juice, canned, unsweetened	250	1 cup	25.0
Bananas, raw	118	1 banana	10.3
Cauliflower, cooked, boiled	124	1 cup	54.9
Broccoli, raw	31	1 spear	27.7
Cabbage, raw	70	1 cup	27.6
Tomatoes, red and ripe	240	1 cup	22.3
Squash	205	1 cup	19.7
Asparagus, cooked, boiled	60	4 spears	14.6
Lime juice, raw	38	Juice of 1 lime	11.4
Potatoes, boiled, cooked	156	1 cup	11.5
Spinach, raw	30	1 cup	8.4

Source: *US Department of Agriculture/US Department of Health and Human Services* [1].

juice can satisfy this requirement. Many fruits and vegetables are rich in β-carotene. These are listed in Table 12.6, whereas foods rich in antioxidant vitamin E are shown in Table 12.7.

12.4.1 Antioxidant Content of Various Foods and Effect of Cooking

In general, antioxidant contents are highest for spices, followed by fruits and vegetables, and usually lower in meat product. Halvorsen *et al.* [26] reported

Table 12.6 β-Carotene Content of Various Fruits and Vegetables as per Common Measure

Food	Weight (g)	Common Measure	β-Carotene (mg)
Carrot juice, canned	236	1 cup	21.9
Carrots, raw	72	1 carrot	9.1
Pumpkin, canned, no salt	245	1 cup	17.0
Sweet potatoes, cooked	156	1 potato	16.8
Spinach, cooked, boiled	180	1 cup	11.3
Winter squash, cooked	205	1 cup	5.7
Pumpkin, cooked	245	1 cup	5.1
Lettuce, butterhead, raw	163	1 head	3.2
Lettuce, green leaf, raw	56	1 cup	2.9
Melon, cantaloupe, raw	160	1 cup	3.2
Pepper, sweet, red, raw	149	1 cup	2.4
Green peas, cooked	160	1 cup	2.0
Mangos, raw	207	1 mango	1.3
Broccoli, cooked	184	1 cup	1.1
Watermelon, raw	286	1 wedge	8.7
Papaya, raw	304	1 papaya	8.3
Tomatoes, red, ripe, raw	180	1 cup	7.7
Cabbage, raw	70	1 cup	4.7
Okra, cooked, boiled	160	1 cup	2.7
Peaches, raw	170	1 cup	2.8

Source: *US Department of Agriculture/US Department of Health and Human Services* [1].

that blackberries have the highest antioxidant content on a per serving basis, followed by walnuts and strawberries. The 25 foods with the highest antioxidant content per serving are summarized in Table 12.8. The authors also studied the effect of cooking on antioxidant content of food compared to uncooked food. Interestingly, for some foods, antioxidant content was increased during cooking because antioxidant compounds became more available due to breakdown of cells. Examples of the effect of cooking on antioxidant capacity of various foods are summarized in Table 12.9.

Cooking methods may positively or negatively affect the concentrations of bioactive components in vegetables. Whereas older publications promoted consuming raw vegetables whenever possible, recent studies have demonstrated that during cooking, availability of beneficial compounds in vegetables may be increased. Miglio *et al.* [27] reported that there was an overall increase in antioxidant capacity of carrots and broccoli after cooking due to softening of matrix and increased extractability of antioxidant compounds. Moreover, some chemical modifications of antioxidant compounds

Table 12.7 Vitamin E Content of Various Foods as per Common Measure[a]

Food	Weight (g)	Common Measure	Vitamin E (mg)
Nuts, almonds	28.3	24 nuts	7.4
Spinach, frozen, chopped, cooked	190	1 cup	6.7
Sunflower oil	13.6	1 tablespoon	5.6
Carrot juice, canned	236	1 cup	2.7
Oil, canola	14	1 tablespoon	2.4
Peanuts, dry roasted with salt	28.35	28 peanuts	2.1
Fish, rainbow trout	85	3 oz	2.4
Fish, swordfish	106	1 piece, 3–4 oz	2.6
Broccoli, chopped, cooked	184	1 cup	2.4
Oil, corn	13.6	1 tablespoon	1.9
Oil, olive	13.6	1 tablespoon	1.9
Peppers, raw, sweet, red	119	1 pepper	1.9
Mangos, raw	207	1 mango	1.9
Blackberries, raw	144	1 cup	1.7
Pinto beans, cooked	171	1 cup	1.6
Peaches, raw	170	1 cup	1.2
Kiwifruit, green, raw	76	1 medium	1.1
Raspberries, raw	123	1 cup	1.1
Fish, salmon, cooked	85	3 oz	1.0
Papaya, raw	304	1 papaya	0.9
Blueberries, raw	145	1 cup	0.83

[a]*Cereals are fortified with vitamin E, and some cereals may contain enough vitamin E required for daily intake from one recommended serving (usually 1 cup).*
Source: *US Department of Agriculture/US Department of Health and Human Services [1].*

also occur during cooking, which increases the antioxidant capacity of the food. The authors concluded that their findings defy the notion that cooked vegetables offer lower nutritional quality. Gliszczynska-Swiglo *et al.* [28] investigated the content of health-promoting and antioxidant compounds in broccoli after water and steam cooking. They observed that steam cooking of broccoli had no effect on vitamin C content, but water cooking significantly reduced the concentration of vitamin C. However, both water and steam cooking of broccoli increased the availability of other beneficial antioxidants, including β-carotene, lutein, and α- and γ-tocopherols, compared to fresh broccoli. Pellegrini *et al.* [29] observed, that based on dry weight, boiling of vegetables usually resulted in a positive change in total antioxidant capacity

Table 12.8 Food with Highest Antioxidant Content per Serving

Rank	Food
1	Blackberries
2	Walnuts
3	Strawberries
4	Antichokes, prepared
5	Cranberries
6	Coffee
7	Raspberries
8	Pecans
9	Blueberries
10	Clove, ground
11	Grape juice
12	Chocolate, baked, unsweetened
13	Cherries, sour
14	Power bar, chocolate flavor
15	Guava nectar
16	Juice drinks (10% juice, blueberry or strawberry, vitamin C-enriched)
17	Cranapple juice
18	Prunes
19	Chocolate, dark, sugar-free
20	Cabbage, red, cooked
21	Orange juice
22	Apple juice with vitamin C
23	Mango nectar
24	Pineapples
25	Oranges
26	Bran Flakes breakfast cereals (Ralston Food, Battle Creek, MI)
27	Plums, black
28	Pinto beans, dried
29	Canned chili with meat, no beans
30	Spinach, frozen

Source: *Data from Halvorsen* et al. [26].

of the vegetables, but a negative change was observed in total antioxidant capacity after pan frying vegetables such as mushroom and onion. Interestingly, for potatoes, artichoke, and aubergine, deep frying increased total antioxidant capacity. Ferrancane *et al.* [30] demonstrated that

Table 12.9 Effect of Cooking on Select Food

Food	Type of Cooking	Antioxidant Content as % of Uncooked Food
Food with increased antioxidant content after cooking		
Carrots	Microwave cooking	113–143
Carrots	Boiling	121–159
Carrots	Cooking by steaming	291
Spinach	Microwave cooking	103–121
Spinach	Boiling	84–114
Mushrooms	Microwave cooking	113
Asparagus	Cooking by steaming	205
Broccoli	Cooking by steaming	122–654
Cabbage	Cooking by steaming	448
Red cabbage	Cooking by steaming	270
Green peppers	Cooking by steaming	467
Red peppers	Cooking by steaming	180
Potatoes	Cooking by steaming	105–242
Tomatoes	Cooking by steaming	112–164
Sweet potatoes	Boiling	413
Wheat bread	Toasting	153–185
Bagels	Toasting	134–367
Pie crust	Baking	311–1450
Food with decreased antioxidant content after cooking		
Spaghetti	Cooking by steaming	42–63
White rice	Cooking by steaming	33–70
Corn grits	Microwave cooking	21–32

Source: *Data from Halvorsen* et al. [26].

antioxidant capacity of artichoke was greatly increased after cooking, particularly after steaming (up to a 15-fold increase) and boiling (up to an 8-fold increase). The observed effect was probably due to matrix softening and higher availability of antioxidant compounds after cooking. The authors concluded that common cooking procedures can enhance the nutritional value of vegetables.

12.4.2 Antioxidant-Rich Fruits

Eating fruits every day is very good for health because fruits are a rich source of vitamins and minerals, as well as polyphenolic antioxidant compounds. Commonly eaten fruits in the United States are apples, apricots, bananas, cherries, grapefruit, grapes, kiwifruit, lemons, limes, mangos, nectarines,

Table 12.10 Phenolic Antioxidant Compounds and Anthocyanins (Antioxidants) of Various Berry Fruits

Berry Fruit	Phenolic Compounds (mg/ 100 g Fresh Weight)	Anthocyanins (mg/ 100 g Fresh Weight)	Glycemic Index
Blackberry	361	135–152	NA
Blueberry	181–473	63–235	53
Chokeberry	662.5	311	NA
Cranberry	120–177	20–66	NA
Raspberry	114–178	38	40
Strawberry	317–443	39	40

NA: The precise value is not available, but it is considered that berry fruits in general have low glycemic index (<56).

oranges, peaches, pears, papaya, pineapple, plums, prunes, raisins, and tangerines. Among berry fruits, strawberries, blueberries, and raspberries are commonly consumed. Among melons, watermelon and cantaloupe are also popular. Commonly consumed juices are orange juice, apple juice, grapefruit juice, and grape juice.

Among all fruits, berry fruits are richest in antioxidants. Anthocyanins are a water-soluble pigment found mostly in the skin of berry fruits. Although berry fruits in general contain small amounts of carotenoids, chokeberry is rich in β-carotene, lycopene, and other carotenoids. In Table 12.10 are listed phenolic antioxidant compounds and anthocyanins present in various berry fruits, as reported by Szajdek and Borowska [31]. Cherries, especially sweet cherries, are dense with nutritional components (some of them are also antioxidants), including anthocyanins, quercetin, hydroxycinnamates, potassium, fiber, vitamin C, carotenoids, and melanin. These bioactive components of cherries reduce the risk of cardiovascular diseases, diabetes, inflammatory diseases, and Alzheimer's disease [32]. Acai berry is the fruit of the acai palm that is traditionally consumed in Brazil but gaining popularity in other countries because the fruit is very high in antioxidant chemicals. In one study, the authors observed that the antioxidant capacity of blood in volunteers was increased two- or threefold after consumption of acai juice and palms [33]. In addition to berries, fruits with high antioxidant capacity include apples, apricots, avocado, cherries, grapefruit, grapes, guava, lemons, oranges, peaches, pears, plums, raisins, and pomegranate.

Total antioxidant content of fruit and vegetables can be measured experimentally using various assays. A common way to measure the total antioxidant capacity of a food is the use of ORAC values expressed as Trolox equivalent. In the past, the USDA provided ORAC values of select common foods consumed by Americans, including spices, fruits, and vegetables. Recently, the USDA nutrient data laboratory removed the ORAC database for select foods due to increasing evidence that values indicating antioxidant capacity have no relevance to the effects of specific bioactive compounds including polyphenols on human health. Moreover, ORAC data generated *in vitro* cannot be extrapolated to its *in vivo* effect. Although there are papers describing total antioxidant capacity of various foods measured *in vitro*, these details were published mostly prior to ORAC values being withdrawn from the USDA. However, for curious readers, total antioxidant capacities of representative fruits, vegetables, and spices are listed in Table 12.11 [26], but data provided in this table are for information purposes only and should be interpreted with caution.

12.4.3 Fruit Versus Fruit Juices

It is always advisable to eat fresh raw fruits rather than fruit juices because raw fruits are rich in fiber, whereas fruit juices are devoid of fiber. Dietary fiber consists of the structural and storage polysaccharides and lignin in plants. Humans cannot digest these fibers. Recommended intake of dietary fibers in adults is 20–35 g/day, and good sources of dietary fiber are fruits, vegetables, and whole and high-fiber grain products. Dietary fiber has many health benefits, including lowering blood cholesterol levels, normalizing blood glucose and insulin levels, and having a laxative effect. A diet rich in fiber is also associated with a lower risk of colon cancer [34]. The USDA Dietary Guidelines for Americans recommends eating fruits and consuming only 100% fruit juices. Drinks that are sweetened with sugar may contain very little fruit juice and may be high in calories.

Crowe and Murray [35] compared the antioxidant capacity of various fruits and fruit juices and observed that antioxidant densities (as measured by ORAC value mmol Trolox/100 g) of fresh apple, orange, and grapefruit were 54%, 23%, and 52% higher, respectively, in fresh fruits compared to commercial name-brand or store-brand fruit juices. These differences may be due to required steps in the production of fruit juice, such as removing certain phenolic compounds, mainly flavonoids, which commonly undergo browning reaction and lead to haze formation in processed fruit juice, especially apple juice. However, antioxidant capacity of grapes and grape juice did not differ significantly. Interestingly, pineapple juice had 56% higher antioxidant density than fresh pineapple fruit. This may be due to the addition of ascorbic acid to pineapple juice to increase its nutrient content as well as prolong shelf life [35].

Although processed fruit juices retain large amounts of the antioxidants present in fresh fruit, processed fruit juices often have added sugar, which increases the calorie values of fruit juices compared to fresh fruit. High fructose intake has been associated with increased lipogenesis in the liver,

Table 12.11 Total Antioxidant Capacity of Representative Fruits and Vegetables

Food	Antioxidant Content (mmol/Serving)
Blackberry	5.746
Artichoke, boiled	3.811
Strawberry	3.584
Cranberry	3.125
Raspberry	2.870
Blueberry	2.680
Sour cherry	2.205
Prune	1.715
Pineapple	1.276
Orange	1.261
Plum, black	1.205
Spinach, chopped, frozen, boiled	1.123
Kiwifruit	0.987
Sweet potato	0.900
Broccoli, cooked	0.823
Grapefruit	0.731
Avocado	0.714
Potato, red, cooked	0.678
Dates	0.503
Apple, red	0.429
Pear	0.358
Cabbage, cooked	0.336
Tomato, cooked	0.273
Honeydew	0.197
Banana	0.192
Onion, yellow	0.189
Peach	0.144
Lettuce, romaine	0.098
Carrot, cooked	0.077
Watermelon	0.064
Mushroom	0.064

Source: *Data from Halvorsen* et al. [26].

increased plasma triglycerides, insulin resistance, and obesity. Vos *et al.* [36] estimated that more than 10% of calories in the American diet derived from fructose, but one-fourth of adolescents consumed 12–15% of all calories from fructose. The major sources of fructose were sugar-sweetened beverages (30%), followed by grains (22%) and fruit or fruit juice (19%) [36].

Wang *et al.* [37] reported that, in the United States, fruit juice drinkers on average consume 148 (age 2–5 years), 136 (age 6–11 years), and 184 (age 12–19 years) kcal/day more than do non-fruit juice drinkers. Consumption of sugar-sweetened beverages (soda, fruit punches, lemonade, and fruit juices) is associated with a significant elevated risk of type 2 diabetes [38]. Chen *et al.* [39] reported that there was no correlation between the risk of developing gestational diabetes and whole fruit consumption, but with fruit juices there were variable responses. Mattei *et al.* [40] observed that homemade fruit juices are healthier than sugar-sweetened beverages. Substituting one serving of homemade fruit juice for a regular soda was associated with a 30% lower chance of developing metabolic syndrome in Hispanic adults. Increasing servings of fruit juice was also positively associated with high-density lipoprotein cholesterol.

12.4.4 Vegetables and Antioxidants

Like fruits, many vegetables are good sources of antioxidants. Dark green vegetables and legumes are probably the richest sources of antioxidants. Among dark leafy vegetables, broccoli, leafy lettuce, kale, and spinach are rich in antioxidants, and regular consumption of these vegetables reduces the risk of various types of cancer [41]. Lycopene, the major pigment in red ripe tomato, has many health benefits, including reducing the risk of both cancer and heart disease. Lycopene is a carotenoid but is not converted to vitamin A. Although the antioxidant property of lycopene was thought to be primarily responsible for its beneficial properties, evidence indicates that it has other beneficial effects, including modulating hormones, the immune system, and several metabolic pathways [42]. Ziegler *et al.* [43] concluded that consumption of dark yellow and orange vegetables was consistently more predictive of reduced risk of cancer than consumption of any other food group. However, the protective effect of these vegetables against cancer risk among smokers was limited. Sweet potato has the highest antioxidant capacity among the red and orange vegetables. Sweet potato is rich in antioxidant compounds, including phenolic compounds, vitamin C, and carotenoids. The β-carotene content of sweet potato after baking or boiling was lower than that of raw sweet potato. In contrast, concentrations of other antioxidant compounds, including vitamin C, were increased after both baking and boiling, and overall antioxidant capacity of sweet potato was increased after cooking [44]. However, after baking, the glycemic index of sweet potato was increased to 82, whereas after boiling it was only 41 (raw sweet potato has a glycemic index of 61). Therefore, boiled sweet potato with skin may be an acceptable food for individuals with diabetes. In one report, the authors concluded that eating sweet potato may even help diabetic patients to control postprandial blood glucose spike [45]. Black beans, kidney beans, lentils, and pinto beans are also rich in antioxidant compounds. Among other vegetables, antichoke, asparagus, beets, and red cabbage are good sources of antioxidants.

Starchy vegetables are relatively low in antioxidant capacity but have other health benefits because these vegetables are a good source of carbohydrates and energy. The common starchy vegetable potato is rich in carbohydrate and is an energy-producing vegetable containing little fat. Potato is also high in vitamin C and is also a good source of several B vitamins and potassium [46]. However, potato has a high glycemic index and may not be a good regular vegetable for diabetics. Halton *et al.* [47] observed a modest positive association between the consumption of potatoes and the risk of type 2 diabetes in women. Glycemic index indicates an incremental increase in blood sugar over 2 h after consumption of a food. The glycemic index of sugar is 100, and, in general, foods with a glycemic index of 55 or less are considered low glycemic index foods, whereas a value greater than 70 is considered a high glycemic number for a food. For diabetic patients, it is better to consume fruits and vegetables with low glycemic index. In general, some fruits and most vegetables have a low glycemic index. For convenience, the glycemic indexes of common fruits are listed in Table 12.12, and those of common vegetables, in Table 12.13. Although individuals without diabetes can consume any fruits and vegetables regardless of glycemic index, sugar-sweetened beverages—which include the full spectrum of soft drinks, fruit drinks, and energy and vitamin drinks—are composed of natural sweeteners such as sucrose, high-fructose corn syrup, or fruit juice concentrate that are high in calories and may cause weigh gain in normal individuals. Foods such as french fries are high in calories. Moreover, acrylamide, a probable carcinogen formed during high-temperature processing of asparagine-rich foods such as potatoes, wheat, and rye product, is a concern with french fries. Acrylamide is formed from reducing sugars and asparagine during the processing and frying process to make french fries. In order to achieve a low concentration of acrylamide in french fries, it is important to control frying temperature and timing [48].

Artichokes, asparagus, and beets are richest in antioxidant compounds. Asparagus is low in calories and has a low glycemic index, and it has a high amount of antioxidant compounds, including carotenes and lutein. It is also a good source of folic acid. Asparagus may have a beneficial effect in patients with diabetes.

12.4.5 Spices, Nuts, and Antioxidants

Spices are the richest source of antioxidants, and historically there is a rich tradition of using spices for not only their flavor-enhancing properties but also for their medicinal properties. Epidemiological research and other

Table 12.12 Glycemic Index of Common Fruits Consumed by Americans

Fruit	Glycemic Index
Acai fruit	24
Apples	38
Apricots, raw	31
Avocado	50
Bananas	52
Cantaloupe, raw	65
Cherries, sweet, raw	22
Dates	103
Grapefruit, raw, pink	25
Grapes, black	46
Grapes, green	46
Guava, common, raw	78
Kiwifruit, fresh, raw	52
Mangoes, raw	55
Nectarines, raw	42
Oranges, raw	42
Papaya, raw	56
Peaches, raw	30
Pears, raw	38
Pineapple, raw	59
Plums, raw	69
Raisins, golden seedless	64
Tangerines (mandarin orange)	42
Watermelon	72

research indicate that although spices are minor dietary constituents, they have antioxidant, antimicrobial, and anticancer properties and are good for health [49]. Cinnamon, a spice with a long history of use, has antioxidant, anti-inflammatory, antitumor, and antibacterial activities. In addition, cinnamon may be useful for lowering cholesterol and controlling blood sugar because it acts as an insulin mimetic [50]. Turmeric (*Curcuma longa*) is a common Indian spice that belongs to the ginger family. In addition to use as a component of popular curry powder, turmeric is also used in Indian Ayurvedic medicine due to its medicinal properties. Curcumin, a hydrophobic polyphenol, is a major constituent of turmeric. Based on more than 60 years of studies, curcumin has been determined to be an antioxidant that also has antibacterial, antifungal, antiviral, anti-inflammatory, antiproliferative, and anti-atherosclerotic activities, thus exerting its medicinal

Table 12.13 Glycemic Index of Common Vegetables

Vegetables	Glycemic Index
Dark green vegetables	
Broccoli, raw	15
Broccoli, boiled	25
Dark green leafy lettuce, raw	<10
Romaine lettuce, raw	<10
Spinach, raw	32
Red and orange vegetables	
Butternut squash	41
Carrots, boiled	47
Pumpkin, raw	75
Red peppers, sweet, raw	32
Sweet potato, baked in skin	82
Sweet potato, cooked, boiled without skin	41
Tomatoes	38
Starchy vegetables	
Corn, sweet, yellow, raw	54
Green peas	48
Lima beans	32
Potatoes, white, flesh and skin baked	85
Potatoes, red, flesh and skin baked	85
Beans and peas	
Black beans, raw	20
Beans, black, boiled	20
Garbanzo beans (chickpeas)	28
Kidney beans	27
Lentils	18–37
Navy beans, raw	38
Pinto beans, raw	45
Pinto beans, boiled	45
Soybeans, raw	18
Split peas	32
Other vegetables	
Artichokes, raw	15
Artichokes, boiled	15
Asparagus, raw	15
Asparagus, cooked, boiled, drained	15
Beets, raw	64

Continued...

Table 12.13 Glycemic Index of Common Vegetables *Continued*

Vegetables	Glycemic Index
Brussels sprouts	15
Cabbage, boiled	15
Cauliflower	15
Celery	15
Cucumbers, raw	32
Eggplant, boiled	15
Green peppers	32
Mushrooms, portabella	15
Mushrooms, shitake	15
Onions, red, raw	10
Onions, white, raw	10
Parsley	15
Yellow squash	32
Zucchini	15

effects on cardiovascular diseases, neurodegenerative diseases, diabetes, arthritis, and a variety of other diseases. Curcumin can interact with and regulate multiple molecular targets, including transcription factors, growth factors, kinases, inflammatory cytokines, adhesion molecules, and apoptosis-related proteins [51]. Iyer *et al.* [52] described the potential health benefits of Indian spices in treating various ailments, such as hypertension, diabetes, and obesity.

Of the various nuts, walnuts, pecans, and chestnuts have the highest antioxidant content. Walnuts contain more than 20 mmol of antioxidant compounds per 100 g. Peanut, although a legume, is also rich in antioxidant. Consuming nuts and peanut butter even once a week can reduce the risk of coronary heart disease, but the best risk reduction was observed in subjects who consumed nuts or peanut butter one to four times a week [53]. Epidemiological studies clearly indicate that consumption of nuts and peanuts can reduce the risk of coronary heart disease by decreasing the incidence of sudden death from heart disease. With higher consumption of various nuts, the risk of cardiovascular disease can be reduced up to 35%. Nuts and peanuts are also full of plant proteins, fibers, potassium, calcium, magnesium, vitamin E, antioxidant phenolic compounds, resveratrol, and arginine [54]. Vinson and Cai [55] noted that daily consumption of nuts provided an average of 158 g of beneficial polyphenolic compounds per day in the European diet and 162 mg of such compounds in the US diet.

These antioxidant compounds in nuts can improve lipid profile, reduce risk of heart diseases, and improve health without weight gain.

12.5 USDA DIETARY GUIDELINES AND AMOUNT OF ANTIOXIDANT PHYTOCHEMICALS CONSUMED BY AMERICANS

According to the 2010 USDA dietary guidelines, adults between ages 19 and 50 years should consume 2 cups of fruits and 3 cups of vegetables each day. This is approximately four servings of fruits and six servings of vegetables (for women 19–30 years old, it is four servings of fruits and five servings of vegetables because one serving is approximately ½ cup). Based on food consumption data from the National Health and Nutrition Examination Survey 2003–2006 and phytonutrient concentration data from the USDA and the literature, Murphy *et al.* [56] noted that only 501 of 8072 subjects surveyed satisfied daily fruit and vegetable consumption recommendations, indicating that the majority of Americans do not eat enough fruits and vegetables daily. The estimated intake of various phytochemicals by Americans based on this report is summarized in Table 12.14.

Eating at fast-food restaurants is associated with higher fat and lower vegetable intake [57]. Larson *et al.* [58] reported that more frequent eating at fast-food restaurants that primarily serve burgers and french fries was associated with a higher risk of being overweight as well as being obese. People who reported eating three or more times per week at restaurants serving burgers and fries also consumed nearly one additional sugar-sweetened beverage per day compared to people who ate at such restaurants only once per week. However, more frequent eating at full-service restaurants where people had a higher intake of vegetables, in contrast to fast-food restaurants, did not cause weight gain [58].

12.6 RELATION BETWEEN FOOD AND VEGETABLE CONSUMPTION AND REDUCED OXIDATIVE STRESS

As expected, eating recommended amounts of fruits and vegetables can significantly reduce oxidative stress experienced by the body. In one study, after 2 weeks of low fruit and vegetable consumption (3.0 servings/day), 246 women were randomly assigned to receive either low (average 3.6) or high (average 9.2) servings of fruits and vegetables per day for 2 weeks. Then, the low fruit and vegetable intake group was switched to high fruit and vegetable intake for the last 2 weeks of the study. The authors observed the

Table 12.14 Estimated Antioxidant Phytochemical Intake by Americans

	Meet Recommendation		Do Not Meet Recommendation	
Phytochemical	**Male**	**Female**	**Male**	**Female**
Carotenoids (μg/day)[a]				
α-Carotene	371	497	148	201
β-Carotene	1706	2678	804	1093
Lutein/zeaxanthin	1150	1746	601	786
Lycopene	4064	4095	2385	2509
Flavonoids (mg/day)[a]				
Anthocyanidin	14.5	20.4	4.5	6.2
Hesperetin	8.9	6.7	3.2	3.7
Quercetin	8.2	9.1	5.3	6.7
Phenolics (mg/day)[a]				
Ellagic acid	6.0	15.1	2.2	3.4

[a]*Values represent mean consumption.*
Source: *Murphy* et al. [56].

greatest reduction in urinary 8-isoprostane $F_{2\alpha}$ in subjects who consumed the highest amounts of fruits and vegetables during the study period, indicating that a diet rich in fruits and vegetables can significantly reduce oxidative stress [59]. In a study of 285 adolescent boys and girls age 13–17 years, Holt *et al.* [60] observed that urinary F_2-isoprostane was inversely correlated with intake of total fruits and vegetables as well as vitamin C, β-carotene, and flavonoids. Serum C-reactive protein concentration was also inversely correlated with intake of fruits, vitamin C, and folate. Serum interleukin-6, a marker of inflammation, was also inversely associated with intake of legumes, vegetables, β-carotene, and vitamin C. Serum tumor necrosis factor-α was inversely associated with β-carotene and luteolin. All these findings demonstrate favorable effects of fruits and vegetables on markers of oxidative stress and inflammation. In a study of 258 healthy premenopausal women aged 18–44 years, Rink *et al.* [61] observed that consuming five servings of any combination of fruits and vegetables per day was associated with significantly lower F_2-isoprostane in plasma. In addition, concentrations of antioxidant vitamins, including α-tocopherol, ascorbic acid, retinal, lutein, β-carotene, and cryptoxanthin, were also higher in women who consumed five servings of fruits and vegetables per day. The authors concluded that five servings of fruits and vegetables per day was associated with lower biomarkers of oxidative stress and improved antioxidant defense in the premenopausal women studied.

12.7 CONCLUSION

Phytochemicals, most of which are antioxidants, are essential components of a daily balanced diet. Therefore, generous amounts of fruits and vegetables must be included in the daily diet. Tea, coffee, and chocolate are also rich in antioxidants. Tea and coffee are devoid of calories and essentially have zero glycemic indexes. Therefore, if milk and sugar are not added, drinking coffee and tea every day is a good habit. The antioxidants present in tea and coffee and the health benefit of drinking tea and coffee are discussed in Chapter 13. Chocolate, especially dark chocolate, is a good source of antioxidants, but it is high in calories and fat (see Chapter 13). Although foods such as fish, meat, and dairy products, as well as grain and rice, are not rich sources of antioxidants, these foods are essential components of a balanced diet.

REFERENCES

[1] US Department of Agriculture/US Department of Health and Human Services. Dietary guidelines for Americans, 7th ed. Washington, DC: US Government Printing Office; 2010. Available at <http://www.health.gov/dietaryguidelines/dga2010/DietaryGuidelines2010.pdf>. Accessed March 2013.

[2] Esfahani A, Wong JM, Truan J, Villa CR, et al. Health effects of mixed fruits and vegetables concentrates: a systematic review of clinical interventions. J Am Coll Nutr 2011;30:285–94.

[3] Khan N, Afaq F, Mukhtar H. Cancer chemoprevention through dietary antioxidants: progress and promise. Antioxid Redox Sig 2008;10:475–510.

[4] Jansen RJ, Robinsin DR, Stolzenbereg-Solomon RZ, Bamlet RZ, et al. Fruits and vegetables consumption is inversely associated with having pancreatic cancer. Cancer Causes Control 2011;22:1613–25.

[5] Yu R, Woo J, Chan R, Sham A, et al. Relationship between dietary intake and the development of type 2 diabetes in a Chinese population: The Hong Kong Dietary Survey. Public Health Nutr 2011;14:1133–41.

[6] Gil A, Ortega RM, Maldonado J. Wholegrain cereals and bread: a duet of the Mediterranean diet for prevention of chronic diseases. Public Health Nutr 2011;14:2316–22.

[7] Li TY, Brennan AM, Wedick NM, Mantzoros C, et al. Regular consumption of nuts is associated with a lower risk of cardiovascular disease in women with type 2 diabetes. J Nutr 2009;139:1333–8.

[8] Pettersen BJ, Anousheh R, Fan J, Jaceldo-Siegl K, et al. Vegetarian diet and blood pressure among white subjects: results from Adventist Health Study-2 (AHS-2). Public Health Nutr 2012;15:1909–16.

[9] Cassidy A, Mukamal KJ, Liu L, Franz M, et al. High anthocyanin intake is associated with a reduced risk of myocardial infarction in young and middle aged women. Circulation 2013;127:188–96.

[10] Finley JW, Kong AN, Hintze KJ, Jeffrey EH, et al. Antioxidants in foods: State of the science important to the food industry. J Agric Food Chem 2011;59:6837–46.

[11] Carlesen MH, Halvorsen BL, Holte K, Bohn SK, et al. The total antioxidant content of more than 3100 foods, beverages, spices, herbs and supplements used worldwide. Nutr J 2010;9:3.

[12] Gulcin I. Antioxidant activity of food constituents: an overview. Arch Toxicol 2012;86:345–91.

[13] Liu RH. Dietary bioactive compounds and their health implications. J Food Sci 2013;78 (Suppl. 1):A18–25.

[14] Pandey KB, Rizvi SY. Plant polyphenols as dietary antioxidants in human health and disease. Oxid Med Cell Longev 2009;2:270–8.

[15] Hallmann E, Rembialkowska E. Characteristics of antioxidant compound in sweet bell pepper (*Capsicum annuum* L) under organic and conventional growing systems. J Sci Food Agric 2012;92:2409–15.

[16] Hallmann E. The influence of organic and conventional cultivation systems on the nutritional value and content of bioactive compounds in selected tomato types. J Sci Food Agric 2012;92:2840–8.

[17] Hunter D, Foster M, McArthur JO, Ojha R, et al. Evaluation of the micronutrients composition of plant foods produced by organic and conventional agricultural methods. Crit Rev Food Sci Nutr 2011;51:571–82.

[18] Raigon MD, Rodriguez-Burruezo A, Prohens J. Effects of organic and conventional cultivation methods on composition of eggplant fruits. J Agric Food Chem 2010;58:6833–40.

[19] Hallmann E, Lipowski J, Marszalek K, Rembialkowska E. The seasonal variation in bioactive compounds content in juice from organic and non-organic tomatoes. Plant Food Hum Nutr 2013;68:171–6.

[20] Juroszek P, Lumpkin HM, Yang RY, Ledesma DR, et al. Fruit quality and bioactive compounds with antioxidant activity of tomatoes grown on farm: comparison of organic and conventional management systems. J Agric Food Chem 2009;57:1188–94.

[21] Briviba K, Stracke BA, Rufer CE, Watzl B, et al. Effect of consumption of organically and conventionally produced apples on antioxidant activity and DNA damage in humans. J Agric Food Chem 2007;55:7716–21.

[22] Fox JT, Reinstein S, Jacob ME, Nagaraja TG. Niche market production practices of beef cattle in the United States and prevalence of foodborne pathogens. Foodborne Pathog Dis 2008;5:559–69.

[23] Jacob ME, Fox JT, Reinstein SL, Nagaraja TG. Antimicrobial susceptibility of foodborne pathogens in organic or natural production systems: an overview. Foodborne Pathog Dis 2008;5:721–30.

[24] Rist L, Mueller A, Barthel C, Snijders B, et al. Influence of organic diets on the amount of conjugated linoleic acid in breast milk of lactating women in the Netherlands. Br J Nutr 2007;97:735–43.

[25] Agte V, Tarwadi K, Mengale S, Hinge A, et al. Vitamin profile of cooked foods: How healthy is the practice of ready-to-eat foods. Int J Food Sci Nutr 2002;53:197–208.

[26] Halvorsen BL, Carlsen MH, Phillips KM, Bohn SK, et al. Content of redox-active compounds (i.e., antioxidants) in foods consumed in the United States. Am J Clin Nutr 2006;84:95–135.

[27] Miglio C, Chiavarao E, Visconti A, Fogliano V, et al. Effects of different cooking methods on nutritional and physiochemical characteristics of selected vegetables. J Agric Food Chem 2008;56:139–47.

[28] Gliszczynska-Swiglo A, Ciska E, Pawlak-Lemanska K, Chmielewski J, et al. Changes in the content of the health-promoting compounds and antioxidant activity of broccoli after domestic processing. Food Addit Contam 2006;23:1088–98.

[29] Pellegrini N, Miglio C, Del Rio D, Salvatore S, et al. Effect of domestic cooking methods on the total antioxidant capacity of vegetables. Int J Food Sci Nutr 2009;60(Suppl. 2):12–22.

[30] Ferrancane R, Pellegrini N, Visconti A, Graziani G, et al. Effect of different cooking methods on antioxidant capacity and physical characteristics of artichoke. J Agric Food Chem 2008;56:8601–8.

[31] Szajdek A, Borowska EJ. Bioactive compounds and health promoting properties of berry fruits: a review. Plant Food Hum Nutr 2008;63:147–56.

[32] McCune LM, Kubota C, Stendell-Hollins NR, Thompson CA. Cherries and health: a review. Crit Rev Food Sci Nutr 2011;51:1–12.

[33] Mertens-Talcott SU, Rios J, Jilma-Stohlawetz P, Pacheco-Palencia LA, et al. Pharmacokinetics of anthocyanins and antioxidant effects after consumption of anthocyanin-rich acai juice and pulp (*Euterpe oleracea* Mart.) in human healthy volunteers. J Agric Food Chem 2008;56:7796–802.

[34] Marlett JA, McBurney MI, Slavin JL. Position of the American Dietetic Association: health implications of dietary fibers. J Am Diet Assoc 2002;102:993–1000.

[35] Crowe KM, Murray E. Deconstructing a fruit serving: comparing the antioxidant density of select whole fruit and 100% fruit juice. J Acad Nutr Diet 2013;113:1354–8.

[36] Vos MB, Kimmons JE, Gillespie C, Welsh J, et al. Dietary fructose consumption among U.S. children and adults: The Third National Health and Nutrition Examination Survey. Medscape J Med 2008;10:160.

[37] Wang YC, Bleich SN, Gotmaker SL. Increasing caloric contribution from sugar-sweetened beverages and 100% fruit juice among U.S. children and adolescents 1988–2004. Pediatrics 2008;121:e1604–14.

[38] De Koning L, Malik VS, Rimm EB, Willett WC, et al. Sugar-sweetened and artificially sweetened beverages consumption and risk of type 2 diabetes in men. Am J Clin Nutr 2011;93:1321–7.

[39] Chen L, Hu FB, Yeung E, Tobias DK, et al. Pregnancy consumption of fruits and fruit juices and the risk of gestational diabetes mellitus: a prospective cohort study. Diabetes Care 2012;35:1079–82.

[40] Mattei J, Malik V, Hu FB, Campos H. Substituting homemade fruit juice for sugar-sweetened beverages is associated with lower odds of metabolic syndrome among Hispanic adults. J Nutr 2012;142:1081–7.

[41] Yang Y, Huang CY, Peng SS, Li J. Carotenoid analysis of several dark green leafy vegetables associated with lower risk of cancer. Biomed Environ Sci 1996;9:386–92.

[42] Rao AV, Agarwal S. Role of antioxidant lycopene in cancer and heart disease. J Am Coll Nutr 2000;19:563–9.

[43] Ziegler RG, Mason TJ, Stemhagen A, Hoover R, et al. Carotenoid intake, vegetables and the risk of lung cancer among white men in New Jersey. Am J Epidemiol 1986;123:1080–93.

[44] Dincer C, Karaoglan M, Erden F, Tetik N, et al. Effects of baking and boiling on the nutritional and antioxidant properties of sweet potato [*Ipomoea batatas* (L.) Lam.] cultivars. Plant Food Hum Nutr 2011;66:341–7.

[45] Bahado-Singh PS, Riley CK, Wheatley AO, Lowe HI. Relationship between processing method and the glycemic indices of ten sweet (*Ipomoea batatas*) cultivars commonly consumed in Jamaica. J Nutr Metab 2011;2011:584832.

[46] Camire ME, Kubow S, Donnelly DJ. Potatoes and human health. Crit Rev Food Sci Nutr 2009;49:823–40.

[47] Halton TL, Willett WC, Liu S, Manson JE, et al. Potato and french fry consumption and risk of type 2 diabetes in women. Am J Clin Nutr 2006;83:284–90.

[48] Sanny M, Luning PA, Jinap S, Bakker EJ, et al. Effect of frying instructions for food handlers on acrylamide concentration in French fries: an explorative study. J Food Prot 2013;76:462–72.

[49] Kaefer CM, Milner JA. The role of herbs and spices in cancer prevention. J Nutr Biochem 2008;19:347–61.

[50] Gruenwald J, Freder J, Armbruster N. Cinnamon and health. Crit Rev Food Sci Nutr 2010;50:822–34.

[51] Zhou H, Beevers CS, Huang S. The targets of curcumin. Curr Drugs Target 2011;12:332–47.

[52] Iyer A, Panchal S, Puudyal H, Brown L. Potential health benefits of Indian spices in the symptoms of the metabolic syndrome: a review. Indian J Biochem Biophys 2009;46:467–81.

[53] Blomhoff R, Carisen MH, Andersen LF, Jacobs Jr D. Health benefits of nuts: potential role of antioxidants. Br J Nutr 2008;96(Suppl. 2):S52–60.

[54] Kris-Etherton PM, Hu FB, Sabete J. The role of tree nuts and peanuts in the prevention of coronary heart disease: multiple potential mechanisms. J Nutr 2008;138:1746S–51S.

[55] Vinson JA, Cai Y. Nuts, especially walnuts, have both antioxidant quantity and efficacy and exhibit significance potential health benefits. Food Funct 2012;3:134–40.

[56] Murphy MM, Barraj LM, Herman D, Bi X, et al. Phytonutrient intake by adults in the United States in relation to fruit and vegetables consumption. J Acad Nutr Diet 2012;112:1626–35.

[57] Satia JA, Galanko JA, Siega-Riz AM. Eating at fast-food restaurants is associated with dietary intake, demographic, psychological, and behavioral factors among African Americans in North Carolina. Public Health Nutr 2004;7:1089–96.

[58] Larson N, Neumark-Sztainer D, Laska MN, Story M. Young adults and eating away from home: associations with dietary intake patterns and weight status differ by choice of restaurant. J Am Diet Assoc 2011;111:1696–703.

[59] Thompson HJ, Heimendinger J, Sedlacek S, Haegele A, et al. 8-Isoprostane $F_{2\alpha}$ excretion is reduced in women by increasing vegetable and fruit intake. Am J Clin Nutr 2005;82:768–76.

[60] Holt EM, Steffen LM, Morgan A, Basu S, et al. Fruit and vegetable consumption and its relation to markers of inflammation and oxidative stress in adolescents. J Am Diet Assoc 2009;109:414–21.

[61] Rink SM, Mendola P, Mumford SL, Poudrier JK, et al. Self-report of fruit and vegetable intake that meets the 5 a day recommendation is associated with reduced levels of oxidative stress biomarkers and increased levels of antioxidant defense in premenopausal women. J Acad Nutr Diet 2013;113:776–85.

CHAPTER 13

Tea, Coffee, and Chocolate: Rich Sources of Antioxidants

CONTENTS

13.1 INTRODUCTION

After water, tea and coffee are the most widely consumed beverages worldwide. Both tea and coffee are rich sources of antioxidants, and there are many health benefits of consuming tea and coffee on a daily basis. Tea is produced by brewing dried leaves of the tea plant (*Camellia sinensis*), which was first cultivated in China. Drinking tea was also an ancient tradition in Japanese culture. Tea reached Great Britain in the 17th century and gained popularity in France, Holland, and Russia. By the mid-18th century, tea was being exported to America. Despite a huge tax levied on tea in 1773 by the British Crown and subsequent dumping of tea in Boston Harbor during the Boston Tea Party, drinking tea remained popular in the United States and is still popular today. The production of 1.8 million tons of dried tea bags provides worldwide per capita consumption of 40 L of tea annually [1].

Coffee plants were first cultivated in Africa, probably in the Ethiopian region, and later introduced into Yemen, Egypt, and Arabia. The name "coffee" was derived from the word *Keffa*, the province of Ethiopia where shepherds discovered coffee beans probably in the 6th century. By the 13th century, coffee consumption was widespread throughout the Muslim civilization, and Muslims imported coffee to India and also to Europe via Italy. Finally, French explorers imported coffee to the Americas probably in the 17th century. Coffee is the third most consumed beverage in the world after water and tea [2]. Coffee berries, which contain the coffee seeds or "beans," are produced by several species of small bushlike plants of the genus *Coffea*. The two most commonly grown coffee plants are *Coffea arabica* and *Coffea robusta*. Coffee prepared from *C. robusta* has more distinct flavor and taste and represents approximately 80% of current coffee production. Once ripe, coffee berries are picked, processed, and dried. The seeds are then roasted to varying degrees, depending on the desired flavor, and roasted coffee has a characteristic black color. Decaffeination of coffee may also be part of the

A. Dasgupta and K. Klein: Antioxidants in Food, Vitamins and Supplements. DOI: http://dx.doi.org/10.1016/B978-0-12-405872-9.00013-6

processing of coffee seeds. Seeds are decaffeinated when they are still green. Many methods can remove caffeine from coffee, but all involve soaking the green beans in hot water or steaming them and then using a solvent to dissolve caffeine-containing oils. However, drinking decaffeinated coffee also has health benefits because it usually retains most antioxidants that are present in caffeinated coffee.

13.2 CONSUMPTION OF TEA AND COFFEE

In 2011, a report by the National Coffee Association stated that 40% of the US population aged 18–24 years reported drinking coffee daily, and that 54% of Americans aged 25–39 years drink coffee on a regular basis. Gourmet coffee represents 37% of all coffee consumption by the US population. Therefore, it is estimated that more than 150 million people in the United States drink coffee on a regular basis [3]. In general, men consume coffee more than women (average 1.9 cups per day by men vs. 1.4 cups per day by women). Tea is a more popular drink than coffee in many areas of the world, such as Middle Eastern countries, Great Britain, and the Indian subcontinent. Only 21.3% of US adults reported drinking tea on a daily basis. Tea drinkers are most likely to be older, female, Caucasian, and have a higher income than non-tea drinkers [4].

In general, three types of tea are consumed throughout the world, but all are produced from young leaves of *Camellia sinensis*. Depending on the processing of tea leaves, tea can be categorized as black tea, green tea, and oolong tea. Green tea is favored in China, Japan, and certain Asian countries, whereas black tea is favored in Western countries and the Indian subcontinent. Of the tea produced worldwide, 78% is black tea, 20% green tea, and 2% oolong tea, which is consumed by people in southern China. Green tea is produced by steaming of tea leaves in Japan and panning (pan-fried) in China so that enzymes are destroyed and no further biotransformation of tea components can take place and the final product is stable. With no fermentation, green tea retains the color of green tea leaves. Black tea is produced by full fermentation (oxidation) of tea leaves, whereas oolong tea is only partially fermented. Therefore, oolong tea has the aroma of black tea and the freshness of green tea [5].

13.2.1 Antioxidants Present in Tea and Coffee

The major polyphenolic antioxidants present in tea are flavonoids and phenolic acid. However, due to different manufacturing processes, different polyphenolic compounds are present in green, black, and oolong tea. Similarly, polyphenolic compounds undergo structural changes as a result of roasting of coffee beans during the manufacturing process.

Polyphenolic compounds in green tea are called *catechins*, which account for 30–42% of water-soluble solids in brewed green tea. The four major compounds are epicatechin, epicatechin gallate, epigallocatechin, and epigallocatechin-3-gallate (Box 13.1). Because black tea is produced by fermentation, catechins undergo chemical transformation by the enzymes polyphenol oxidase and peroxidase, which are present in tea leaves. Usually, 75% of catechins present in green leaves undergo oxidation to produce theaflavins and thearubigins. The four major theaflavins are theaflavin, theaflavin-3-gallate, theaflavin-3′-gallate, and theaflavin-3-3′-digallate. Thearubigins represent 10–20% of the dry weight of black tea [6]. Antioxidant components of black tea are also summarized in Box 13.1. Usually, these antioxidants account for approximately 30% of dry weight of tea leaves, indicating that both green and black tea are good sources of natural antioxidants. Green tea and black tea polyphenols undergo epimerization and oxidation during the brewing process. Oxidation of polyphenolic compounds of green tea is limited because enzymes are inactivated during the processing of green tea. Theanaphthoquinone has been identified as a major oxidation product of theaflavin in black tea [6].

Coffee also contains many chemicals that exhibit antioxidant properties. However, phenolic compounds and phenolic melanoidins are major antioxidant compounds in coffee. Caffeic acid and structurally related chlorogenic

Box 13.1 MAJOR ANTIOXIDANTS IN BREWED GREEN TEA AND BLACK TEA

Green Tea—Brewed

- Epicatechin
- Epicatechin-3-gallate
- Epigallocatechin
- Epigallocatechin-3-gallate
- Catechin
- Gallocatechin
- Theaflavin
- Theaflavin-3,3′-digallate
- Theaflavin-3-gallate
- Thearubigins

Black Tea—Brewed

- Epicatechin
- Epicatechin-3-gallate
- Epigallocatechin
- Epigallocatechin-3-gallate
- Catechin
- Gallocatechin
- Theaflavin
- Theaflavin-3,3′-digallate
- Theaflavin-3-gallate
- Theaflavin-3′-gallate
- Thearubigins
- Kaempferol

acids (esterification of caffeic acid and quinic acid produces chlorogenic acid) are the major polyphenolic compounds in green coffee beans. However, during coffee roasting, these compounds undergo oxidation and polymerization into chlorogenic acid lactones and high-molecular-weight melanoidins with excellent antioxidant properties. The antioxidant capacity of roasted coffee is generally 30-fold higher than that of green coffee [7]. In general, melanoidins are defined as brown nitrogenous macromolecular compounds that absorb light at 405 nm, allowing their quantification in coffee fraction. These compounds are of interest not only because they produce the characteristic brown color of brewed coffee but also because of their flavor-binding properties, antioxidant activity, and metal-chelating capacity. Interestingly, dark roasted coffee contains more melanoidins than light roasted coffee [8].

Caffeine is the natural constituent of coffee. Coffee has the highest and most variable amounts of caffeine among brewed beverages from natural sources. Caffeine present in coffee can produce addiction and stimulation of the central nervous system, thus slightly increasing blood pressure. However, consuming coffee in moderation has many beneficial health effects because it is rich in various antioxidants. A single cup of coffee contains 70–350 mg of chlorogenic acid [9]. Bonita *et al.* [10] estimated that a 180-mL cup of brewed coffee provides approximately 396 mg of polyphenolic compounds, whereas instant coffee contains 316 mg of polyphenolic compounds. Combining this data with US per capita consumption of coffee, the authors speculated that coffee is the number one source of antioxidants in the US diet.

It is advised to drink filtered coffee rather than boiled coffee or espresso because two components of coffee, cafestol and kahweol, are only present in unfiltered coffee. Cafestol, a diterpene, is the most potent cholesterol-elevating compound present in the human diet, although it also has a hepatoprotective effect [11]. However, during filtration, these components are trapped in the filter paper, and an almost negligible amount is found in filtered coffee (Table 13.1) [12].

Table 13.1 Amounts of Harmful Cholesterol-Elevating Components in Boiled Coffee, Espresso, and Filtered Coffee

Type of Coffee	Kahweol (mg/100 mL Coffee)	Cafestol (mg/100 mL of Coffee)
Boiled coffee	0.7–10	0.5–8
Espresso	0.1–2.6	0.1–1.9
Filtered coffee	0–0.1	0–0.1

Source: *Ranheim and Halvorsen* [12].

13.2.2 Tea or Coffee Consumption and the Antioxidant Capacity of Blood

The antioxidant capacity of blood increases after consuming coffee and tea because antioxidants present in both are absorbed from the gastrointestinal tract and such compounds can be detected in blood. In one study, volunteers consumed a single dose of green tea (0.9 g catechins), black tea (0.3 g total catechins), or black tea with milk. The authors observed rapid increases in catechin levels after consumption of both green tea and black tea, with maximum increases observed after 2.3 h for green tea and 2.2 h for black tea. Then blood levels of catechins declined rapidly, with a half-life of 4.8 h for green tea and 6.9 h for black tea. Drinking black tea with milk did not affect blood levels of catechins. The authors concluded that catechins from both green tea and black tea are rapidly absorbed after consumption and milk does not impair bioavailability of catechins [13]. In a study of 21 healthy volunteers (10 male and 11 female), Leenen *et al.* [14] examined the effect of plasma antioxidant capacity after consumption of green tea or black tea without or with milk. The authors observed significant increases in plasma antioxidant capacity after consumption of both green tea and black tea, but larger increases in antioxidant capacity of plasma were observed with green tea compared to black tea. This was due to the higher catechin content of green tea because concentrations of plasma catechins were higher after consumption of green tea compared to black tea. The addition of milk to the tea did not alter these increased concentrations. The authors concluded that consumption of a single dose of green or black tea increases plasma antioxidant capacity of humans *in vivo*.

Pecorari *et al.* [15] studied biomarkers of antioxidant status in 15 healthy volunteers after each volunteer consumed 500 mL of green tea with different solid counts (1.4, 1.6, 1.8, and 2.0 g/L). They observed that after consuming green tea with a solid count of 2.0 g/L, the total radical trapping antioxidant parameter of plasma was increased by 8.4% after 1 h of consumption of green tea; even after 4 h, the plasma antioxidant potential was still increased by 5.9% compared to the baseline. The effects were less significant when volunteers consumed green tea with lower solid content. In a study of 34 people who drank 1 L of green tea daily, Coimbra *et al.* [16] observed lower concentrations of malondialdehyde and 4-hydroxy-2-nonenal in serum 3 weeks after tea drinking, indicating significant increases in antioxidant capacity of serum following tea consumption. In addition, a lower level of membrane-bound hemoglobin in erythrocytes after tea consumption indicated reduced oxidative stress experienced by erythrocyte membranes in these subjects. The authors concluded that green tea consumption is beneficial because it reduces oxidative stress, thus protecting individuals from diseases caused by excessive oxidative stress. In a study of 12 healthy male volunteers who consumed 600 mL of green tea daily for 4 weeks, Sung *et al.*

[17] reported a significant reduction in oxidized low-density lipoprotein (ox-LDL) and soluble vascular adhesion molecule-1 (sVCAM-1) following consumption. However, lipid profile did not change in these individuals. The authors concluded that the *in vivo* antioxidant effect of green tea and the effect of green tea on reducing ox-LDL (elevated ox-LDL increases the risk of atherosclerosis) as well as sVCAM-1 may provide a potential mechanism for the known benefit of tea consumption in reducing the risk of cardiovascular disease. Inami *et al.* [18] demonstrated that catechins, the major antioxidant compounds present in green tea, are responsible for a reduction in the concentration of ox-LDL. The authors studied 40 healthy volunteers who orally consumed 500 mg of catechin (extracted from green tea) daily for 4 weeks. After the study period, the concentrations of ox-LDL in volunteers who consumed catechin were significantly reduced compared to those of controls (9.56 ± 9.2 vs. 7.76 ± 7.7 U/mL, respectively). The authors concluded that catechin decreased the plasma ox-LDL concentration without affecting the plasma LDL concentration, and the beneficial effect of green tea in reducing the risk of coronary artery disease might result from a decrease in plasma ox-LDL level. In contrast, consumption of black tea reduced total and LDL cholesterol significantly without affecting the concentration of ox-LDL [19]. However, in a study of 46 patients with type 2 diabetes, Neyestani *et al.* [20] showed that regular intake of black tea extract reduced serum malondialdehyde concentrations as well as C-reactive protein concentrations, indicating both antioxidant and anti-inflammatory effects of black tea.

Total amounts of antioxidant polyphenolic compounds are similar in both black tea and green tea (15–25% dry weight for both), but catechins are present in much higher amounts in green tea (12–18% dry weight) than in black tea (2 or 3% dry weight). However, black tea is richer than green tea (2–4% dry weight) in other polyphenolic antioxidants (7-15% dry weight). Some authors report that after consumption of black or green tea, antioxidant capacity of human plasma is increased similarly, although other investigators claim that green tea has better antioxidant potential than black tea, and this may be related to higher levels of catechins in plasma after drinking green tea compared to black tea [3]. The total oxygen radical absorbance capacity (ORAC) is similar for both brewed green tea and brewed black tea (1253 vs. 1153, respectively). However, ready-to-drink green tea (usually ice tea) has a much lower ORAC of 520, whereas ready-to-drink black tea has an ORAC of 313.

Chlorogenic acids found in coffee are also absorbed after coffee intake, and these compounds are found in human plasma [21]. Therefore, plasma antioxidant capacity is increased following consumption of coffee. Natella *et al.* [22] compared the antioxidant capacity of human plasma in subjects after they drank tea or coffee. Using the TRAP test (total radical trapping antioxidant parameter), which measures total antioxidant capacity of the plasma, the authors observed that after drinking 200 mL of coffee, the antioxidant capacity of plasma was increased an average of 5.5% after 1 h and a 4% increase was maintained 2 h after drinking coffee. However, after drinking 200 mL of tea, the antioxidant capacity of the plasma was increased by 4.7% 1 h after ingestion, indicating that antioxidant capacity of plasma was increased more by coffee than by tea. These authors further investigated the total antioxidant capacity of both coffee and tea and observed that, in general, coffee had an antioxidant capacity five to eight times higher than that of tea. However, coffee contains more caffeine than tea. A cup of coffee (approximately 200 mL in this study but usually in the United States one coffee cup is 8 oz or 240 mL) contains about 181 mg of caffeine versus an average of 130 mg of caffeine in a cup of tea. The amount of theobromine in a cup of coffee was 28.9 mg, whereas the amount of theobromine in a cup of tea was 5.9 mg. However, both caffeine and theobromine, which are found in a relatively higher concentration in coffee than in tea, have negligible antioxidant capacity. Therefore, a higher concentration of these compounds in coffee compared to tea does not explain the higher antioxidant capacity of coffee.

13.2.3 Health Benefits of Drinking Tea

Drinking tea on a regular basis has many health benefits, the most important of which is reducing the risk of cardiovascular diseases. The major health benefit of drinking tea comes from its rich antioxidant properties. Although more research has been done on green tea than black tea and there is a general perception that green tea has more health benefits than black tea, black tea has similar health benefits. Health benefits of drinking tea (both black and green) are summarized in Box 13.2.

Box 13.2 HEALTH BENEFITS OF DRINKING TEA

- Improved antioxidant defense of the body (both green and black tea)
- Reduced risk of heart disease (both green and black tea)
- Reduced risk of stroke (both green and black tea)
- Preventing various types of cancer (green tea and black tea may prevent different types of cancer)
- Boosting immune system (both green and black tea)
- Protecting bone health (both green and black tea)
- Less chance of cognitive impairment (green tea may be more effective)
- Green tea may be effective in reducing obesity
- Green tea may help with bone health
- Green tea has antidepressant properties
- Green tea reduces risk of death from pneumonia in Japanese women
- Black tea may lower the risk of type 2 diabetes

Stangl *et al.* [23] reported that consumption of either black or green tea can reduce the risk of cardiovascular diseases, especially atherosclerosis. Regular consumption of tea among the elderly population has been associated with a 40–50% reduction in mortality from cardiovascular disease. The most likely mechanism by which tea consumption reduces heart disease is through its antioxidant effect. A 6-year study of Dutch men and women found that those who consumed three or more cups of black tea per day had significantly lower risk of myocardial infarction than those who did not drink tea [24]. Current research indicates that consumption of three or more cups of black tea per day can reduce the risk of myocardial infarction. Consumption of a similar amount of green tea per day also provides a similar benefit, but people may not obtain this beneficial effect if they consume one cup of tea or less per day [25].

Tea can also reduce the risk of stroke. Arab and Lieberskind [26] noted that based on current research with both human and animal models as well as epidemiological studies, the evidence is strong that daily consumption of tea is beneficial in reducing the risk of stroke. Moderate consumption of black or green tea (three cups per day) may prevent the onset of stroke. Other investigators have also demonstrated that moderate consumption of tea (two or three cups per day; 500 mL) enhances endothelial-dependent vasodilation, thus reducing the risk of stroke. In a study of 74,961 Swedish men and women who were free from cardiovascular disease and cancer at baseline in 1997 and followed until December 2008, Larsson *et al.* [27] observed that consuming four or five cups of black tea each day significantly reduced the risk of stroke, including cerebral infarction and hemorrhagic stroke, compared to no tea consumption. Green tea is also effective at reducing the risk of both cardiovascular disease and stroke [28]. Based on a meta-analysis of nine studies, Ras *et al.* [29] concluded that vasodilation may be the mechanism by which tea reduces the risk of both stroke and heart attack in individuals who drink tea on a regular basis.

Drinking tea effectively protects an individual against cancer as well as slows the progression of disease in people already diagnosed with cancer. Animal experiments clearly demonstrate that components of green tea are very effective at preventing cancer induced by carcinogens. However, there is controversy regarding the efficacy of tea (both black and green tea) in preventing various types of cancer in humans. Although many studies have clearly demonstrated that tea can prevent various types of cancer as well as prolong life in cancer patients, other studies have failed to clearly establish the link between preventing cancer and moderate consumption of tea. It has been postulated that epigallocatechin gallate, the major antioxidant component of green tea, is responsible for the protective effect of green tea against cancer. Butt and Sultan [30] noted that green tea is nature's defense against malignancies, especially colon, skin, lung, prostate, and breast cancer. Six epidemiological studies that investigated the efficacy of green tea in preventing gastric cancer concluded that green tea is effective at preventing gastric cancer. A large study of 501 breast cancer patients and 594 women who did not have breast cancer concluded that green tea is effective at reducing the risk of breast cancer, but no such protective effect was observed among drinkers of black tea. Another study found that regular consumption of green tea among nonsmoking women was associated with an overall reduced risk of lung cancer. In addition, studies indicated that consumption of green tea was associated with slowing the progression of ovarian cancer in Chinese women already diagnosed with ovarian cancer. Green tea was also associated with slowing the progression of breast cancer. However, the role of green tea in preventing prostate cancer, esophageal cancer, colon cancer, rectal cancer, and pancreatic cancer is controversial, with some studies showing a beneficial effect and others failing to demonstrate any protective effect of green tea against these cancers [31].

Based on a study of 41,761 Japanese adults, Naganuma *et al.* [32] concluded that consuming five or more cups of green tea every day reduced the risk of hematological malignancies such as lymphoid neoplasm and myeloid neoplasm. Tang *et al.* [33] reported an overall 18% decrease in the risk of developing lung cancer with the consumption of at least two cups of green tea each day, but no such benefit was observed in people who drank black tea. However, other studies have indicated a protective effect of black tea against certain types of cancer. Polyphenolic compounds of black tea are potent antioxidants and have anticancer activities. Black tea extract is enriched in theaflavins, which can inhibit the chymotrypsin-like activity of the proteasome and proliferation of human multiple myeloma cells in a dose dependent manner. In addition, isolated theaflavin from black tea can bind to and inhibit the purified 20S proteasome, accompanied by suppression of proliferation of tumor cells, indicating that tumor proteasome may be an important target for theaflavin by which black tea may exert its anticancer effects [34]. Lee *et al.* [35] observed that regular consumption of green tea or black tea could reduce the risk of developing ovarian cancer.

Other health benefits of drinking tea include boosting the immune system, improving mental alertness, reducing the risk of cognitive impairment in old age, reducing the risk of depression, and possibly lowering blood pressure and blood cholesterol. Theanine, which is more abundant in green tea than in black tea, is responsible for boosting the immune system. Regular tea consumption has a beneficial effect on cognitive function of the elderly, and theanine and other components of tea may play an important role in improving cognitive function late in life [36]. Catechins in green tea can penetrate the blood–brain barrier. In addition to their antioxidant effects, these compounds can chelate metal ions, thus protecting neurons from death in a variety of neurodegenerative diseases, including Parkinson's disease and Alzheimer's disease [37]. Although caffeine may elevate blood pressure, theanine acts to neutralize this effect of caffeine. However, both caffeine and theanine play important roles in improving mental alertness after drinking tea and improving performance requiring sustained attention [38].

Drinking tea on a regular basis can reduce stress by reducing the level of the stress-producing chemical cortisol. It may also have anti-inflammatory effects. In addition, regular consumption of tea is beneficial in lowering the prevalence of depression in the community-dwelling older population. Usually, drinking three or four cups of green tea per day is effective at preventing depression [39]. A study on the Finnish general population found that none of the subjects who drank five cups of tea or more suffered from depression. In general, those who drank an amount of tea on a regular basis were less depressed than those who did not drink tea [40]. Green tea, but not black tea, has an anti-obesity effect. Catechins which are green tea antioxidants are probably responsible for preventing obesity in regular tea drinkers [41]. Based on a study of 35 obese patients with metabolic syndrome, Basu *et al.* [42] observed that green tea consumption (four cups per day) or green tea extract supplementation (two capsules per day) for 8 weeks significantly decreased body weight and body mass index in these patients. In addition, patients who consumed green tea showed a decreasing trend in LDL cholesterol as well as significantly decreased levels of both malondialdehyde and 4-hydroxy-2-nonenal (markers of lipid peroxidation) compared to controls. Therefore, drinking green tea has favorable effects in obese patients with metabolic syndrome. Watanabe *et al.* [43] reported that green tea consumption was associated with a lower risk of death from pneumonia in Japanese women. Tomata *et al.* [44] concluded that green tea consumption was significantly associated with a lower risk of functional disability (disability stroke, cognitive impairment, and osteoporosis) even after adjustment for possible confounding factors. Another study demonstrated that both green tea and black tea can also improve bone health by protecting individuals from osteoporosis [45]. Beresniak *et al.* [46] observed a statistical correlation between high black tea consumption and low diabetes prevalence. However, the role of green tea in reducing the risk of type 2 diabetes is controversial because some studies have shown a beneficial effect but other studies have shown no significant effect.

13.2.4 Health Benefits of Drinking Coffee

In general, coffee is richer than tea in antioxidants, but coffee also contains approximately double the amount of caffeine compared to tea. Depending on the type of coffee bean, the ORAC value of coffee can be as high as

15,264 [47]. Excess consumption of caffeine may elevate blood pressure, but such effect is not observed if a person drinks 4 or fewer cups of coffee per day. Consumption of 250 mg or more caffeine per day (2 cups of coffee or more) may cause caffeine dependence. One way to avoid caffeine intake is to drink decaffeinated coffee; however, it should be borne in mind that decaffeinated coffee contains a very small amount of caffeine (approximately 3 mg per cup). Death from caffeine overdose is impossible from coffee consumption alone because one needs to consume at least 50 cups of coffee during a single occasion. Deaths from caffeine overdose have been reported from ingesting large amounts of caffeine tablets. A 21-year-old woman who ingested 100 caffeine tablets (100 mg caffeine in each tablet; total = 10,000 mg of caffeine) died from caffeine overdose [48].

Drinking both caffeinated and decaffeinated coffee is effective at preventing type 2 diabetes. It is postulated that chlorogenic acid in coffee reduces plasma glucose concentration. A study in the Netherlands reported that people who drank seven cups of coffee per day were half as likely as non-coffee drinkers to develop type 2 diabetes. A Swedish study reported that drinking slightly more than two cups of regular coffee per day is effective at reducing the risk of developing type 2 diabetes. In a study involving the Scottish population, the authors observed that drinking coffee had a beneficial effect in reducing the risk of heart disease [12]. Based on a review of 13 cohort studies, Muley *et al.* [49] concluded that drinking coffee reduced the risk of type 2 diabetes, especially for people who drank four to six cups of coffee per day compared to people who consumed less than two cups per day. Interestingly, filtered coffee was superior to pot-boiled coffee for reducing the risk of type 2 diabetes, and decaffeinated coffee was also superior to caffeinated coffee. A stronger correlation was also observed for people younger than age 60 years [49].

Mineharu *et al.* [50] compared coffee, black tea, green tea, and oolong tea consumption with risk of mortality from cardiovascular disease in 76,979 Japanese men and women who were free from stroke, cancer, and cardiovascular disease during enrollment in the study. The authors observed that consumption of coffee (three or more cups per day), green tea (three to five cups per day), and oolong tea (more than one cup per day) was associated with a reduced risk of mortality from cardiovascular disease. Based on a meta-analysis of five prospective studies of coffee consumption and risk of heart failure that included a total of 140,222 participants, Mostofsky *et al.* [51] concluded that moderate coffee consumption was inversely associated with the risk of heart failure, and the largest inverse relation was observed with consumption of four cups per day. Although coffee consumption increases homocysteine in blood, which is associated with an increased risk of coronary heart disease, it is currently believed that, mostly due to the antioxidant effect of coffee, consumption in moderation (two or three cups per day) may reduce the risk of coronary heart disease [52]. Even drinking up to five cups of regular caffeinated coffee does not increase the risk of coronary heart disease [53]. Based on a meta-analysis of nine cohort studies, Kim *et al.* [54] concluded that drinking four cups or more of coffee per day has a preventive effect on stroke.

Drinking regular coffee can also reduce the risk of Parkinson's disease. In a study of 8000 Japanese American men, it was found that people who did not drink coffee were three to five times more likely to develop Parkinson's disease during the next 24–30 years than those who drank at least 28 oz of coffee daily (830 mL; approximately four cups of coffee per day). A large study of 86,000 women found that drinking at least two cups of coffee per day reduced the risk of suicide by 50%. Other studies have also found that drinking coffee on a regular basis can reduce the risk of suicide. Drinking coffee also reduces the risk of colon and liver cancer (hepatocellular carcinoma). Coffee drinking may also reduce the risk of developing liver cirrhosis. Drinking coffee can also reduce the risk of liver and kidney cancer [55]. Hildebrand *et al.* [56] observed that intake of four or more cups of caffeinated coffee per day was associated with a 49% lower risk of death from oral/pharyngeal cancer compared to nondrinkers or occasional drinkers. However, no such association was found with tea drinking. Sinha *et al.* [57] observed that, compared to nondrinkers, drinkers of four to six cups of coffee, predominantly caffeinated coffee, per day had a lower risk of colon cancer. No such association was found with tea drinkers. In a large prospective study, Freedman *et al.* [58] found an inverse relationship between coffee drinking and total and cause-specific mortality (death due to heart disease, respiratory disease, stroke, injuries, accidents, diabetes, and infection) but did not observe reduced mortality from cancers due to coffee drinking.

Drinking coffee on a regular basis can reduce the risk of developing depression. Hermansen *et al.* [59] concluded that moderate coffee drinking (three or four cups per day) reduces the risk of Parkinson's disease, Alzheimer's disease, suicide, and depression. Coffee has some modest laxative effect, and it can also prevent the formation of gallstones and gout. Usually, a high concentration of uric acid in blood is responsible for gout attack. Drinking coffee on a regular basis reduces the level of uric acid in blood more significantly in men than in women, thus reducing the risk of gout [60]. Various health benefits of drinking coffee are listed in Box 13.3. However, drinking excessive amounts of coffee may impair mineral and iron absorption. Individuals such as the elderly, people with hypertension, children, and adolescents may be more vulnerable to the adverse effects of excessive caffeine consumption. In addition, it is recommended that pregnant women limit coffee drinking to not more than two cups per day (no more than 300 mg of caffeine) to exclude the possibility of spontaneous abortion or impaired fetal growth [61].

Box 13.3 HEALTH BENEFITS OF DRINKING COFFEE

- Reduced risk of developing type 2 diabetes
- Improved alertness and cognitive function
- Reduced risk of stroke
- May prevent cardiovascular disease
- Reduced risk of suicide
- Preventing Parkinson's disease
- Preventing various types of cancers
- Lower mortality from oral/pharyngeal cancer
- Preventing formation of gallstones and disease of gallbladder
- Reduced chance of gout
- Reduced chance of depression
- Laxative effect of coffee

13.2.5 How Much Tea or Coffee per Day?

There is more caffeine in coffee than in tea. Therefore, more caution should be exercised when drinking coffee compared to tea. However, more coffee can be consumed if a person drinks mostly decaffeinated coffee because many adverse effects of coffee are related to caffeine. In general, drinking three to five cups of tea per day (green tea or a combination of green and black tea) is sufficient to obtain the health benefits from tea. The current recommendation for tea and coffee consumption is up to 40 oz (approximately 1200 mL or five cups total) of tea or coffee or a combination per day. For drinking regular coffee alone, the suggested limit is 400 mg of total caffeine per day, which is 32 oz of coffee (960 mL or approximately four cups per day) [62].

13.3 CHOCOLATE: A HISTORICAL PERSPECTIVE

The cacao tree (*Theobroma cacao*) was probably first cultivated in approximately AD 250–900 by the ancient Maya civilization in Mexico and Central America. During the 12th through the 16th century, the elites of Aztec civilization drank chocolate derived from cocoa beans and also offered chocolate to Aztec gods. The beans were also used as both currency and medicine (to fight fatigue). Following the Spanish conquest of Mexico, cocoa beans were brought to Spain in 1528, and during the next 100 years cane sugar, vanilla, cinnamon, etc. were added to bitter-tasting chocolate rather than peppers and natural herbs used traditionally in chocolate production. In 1657, the first chocolate house was opened in London, and chocolate was introduced in North America in approximately the mid-1800s. During World War I, chocolate was provided as a ration to US soldiers in Europe. Currently, chocolate is a favorite food product throughout the world [63]. The word "chocolate" was probably derived from the word *xocolati*, meaning "bitter water" in the Aztec language. The seeds of the cacao tree have a bitter taste and must be fermented, dried, and roasted in order to prepare chocolate. Then shells are removed to produce cacao nibs. The nibs are ground to cocoa mass, which is pure chocolate in rough form. Because this cocoa mass is usually liquefied and then molded with or without other ingredients, it is called chocolate liquor, which contains cocoa solid and cocoa butter. Much of the chocolate consumed today is in the form of sweet chocolate, combining cocoa solids and cocoa butter or other fat with sugar. Milk chocolate is sweet chocolate that additionally contains milk powder or condensed milk. White chocolate contains cocoa butter, sugar, and milk but no cocoa solids.

13.3.1 Antioxidants Present in Chocolate: Health Benefits

Chocolate has excellent antioxidant properties, and the ORAC values of various types of chocolate are given in Table 13.2. Chocolate is rich in flavonoids, which are natural antioxidants. The primary antioxidants present in chocolate are flavonoids including catechin, epicatechin, and proanthocyanidins. These antioxidants not only can scavenge free radicals due to their tricyclic structures but also can chelate iron and copper ions. Dark chocolate is formulated with a higher percentage of cocoa bean liquor and contains more antioxidants than does milk chocolate. In addition to a strong antioxidant effect, chocolate can reduce platelet aggregation, particularly due to its epicatechin content. Reducing platelet aggregation may reduce the risk of heart attack or stroke [64]. Studies of the Kuna Indians in Panama, who consume large amounts of natural cocoa beverages, indicate that these people have lower blood pressure, better renal function, and lower risk of cardiovascular diseases as well as cardiovascular mortality compared to a Panamanian control population. In addition, the antioxidant effect of cocoa may directly

Table 13.2 Oxygen Radical Absorbance Capacity (ORAC) of Various Types of Chocolate

Type of Chocolate	ORAC Value (μmol per Trolox Equivalent/100 g)
Baking chocolate, unsweetened square	49,093
Dark chocolate	20,816
Milk chocolate	7,519
Semisweet chocolate	18,053
Cocoa, dry powder unsweetened	55,653
Chocolate syrup	6,330

Source: *Old USDA website. This information was deleted from the website.*

influence insulin resistance, thus reducing the risk of developing type 2 diabetes. Furthermore, cocoa consumption may stimulate changes in redox-sensitive signaling pathways involved in gene expression and the immune system, thus boosting the immune system. Cocoa can also protect nerves from injury and inflammation, protect skin from ultraviolet-induced oxidative damage, and may improve cognitive function and mood [65]. It has been accepted that the major benefit of eating chocolate is to lower the risk of cardiovascular diseases, including myocardial infarction. In patients who already suffered from myocardial infarction, eating chocolate twice a week reduced the risk of mortality from heart diseases by 66% compared to the group that never consumed chocolate. Chocolate may also be effective in reducing blood pressure as well as improving flow of blood through the arteries, thus lowering the risk of myocardial infarction. Chocolate consumption on a regular basis may also lower triglycerides and improve blood level of high-density lipoprotein (HDL) cholesterol. Chocolate may also modulate insulin resistance, thus improving the glycemic control in diabetic patients, especially those with type 2 diabetes [66].

Preeclampsia is a major complication of pregnancy with cardiovascular manifestation. Triche *et al.* [67] studied the association between chocolate consumption and preeclampsia in 2291 pregnant women and observed that women who consumed five or more servings of chocolate per week had a reduced risk of preeclampsia compared to pregnant women who consumed less than one serving of chocolate per week. In addition, there was an inverse relation between cord serum theobromine level (theobromine is a component of chocolate) and the risk of developing preeclampsia. Eating chocolate in the third trimester was more beneficial than eating chocolate in the first trimester. Chocolates rich in polyphenolic compounds may help improve symptoms in patients with chronic fatigue syndrome [68]. However, chocolate is full of calories, a significant amount of fat being present in cocoa butter—mostly stearic acid but also oleic acid, palmitic acid, and other fatty acids. A major problem with consuming too much chocolate is its high calorie content: 100 g of chocolate may produce 500–600 calories or more. It has been well-established that eating dark chocolate has more health benefits than eating milk chocolate, and white chocolate has only a few benefits; however, chocolate must be consumed in moderation (Box 13.4). Currently, consumption of not more than 25 g of dark chocolate per day is recommended for prevention of heart diseases and to obtain optimal health benefits of dark chocolate [69]. Milk chocolate is rich in calcium and may have health benefits, but eating white chocolate other than for enjoyment and taste provides only little benefit. It is recommended that chocolate not be consumed with milk because milk proteins may inhibit absorption of antioxidant flavonoids present in chocolate [70].

Box 13.4 BENEFICIAL EFFECTS OF EATING CHOCOLATE IN MODERATION AND PROBLEM WITH EATING TOO MUCH CHOCOLATE

Beneficial Effects of Eating Chocolate in Moderation

- Lower risk of various cardiovascular diseases, including heart attack
- Inhibit platelet aggregation
- Lowering blood pressure and better flow of blood through arteries
- May lower triglycerides
- May improve concentration of HDL cholesterol
- May reduce the risk of developing type 2 diabetes
- Better glycemic control in diabetic patients
- May reduce the likelihood of preeclampsia
- May reduce the burden of symptoms in chronic fatigue syndrome

Adverse Effect of Eating Too Much Chocolate

- Possible weight gain

13.3.2 Can Chocolate be Addictive?

Craving chocolate is more common in females; almost half of all females surveyed in a study craved chocolate. In general, craving for chocolate in females is highest a few days before menses and extends into the first few days of menses. There is no significant correlation between craving chocolate by parents and craving by their children. A liking for chocolates also correlates with a liking for sweets and white chocolate [71]. Chocolate contains several biologically active compounds (methylxanthine, biogenic amines, and cannabinoid-like fatty acids), all of which may cause psychological sensations that biochemically mimic the biological process involved in the brain after ingesting addictive substances. Most likely, a combination of sensory characteristics, psychoactive ingredients, and monthly hormonal fluctuations and mood swings among women contributes to chocolate craving [72]. However, regular exercise may lower craving for chocolate.

13.3.3 Adverse Effect of Eating Too Much Chocolate

Americans consume much less chocolate than Europeans (approximate per capita consumption of 4.6–4.8 kg/year in the United States vs. 7–10 kg/year in Europe), and only 1% of the US population consumes chocolate daily. Mean chocolate consumption in the United States is 30–90 g/day, and chocolate consumption is a minor contributor to total calorie intake in the daily US diet [73]. Desch *et al.* [74] observed that consumption of 6 g of chocolate per day had the same beneficial effect on lowering blood pressure as consumption of 25 g of chocolate per day. There was no weight gain in the group of participants who consumed 6 g of chocolate per day for 3 months, but there was a slight weight gain (an average of 0.8 kg or 1.8 lb) in participants who consumed 25 g of chocolate per day. However, no study has directly linked weight gain to a ceratin amount of chocolate consumption because obesity caused by high calorie intake is usually a result of the consumption of various foods.

Mood alternation by chocolate is a controversial issue. Some researchers have reported that depressed people crave chocolate more than people who are not depressed. In contrast, one publication indicated that both men and women who consumed more chocolate were more susceptible to depression [75]. Currently, there is no consensus that chocolate is a major factor in causing either depression or euphoria. Therefore, chocolate should not be considered a mood-altering substance.

13.4 TEA, COFFEE, AND CHOCOLATE: GLYCEMIC INDEX

If no sugar or milk is added to tea or coffee, the glycemic index is essentially zero. Therefore, drinking tea and coffee is safe for patients with diabetes. In fact, drinking tea may help control blood sugar in these patients. Dark unsweetened chocolate has a low glycemic index (20–30; a glycemic index <55 is considered low). Milk chocolate has a higher glycemic index [34–50] because milk and sugar are added. However, a diabetic patient may consume dark, unsweetened chocolate in moderation.

13.5 CONCLUSION

Tea and coffee are rich in antioxidants and may be consumed on a daily basis for good health. However, when drinking regular coffee, caffeine intake should not exceed 300–400 mg daily (up to three cups), but more decaffeinated coffee may be consumed. Both green and black tea have health benefits, and consuming three to five cups per day of a combination of black and green tea may also be beneficial. Dark chocolate is also good for health, but milk chocolate and white chocolate provide less benefit than dark chocolate. Moreover, milk chocolate has more calories than dark chocolate.

REFERENCES

[1] Wang Y, Ho CT. Polyphenolic chemistry of tea and coffee: a century of progress. J Agric Food Chem 2009;57:8109–14.

[2] Villanueva CM, Cantor KP, King WD, Jaakkola J, et al. Total and specific fluid consumption as determinants of bladder cancer risk. Cancer 2006;118:2040–7.

[3] Patil H, Lavie CJ, O'Keefe JH. Cuppa joe: Friend or foe? Effects of chronic coffee consumption on cardiovascular and brain health. Mol Med 2011;108:431–8.

[4] Song WO, Chun OK. Tea is the major source of flavan-3-ol and flavonol in the U.S. diet. J Nutr 2008;138: 1543S–7S.

[5] Chan EW, Soh EY, Tie PP, Law YP. Antioxidant and antibacterial properties of green, black and herbal teas of *Camellia sinensis*. Pharmacognosy Res 2011;3:266–72.

[6] Wang Y, Ho CT. Polyphenolic chemistry of tea and coffee: a century of progress. J Agric Food Chem 2009;57:8109–14.

[7] Chu YF, Brown PH, Lyle BJ, Chen Y, et al. Roasted coffee high in lipophilic antioxidants and chlorogenic acid lactones are more neuroprotective than green coffee. J Agric Food Chem 2009;57:9801–8.

[8] Bekedam E, Loots MJ, Shols HA, Van Boekel MA, et al. Roasting effects on formation mechanisms of coffee brew melanoidins. J Agric Food Chem 2008;56:7138–45.

[9] Stark T, Justus H, Hofmann T. Quantitative analysis of N-phenylpropenoyl L-amino acids in roasted coffee and cocoa powder by means of stable isotope dilution assay. J Agric Food Chem 2006;54:2859–67.

[10] Bonita J, Mandarano M, Shuta D, Vinson J. Coffee and cardiovascular disease: In vitro, cellular, animal and human studies. Pharmacol Res 2007;55:187–98.

[11] Cruchten ST, de Waart DR, Kunne C, Hooiveld GJ, et al. Absorption, distribution and biliary extraction of cafestol, a potent cholesterol elevating compound in unfiltered coffee in mice. Drug Metab Dispos 2010;38:635–40.

[12] Ranheim T, Halvorsen B. Coffee consumption and human health: Beneficial or detrimental? Mechanism for effects of coffee consumption on different risk factors for cardiovascular and type 2 diabetes mellitus. Mol Nutr Food Res 2005;49:274–84.

[13] Van het Hof KH, Kivits GA, Weststrate JA, Tijburg LB. Bioavailability of catechins from tea: the effect of milk. Eur J Clin Nutr 1998;52:356–9.

[14] Leenen R, Roodenburg AJ, Tijburg LB, Wiseman SA. A single dose of tea with or without milk increases plasma antioxidant activity in humans. Eur J Clin Nutr 2000;54:87–92.

[15] Pecorari M, Villano D, Testa MF, Schmid M, et al. Biomarkers of antioxidant status following ingestion of green tea at different polyphenol concentrations and antioxidant capacity in human volunteers. Mol Nutr Food Res 2010;54(Suppl. 2):S278–83.

[16] Coimbra S, Castro E, Rocha-Pereira P, Rebelo I, et al. The effect of green tea in oxidative stress. Clin Nutr 2006;25:790–6.

[17] Sung H, Min WK, Lee W, Chun S, et al. The effects of green tea ingestion over four weeks on atherosclerotic markers. Ann Clin Biochem 2005;42:292–7.

[18] Inami S, Takano M, Yamamoto M, Murakami D, et al. Tea catechin consumption reduces circulating oxidized low density lipoprotein. Int Heart J 2007;48:725–32.

[19] Davies MJ, Judd JT, Baer DJ, Clevidence BA, et al. Black tea consumption reduces total and LDL cholesterol mildly in hypercholesterolemic adults. J Nutr 2003;133:3298S–302S.

[20] Neyestani TR, Shariatzade N, Kalayi A, Gharavi A, et al. Regular daily intake of black tea improves oxidative stress biomarkers and decreases serum C-reactive protein levels in type 2 diabetes. Ann Nutr Metab 2010;57:40–9.

[21] Monteiro M, Farah A, Perrone D, Trugo LC, et al. Chlorogenic acid compounds from coffee are differentially absorbed and metabolized in humans. J Nutr 2007;137:196–201.

[22] Natella F, Nardini M, Giannetti I, Dattilo C, et al. Coffee drinking influences plasma antioxidant capacity in humans. J Agric Food Chem 2002;50:6211–66.

[23] Stangl V, Lorenz M, Stangl K. The role of tea and tea flavonoids in cardiovascular health. Mol Nutr Food Res 2006;50:218–28.

[24] Geleijnse JM, Launer LJ, Van der Kuip DA, Hofman A, et al. Inverse association of tea and flavonoid intakes with incidence of myocardial infarction: the Rotterdam study. Am J Clin Nutr 2002;75:880–6.

[25] Popkin MB, Armstrong LE, Bray GM, Caballero B, et al. A new proposed guidance system for beverage consumption in the Unites States. Am J Clin Nutr 2006;83:529–42.

[26] Arab L, Lieberskind DS. Tea, flavonoids and stroke in man and mouse. Arch Biochem Biophys 2010;50:31–6.

[27] Larsson SC, Virtamo J, Wolk A. Black tea consumption and risk of stroke in women and men. Ann Epidemiol 2013;23:157–60.

[28] Clement Y. Can green tea do that? A literature review of clinical evidence. Prev Med 2009;49:83–7.

[29] Ras RT, Zock PL, Draijer R. Tea consumption enhances endothelial-dependent vasodilation; a meta-analysis. PLoS One 2011;6:e16974.

[30] Butt MS, Sultan MT. Green tea: Nature's defense against malignancies. Crit Rev Food Sci Nutr 2009;49:463–73.

[31] Sturgeon JL, Williams M, van Servellen G. Efficacy of green tea in the prevention of cancer. Nurs Health Sci 2009;11:436–46.

[32] Naganuma T, Kuriyama S, Kakizaki M, Sone T, et al. Green tea consumption and hematological malignancies in Japan: the Ohsaki study. Am J Epidemiol 2009;170:730–8.

[33] Tang N, Wu Y, Zhou B, Wang B, et al. Green tea, black tea consumption and risk of lung cancer: a meta-analysis. Lung Cancer 2009;65:274–83.

[34] Mujtaba T, Dou QP. Black tea polyphenols inhibit tumor proteasome. *In Vivo* 2012;26:197–202.

[35] Lee AH, Su D, Pasalich M, Binns CW. Tea consumption reduces ovarian cancer risk. Cancer Epidemiol 2013;37:54–9.

[36] Song J, Xu H, Liu F, Feng L. Tea and cognitive health in late life: current evidence and future directions. J Nutr Health Aging 2012;16:31–4.

[37] Mandel S, Amit T, Reznichenko L, Weinreb O, et al. Green tea catechins as brain permeable natural ion chelator–antioxidants for the treatment of neurodegenerative diseases. Mol Nutr Food Res 2006;50:229–34.

[38] Foxe JJ, Morie KP, Laud PJ, Rowson MJ, et al. Assessing the effects of caffeine and theanine on the maintenance of vigilance during a sustained attention task. Neuropharmacology 2012;62:2320–7.

[39] Niu K, Hozawa A, Kuriyama S, Ebihara S, et al. Green tea consumption is associated with depressive symptoms in the elderly. Am J Clin Nutr 2009;90:1615–22.

[40] Hintikka J, Tolmunen T, Honkalampki K, Haatainen K, et al. Daily tea drinking is associated with a low level of depressive symptoms in the Finnish general population. Eur J Epidemiol 2005;20:359–63.

[41] Rains TM, Agarwal S, Maki KC. Antiobesity effects of green tea catechins: a mechanistic review. J Nutr Biochem 2011;22:1–7.

[42] Basu A, Sanchez K, Leyva MJ, Wu M, et al. Green tea supplementation affects body weight, lipids, and lipid peroxidation in obese subjects with metabolic syndrome. J Am Coll Nutr 2010;29:31–40.

[43] Watanabe I, Kuriyama S, Kakizaki M, Sone T, et al. Green tea and death from pneumonia in Japan: the Ohsaki cohort report. Am J Clin Nutr 2009;90:672–9.

[44] Tomata Y, Kakizaki M, Nakaya N, Tsuboya T, et al. Green tea consumption and risk of incident functional disability in elderly Japanese: the Ohsaki 2006 study. Am J Clin Nutr 2012;95:732–9.

[45] Shen CL, Yeh JK, Cao JJ, Chyu MC, et al. Green tea and bone health: Evidence from laboratory studies. Pharmacol Res 2011;64:155–61.

[46] Beresniak A, Duru G, Berger G, Bremond-Gignac D. Relationships between black tea consumption and key health indicators in the world: an ecological study. BMJ Open 2012;8 (2):e000648.

[47] Mullen W, Nemzer B, Ou B, Stalmach A, et al. The antioxidant and chlorogenic acid profiles of whole coffee fruits are influenced by the extraction procedures. J Agric Food Chem 2011;59:3754–62.

[48] Rudolph T, Knudsen K. A case of fatal caffeine poisoning. Acta Anaesthesiol Scand 2010;54:521–3.

[49] Muley A, Muley P, Shah M. Coffee to reduce the risk of type 2 diabetes: a systematic review. Curr Diabetes Res 2012;8:162–8.

[50] Mineharu Y, Koizumi A, Wada Y, Iso H, et al. Coffee, green tea, black tea and oolong tea consumption and risk of mortality from cardiovascular disease in Japanese men and women. J Epidemiol Community Health 2011;65:230–40.

[51] Mostofsky E, Rice MS, Levitan EB, Mittleman MA. Habitual coffee consumption and risk of heart failure: a dose–response meta-analysis. Circ Heart Fail 2012;5:401–5.

[52] Cornelis MC, El-Sohemy A. Coffee, caffeine and coronary heart disease. Curr Opin Clin Nutr Metab Care 2007;10:745–51.

[53] Gensini GF, Conti AA. Does coffee consumption represent a coronary risk factor? Recenti Prog Med 2004;95:563–5 [in Italian].

[54] Kim B, Nam Y, Kim J, Choi H, et al. Coffee consumption and stroke risk: a meta-analysis of epidemiological studies. Korean J Fam Med 2012;33:356–65.

[55] Nkondjock A. Coffee consumption and the risk of cancer: An overview. Cancer Lett 2009;277:121–5.

[56] Hildebrand JS, Patel AV, McCullough ML, Gaudet MM, et al. Coffee, tea and fatal oral/pharyngeal cancer in a large cohort study. Am J Epidemiol 2013;177:50–8.

[57] Sinha R, Cross AJ, Daniel CR, Grauband BI, et al. Caffeinated and decaffeinated coffee and tea intakes and risk of colorectal cancer in a large prospective study. Am J Clin Nutr 2012;96:374–81.

[58] Freedman ND, Park Y, Abnet CC, Hollenbeck AR, et al. Association of coffee drinking with total and cause specific mortality. N Engl J Med 2012;366:1891–904.

[59] Hermansen K, Krogholm KS, Bech KS, Dragsted LO, et al. Coffee can protect against disease. Ugeskr Laeger 2012;174:2293–7 [in Danish].

[60] Pham NM, Yoshida D, Morita M, Yin G, et al. The relation of coffee consumption to serum uric acid in Japanese men and women aged 49–76 years. J Nutr Metab 2010;2010:930757.

[61] Higdon JV, Frei B. Coffee and health: a review of recent human research. Crit Rev Food Sci Nutr 2006;46:101–23.

[62] Popkin MB, Armstrong LE, Bray GM, Caballero B, et al. A new proposed guidance system for beverage consumption in the Unites States. Am J Clin Nutr 2006;83:529–42.

[63] Lopez R. Chocolate: the nature of indulgences. New York: Harry N. Abrams; 2002.

[64] Steinberg FM, Bearden MM, Keen CL. Cocoa and chocolate flavonoids: implication for cardiovascular help. J Am Diabetic Assoc 2003;103:215–23.

[65] Katz DL, Doughty K, Ali A. Cocoa and chocolate in human health. Antioxid Redox Signal 2011;15:2779–811.

[66] Hooper L, Kay C, Abdelhamid A, Kroon PA, et al. Effects of chocolate, cocoa, and flavan-3-ols on cardiovascular health: A systematic review and meta-analysis of random trials. Am J Clin Nutr 2012;95:740–51.

[67] Triche EW, Grosso LM, Belanger K, Darefsky AS, et al. Chocolate consumption in pregnancy and reduced likelihood of preeclampsia. Epidemiology 2008;19:459–64.

[68] Sathyapalan T, Beckett S, Rigby AS, Melloe DD, et al. High cocoa polyphenol rich chocolate may reduce the burden of the symptoms in chronic fatigue syndrome. Nutr J 2010;9:55.

[69] Belz GG, Mohr-Kahaly S. Cocoa and dark chocolate in cardiovascular prevention? Dtsch Med Wochenschr 2011;136:2657–63 [in German].

[70] Lippi G, Franchini M, Montagnana M, Favaloro EJ, et al. Dark chocolate: consumption for pleasure or therapy? J Thromb Thrombolysis 2009;28:482–8.

[71] Rozin P, Levine E, Stoess C. Chocolate craving and liking. Appetite 1991;17:199–212.

[72] Bruinsma K, Taren DL. Chocolate: food or drug? J Am Dietetic Assoc 1999;99:1249–56.

[73] Seligson FH, Krummel DA, Apgar JL. Patterns of chocolate consumption. Am J Clin Nutr 1994;60(6 Suppl): 1060S–4S.

[74] Desch S, Kobler D, Schmidt M, Adams V, et al. Low vs. higher dose dark chocolate and blood pressure in cardiovascular high-risk patients. Am J Hypertens 2010;23:694–700.

[75] Rose N, Koperski S, Golomb BA. Mood food: chocolate and depressive symptoms in a cross-sectional analysis. Arch Intern Med 2010;170:6990703.

Alcoholic Beverages: Antioxidant and Other Health Benefits of Moderate Consumption

14.1 INTRODUCTION

The history of alcohol use by mankind can be traced back to 10,000 BC, but the origins of drinking may be traced back to ape ancestors 40 million years ago. Originally proposed by Professor Robert Dudley of the University of California at Berkeley, the drunken monkey hypothesis speculates that the human attraction to alcohol may have a genetic basis due to the high dependence of primate ancestors for millions of years on fruits as primary diet. Yeasts on fruit skin and within fruit convert sugars into ethanol, which diffuses out of the fruit and the alcoholic smell indicates that the fruit is ripe and ready for consumption. In tropical forests where monkeys lived, competition for ripe fruit was intense, and hungry monkeys capable of following the smell of alcohol to identify ripe foods and consuming them rapidly survived better than others. Eventually, "natural selection" favored monkeys with a keen appreciation for the smell and taste of alcohol, and it is possible that the human taste for alcohol originated from shared ancestry of humans with primates [1].

Alcoholic beverages can be classified into three broad categories: beer, wine, and distilled liquor (spirits). Beer and wine are fermented beverages produced from sugar- or starch-containing plant materials. The normal fermentation process, which uses yeast, cannot produce alcoholic beverages with alcohol content greater than 14%. Therefore, hard liquors or spirits are produced using fermentation followed by distillation. Alcoholic drinks primarily consist of water, alcohol, and variable amounts of sugars and carbohydrates (residual sugar and starch left after fermentation), but distilled alcoholic beverages such as spirits are devoid of any residual sugar. There are many benefits to consuming alcohol in moderation, but these benefits disappear with excessive drinking and binge drinking. Therefore, there is a "J"-shaped curve for alcohol consumption and effects on the human body, with beneficial effects with alcohol consumption in moderation and detrimental effects with excessive alcohol consumption.

CONTENTS

A. Dasgupta and K. Klein: Antioxidants in Food, Vitamins and Supplements. DOI: http://dx.doi.org/10.1016/B978-0-12-405872-9.00014-8

14.2 ALCOHOLIC BEVERAGES: ALCOHOL CONTENT

The alcohol content of various alcoholic beverages varies widely. Beer and wine, which are prepared by fermentation, contain much less alcohol per volume than distilled liquors such as whiskey and vodka. In the United States, a standard drink is defined as: a bottle of beer (12 oz) containing approximately 5% alcohol; 8.5 oz of malt liquor with approximately 7% alcohol; a 5-oz glass of wine with approximately 12% alcohol; 3.5 oz of fortified wine such as sherry or port with approximately 17% alcohol; or one shot of distilled spirits such as gin, rum, vodka, or whiskey (1.5 oz) containing approximately 40–50% alcohol. Although alcohol content varies widely among alcoholic beverages due to differences in serving size (the higher the alcohol content, the smaller the serving size), one standard drink contains approximately 0.6 oz (18 mL) of pure alcohol, which is the equivalent of 14 g of pure alcohol because the specific gravity of alcohol is 0.78 [2]. Logan *et al.* [3] provide detailed alcohol content of various beers and malt beverages commercially available in the United States.

In a mixed drink or cocktail, if one shot of distilled spirits is used, the alcohol content should be 0.6 oz. In a bar, a bartender uses a "jigger" for preparing mixed drinks. A jigger is a measuring device used to pour precisely 1.5 oz of alcohol. For simple mixed drinks such as rum and Coke or gin and tonic, each cocktail has one shot of liquor. Regarding the blood alcohol level, consuming one such drink is the equivalent of drinking one bottle of beer. However, in some mixed drinks, two shots are used, and thus the alcohol content in these drinks is the equivalent of that in two standard drinks. A cocktail usually contains one or more types of distilled spirits, fruit juice, honey, milk, or other ingredients. If a mixed drink is prepared using more than one distilled spirit, then one mixed drink may be equivalent to two to three standard drinks.

Historically, the alcohol content of various drinks was expressed as "proof." The term originated in the 18th century when British sailors were paid with money as well as rum. To ensure that the rum was not diluted with water, it was "proofed" by dousing gunpowder with it and setting it on fire. If the gunpowder failed to ignite, this indicated that the rum had too much water and was considered "under proof." A sample of rum that was 100 proof contained approximately 57% alcohol by volume (43% water). In the United States, proof to alcohol by volume is defined as a ratio of 1:2. Therefore, a beer that has 5% alcohol by volume is defined as 10 proof. On the other hand, rum containing 50% alcohol by volume is 100 proof. In the United States, the alcohol content of an alcoholic beverage is measured by the percentage of alcohol by the volume. The Code of Federal Regulations (27 CFR

[4-1-03 edition] §5.37 Alcohol Content) requires that labels for alcoholic beverages must state the alcohol content by volume. The regulation permits but does not require the "proof" of the drink to be printed.

14.3 ANTIOXIDANTS PRESENT IN BEER

Fermented alcoholic beverages such as beer and wine are full of beneficial polyphenolic compounds that are potent antioxidants. Because polyphenolic compounds are not volatile, they are essentially absent in distilled liquors such as rum and whiskey. However, alcoholic fruit drinks contain polyphenolic compounds that originate from fruit juice.

More than 2000 compounds have been identified in various beers, including more than 50 polyphenolic compounds, because barley and hops are rich in many organic compounds, including antioxidant polyphenolic compounds. Beer is also rich in nutrients such as carbohydrates, amino acids, minerals, and vitamins, as well as polyphenolic antioxidants. Beer is a good source of dietary antioxidants because 1 L of beer contains 366–875 mg of polyphenolic compounds [4]. Therefore, one bottle of beer (355 mL) contains approximately 130–311 mg of polyphenolic compounds. The total amount of polyphenolic compounds varies widely among different types of beer, but a substantial amount of these polyphenolic compounds are also present in nonalcoholic beer, thus making beer of any provenance a good source of antioxidants. Granato *et al.* [5] reported that brown ale beer manufactured in Brazil had higher amounts of flavonoids and a higher oxygen radical absorption capacity (ORAC) value than other beer tested. Structural classes of polyphenolic compounds in beer include simple phenols, benzoic acid derivatives, cinnamic acid derivatives, coumarines, catechins, di- and tri-oligomeric proanthracyanidins, and prenylated chalcones. Xanthohumol, a principal prenylated chalcone of hops, is a potent antioxidant that is also present in beer. Human exposure to xanthohumol and related compounds such as isoxanthohumol is primarily through beer consumption. *In vitro* studies have indicated that these compounds are broad-spectrum cancer chemopreventive agents [6]. Based on the health-promoting properties of xanthohumol, production of beer enriched with this compound is of interest to brewing companies in order to provide consumers with a better product [7]. Major polyphenolic compounds in beer are listed in Box 14.1.

Many of the beneficial polyphenolic antioxidants present in beer are well absorbed after beer consumption and can be detected in plasma. Therefore, after drinking beer, the antioxidant capacity of plasma is increased significantly. Nardini *et al.* [8] reported two- to fourfold increases in plasma levels of the antioxidant phenolic acid 30 min after beer consumption.

Box 14.1 MAJOR POLYPHENOLIC ANTIOXIDANTS IN BEER

- Ferulic acid
- Catechin
- Catechin gallate
- Epicatechin
- Epicatechin gallate
- Caffeic acid
- Gallic acid
- Vanillic acid
- Sinapic acid
- 4-Hydroxyphenylacetic acid
- 4-Hydroxybenzoic acid
- 2,6-Dihydroxybenzoic acid
- *p*-Coumaric acid
- Syringic acid
- Phenolic acid
- Kanpherol
- Quercetin
- Xanthohumol

4-Hydroxyphenylacetic acid was present in plasma mostly in nonconjugated form, but *p*-coumaric acid was present in both conjugated and nonconjugated form. Ferulic acid, vanillic acid, and caffeic acid were present in plasma mostly in conjugated form (both sulfate and glucuronide). The authors concluded that phenolic acids present in beer are well absorbed from the gastrointestinal tract and are present in plasma. Ghiselli *et al.* [9] observed significant increases in plasma antioxidant capacity in humans after consumption of beer, with maximum antioxidant capacity observed 1 h after consumption. The antioxidant capacity of plasma returned to baseline 2 h after consumption of beer. All antioxidant phenolic acids measured in plasma were also increased after consumption of beer. Alcohol present in beer enhanced the absorbance of these phenolic compounds from the gastrointestinal tract because plasma antioxidant capacity did not increase significantly after consumption of dealcoholized beer. Consumption of a similar amount of ethanol alone (4.5% by volume) did not affect plasma antioxidant capacity, indicating that phenolic compounds present in beer are responsible for increased antioxidant capacity of human plasma after beer consumption.

14.4 ANTIOXIDANTS PRESENT IN WINE

Wines are prepared from grapes, which are rich in antioxidant polyphenolic compounds. After harvest, the grapes are crushed and allowed to ferment. Red wine is made from the must (pulp) of red or black grapes together with the grape skin. White wine is usually made by fermenting juice pressed from white grapes, but it can also be made from must extracted from red grapes that contain very little or no grape skin. Grape skin is full of polyphenolic compounds and anthocyanins, which are capable of destroying free radicals. Concentrations of these compounds in grapes depend on the geographic location where grapes are cultivated, soil type, climate, vine cultivation practice, harvesting time, the production process, and aging of wine. Grape seeds, grape skin, and must contain many phenolic compounds, including catechin; epicatechin; dimeric, trimeric, and polymeric proanthocyanidin; phenolic acids; and anthocyanins. Soleas *et al.* [10] demonstrated that these compounds are more abundant in red wine compared to white wine. For example, the average catechin level in red wine was 191 mg/L, whereas it was 35 mg/L in white wine. Similarly, average gallic acid and rutin contents of red wine were 95 and 82 mg/L, respectively, whereas those of white wine were 7 and 21 mg/L, respectively. Major antioxidant compounds in red wine are listed in Box 14.2. Sanchez-Moreno *et al.* [11] directly tested the antioxidant capacity of various brands of white and red wine by measuring their ORAC using Trolox (TE), a water-soluble analog of vitamin E, as the control. The authors observed significantly higher antioxidant capacity of red wines compared to white wines. Jakubec *et al.* [12] measured total antioxidant capacity and total polyphenol content of various brands of red, rosé, and white wines and observed highest concentrations of polyphenolic compounds in red wines compared to other wines. Antioxidant activity was highest in red wine, with one glass of red wine being equivalent to two glasses of white wine or two cups of black tea. In addition, red wine contained more antioxidants than beer: one glass of red wine (150 mL) was equivalent to 1750 mL of beer. Alcohol content, total radical trapping antioxidant parameter (TRAP; luminol-enhanced chemiluminescence was used for measurement), and ORAC (measured using fluorescein probe) values of various wines as measured by these authors are summarized in Table 14.1. As expected, both TRAP and ORAC values were significantly higher in red wines compared to rosé or white wines.

Box 14.2 MAJOR POLYPHENOLIC ANTIOXIDANTS IN RED WINE

- Resveratrol
- Caffeic acid
- Gallic acid
- Gallic acid ethyl ester
- Catechin
- Epicatechin
- Epicatechin gallate
- Cyanidin-3-*O*-glucoside
- Delphinidin-3-*O*-glucoside
- Melvidin-3-*O*-glucoside
- Peonidin-3-*O*-glucosdie
- Petunidin-3-*O*-glucoside
- Procyanidin dimer and trimer
- Gentisic acid
- Vanillic acid
- Syringic acid
- 4-Hydroxybenzoic acid
- Myricetin
- *p*-Coumaric acid
- Phenolic acid
- Quercetin

Similarly to beer, polyphenolic antioxidants are rapidly absorbed after consumption of wine, thus increasing antioxidant capacity of blood. In one report, the authors measured the antioxidant capacity of blood after each healthy volunteer (a total of eight) consumed 300 mL of red wine. The authors observed a maximum increase in the antioxidant capacity of blood 55 min after drinking wine. Increased levels of uric acid in blood after drinking wine may provide blood with additional antioxidant capacity [13]. Modun *et al.* [14] noted that increased plasma antioxidant capacity after consumption of red wine is due to absorption of polyphenolic antioxidants from the gut as well as increased uric acid in plasma after consumption.

Resveratrol, which is found in abundance in grape skin, has been described as a strong antioxidant compound. In one study, the authors found approximately 73 μg of resveratrol in 1 oz of grape juice. In contrast, 1 oz of red wine contained an average 160 μg of resveratrol. In addition, certain antioxidant polyphenolic compounds were also produced during the aging process of red wine. Structural changes in some of these compounds also occurred

Table 14.1 Antioxidant Capacity of Various Wines

Wine	Alcohol Content	Antioxidant Capacity	
		TRAP (μmol TE/L)	ORAC (μmol TE/L)
Red Wine			
Ruby Cabernet dry, California	13.0	25,286	25,953
Cabernet Sauvignon dry, Chile	12.5	16,549	22,446
Cabernet Sauvignon dry, Italy	11.5	25,831	20,421
Rosé Wine			
Zinfandel rosé, California	10.5	4699	4567
Cinsault–Grenache, France	12.0	3973	4567
White Wine			
Chardonnay, Slovakia	11.0	2596	2342
Chardonnay, Chile	14.0	5063	5037
Chardonnay, California	13.0	4707	3909
Chardonnay, Australia	12.5	1916	4043

ORAC, oxygen radical absorption capacity measured as Trolox (TE) equivalent/L; TRAP, total radical trapping antioxidant parameter measured as TE equivalent/L.
Source: *Data from Jakubec* et al. [12].

during aging of wine as these pigments changed to more stable structural configurations that are deep red in color, thus explaining darkening of the red color of the wine during the aging process. Therefore, eating grapes may have some benefit, but most epidemiological studies have focused on consuming wine, especially red wine, which could certainly boost the antioxidant defense of the body [15]. Whereas many authors have focused on the beneficial effects of drinking red wine, Dudley *et al.* [16], using a rat model, demonstrated that white wine is also capable of protecting the heart, similarly to red wine. White wine contains tyrosol and hydroxytyrosol, which have a cardioprotective effect similar to that of resveratrol, which is found in higher concentration in red wine.

14.5 ALCOHOL INCREASES ANTIOXIDANT LEVELS IN STRAWBERRIES AND BLACKBERRIES

Strawberries and blackberries are rich in oxidants, but treating them with alcohol boosts their antioxidant power (free radical scavenging capacity), as reported by researchers in Thailand and by the US Department of Agriculture. Therefore, serving strawberries in daiquiri form may be even better for health. People who do not like strawberries may consider blackberry champagne cocktail because this is also a good antioxidant-containing drink. In addition, storing strawberries and blackberries for 1 or 2 weeks in alcohol keeps them relatively fresh and prevents decay of antioxidants present in these fruits. Strawberries and blackberries contain polyphenolic compounds and anthocyanins, which are strong antioxidants, and storing these fruits in alcohol or serving them in alcohol increases their antioxidant potential [17].

14.6 GUIDELINES FOR DRINKING IN MODERATION

Alcohol has many beneficial effects for individuals who drink in moderation, but heavy drinking is hazardous to health because most effects of alcohol follow a "J"-shaped curve, with benefits from moderation and health hazards from excess consumption. Guidelines for moderate alcohol consumption are provided by the US Government [18]:

- Men: No more than two standard alcoholic drinks per day
- Women: No more than one standard alcoholic drink per day
- Adults older than 65 years (both male and female): No more than one drink per day
- Pregnant women or women who want to become pregnant: Should not drink alcohol

Consuming more than 14 drinks per week for men (or more than 4 drinks per occasion) and more than 7 drinks per week for women (or more than 3 drinks per occasion) may cause harm. Hazardous drinking is defined as 21 or more drinks per week by men or more than 7 drinks per occasion at least three times a week. For women, consuming more than 14 drinks per week or more than 5 drinks in one occasion at least three times a week is considered hazardous drinking. Risk drinking (also known as binge drinking) is defined as consumption of 5 or more drinks in a single occasion or day for men and 4 or more drinks for women. There is a misconception that if the number of total drinks in a week does not exceed 14 for men and 7 for women, then the drinking practice is safe. In reality, frequency of drinking is also important to define moderate drinking. The London-based Whitehall II Cohort Study, which included 10,308 government employees between the ages of 35 and 55 years with an 11-year follow-up, concluded that optimal drinking is once or twice a week, with a maximum daily consumption of one drink or less. People who consumed alcohol even twice a day had an increased risk of mortality compared to those who drank once or twice a week [19]. It is also better to consume alcohol with food because food can reduce absorption of alcohol from the gut, causing lower peak blood alcohol concentration.

14.7 THE FRENCH PARADOX AND HEALTH BENEFITS OF DRINKING IN MODERATION

French people favor drinking red wine with their meals. Although in most countries intake of fatty meals is positively related to high mortality from coronary heart disease, in France mortality from coronary heart disease is relatively low despite the consumption of fatty meals. Epidemiological studies indicate that consumption of alcohol at a level of intake in France (20–30 g/day; one drink contains 14 g of alcohol) can reduce the risk of coronary heart disease by 40% [20].

Other than the French paradox, the relationship between alcohol consumption and coronary heart disease was examined in the original Framingham Heart Study initiated in 1948 (the study originated in Framingham, Massachusetts, and is the basis of our current understanding of various factors that lead to heart disease, including cholesterol). The Framingham study indicated that drinking in moderation reduced the risk of heart disease and heart attack [21]. It was found that it is beneficial to consume one drink per day or at least six drinks per week to reduce the risk of coronary heart disease for men, but women may require a lower amount of alcohol intake to obtain benefits. Those women who consumed alcohol at least one day per week had a lower risk of coronary heart disease than those who did not drink at all. However, little difference was found between women who consumed at least one drink per week and those who consumed two to six drinks per week [22]. Based on a study of 1154 participants (580 men and 574 women) in Winnipeg, Manitoba, Canada, Snow *et al.* [23] reported that the well-established relationship between reduced risk of cardiovascular disease and moderate consumption of alcohol may not be evident until middle age (35–49 years) or older (50–64 years) in men. However, women may benefit from moderate consumption of alcohol at a much younger age (18–34 years). The beneficial effects of alcohol consumption are negated when alcohol is consumed in heavy episodic drinking patterns (eight or more drinks per occasion), especially for middle-aged and older men.

Another beneficial effect of consuming alcohol in moderation is a significant reduction in the risk of stroke among both men and women, regardless of age or ethnicity. The Copenhagen City Heart Study found that people who consumed low to moderate alcohol experienced a protective effect of alcohol against stroke [24]. The Rotterdam Study investigated the effect of alcohol consumption in subjects 55 years old or older in preventing age-related dementia and concluded that light to moderate alcohol consumption was associated with a lower risk of developing any dementia [25]. It has also been demonstrated that moderate alcohol consumption, especially red wine, may lower the risk of developing Alzheimer's disease [26].

Moderate consumption of alcohol may reduce the risk of certain types of cancer. It has been suggested that moderate alcohol consumption facilitates the elimination of *Helicobacter pylori*, a bacterium found in the gut that causes chronic atrophic gastritis (CAG) and gastric cancer. In a study of 9444 subjects aged 50–74 years, Gao *et al.* [27] observed that moderate drinkers (<60 g of alcohol per week or four drinks per week) had a 29% lower chance of developing CAG compared to nondrinkers. Both beer and wine consumption provided protection against CAG. In the California Men's Health Study of 84,170 men between ages 45 and 69 years, consumption of one or more drinks per day was associated with an approximately 60% reduction in lung cancer risk, ever in smokers. Interestingly, heavy smokers also benefited from consuming red wine in moderation [28]. In another study, Chao [29] observed that although moderate consumption of wine (one drink or less per day) was associated with an approximately 23% reduced risk of developing lung cancer, moderate consumption of beer (one or more per day) increased the risk of developing lung cancer only in men but not in women. Jiang *et al.* [30] reported that, compared to nondrinkers, the risk of bladder cancer was reduced by up to 32% in those who consumed beer and wine, but not sprits (hard liquor). Consumption of up to one drink per day reduced the risk of head and neck cancer in both men and women, but consuming more than three alcoholic beverages increased the risk of developing cancer [31]. Moderate alcohol consumption also reduced the risk of developing renal cell carcinoma in both males and females [32].

Moderate alcohol consumption reduces the risk of developing rheumatoid arthritis by 40–50% [33]. Moderate alcohol intake also reduces the risk of developing type 2 diabetes, but excessive alcohol intake increases the risk of developing diabetes. Based on a review of 20 studies, Baliunas *et al.* [34] observed a U-shaped relationship between alcohol consumption and the risk of developing type 2 diabetes: moderate alcohol consumption decreased the risk, but heavy alcohol consumption increased the risk. Even individuals with type 2 diabetes can obtain better glycemic control by consuming alcohol in moderation. Shai *et al.* [35] investigated the effect of daily consumption of alcohol on glycemic control in patients with type 2 diabetes. The authors divided 109 patients (41–74 years old) who had previously abstained from alcohol into two groups. One group received 150 mL of wine daily (one glass of wine) and another group received nonalcoholic beer at dinner, over a period of 3 months. The fasting glucose was significantly reduced from an average of 139.6 mg/dL initially to 118.0 mg/dL after 3 months in patients with diabetes who had consumed one alcoholic drink during dinner. In contrast, the mean fasting glucose value had not changed in those diabetic patients who had abstained from alcohol during the 3-month period (136.7 mg/dL at the beginning and 138.6 mg/dL at the end of the study). Participants in the alcohol group also reported improvement in their ability to fall asleep.

Drinking in moderation may also offer protection from macular degeneration, osteoporosis, and the development of gallstones. Moderate drinking can also reduce stress levels and elevate mood [36]. Although beer and spirits have no protective effect against the common cold, moderate wine consumption, especially red wine, can reduce the risk of getting the common cold [37]. Because moderate consumption of alcohol can prevent many diseases, it is expected that moderate drinkers may live longer than lifetime abstainers from alcohol. Klatsky *et al.* [38] studied 10-year mortality in relation to alcohol in 8060 subjects and observed that persons who consumed two drinks or fewer daily fared best and had a 50% reduction in mortality rate compared to nondrinkers. The mortality rate was doubled in the heaviest drinkers (six or more drinks per day) compared to moderate drinkers. In the Physician's Health Study, which included 22,071 male physicians in the United States between ages 40 and 84 years with no history of myocardial infarction, stroke, or cancer, and which had a 10-year follow-up, the authors observed that men who consumed two to six drinks per week had the most favorable results with a lower mortality rate than that of people who consumed one drink per week. In contrast, people who consumed more than two drinks per day had a higher mortality rate than people who consumed one drink per week [39]. The health benefits of drinking are summarized in Box 14.3, and the health hazards of drinking in excess are summarized in Box 14.4.

Box 14.3 BENEFITS OF DRINKING IN MODERATION

- Reduced risk of cardiovascular disease
- Reduced risk of myocardial infarction
- Reduced risk of congestive heart failure
- Reduced risk of stroke
- Reduced risk of certain types of cancer
- Reduced risk of type 2 diabetes
- Reduced risk of forming gallstones
- Reduced risk of developing rheumatoid arthritis
- Reduced risk of dementia/Alzheimer's disease
- Reduced risk of Parkinson's disease
- Reduction of stress level and depression
- Increased longevity
- Reduced risk of common cold

Box 14.4 HARMFUL EFFECTS OF DRINKING IN EXCESS

- Decreased life span
- Proneness to substance abuse
- Alcoholic liver disease
- Anxiety and depression
- Increased risk of stroke
- Damage to endocrine/reproductive system
- Increased risk of various cancers
- Fetal alcohol syndrome
- Brain damage in teenagers

14.8 ANTIOXIDANT AND PRO-OXIDANT EFFECTS OF ALCOHOLIC BEVERAGES

Moderate consumption of alcohol increases the antioxidant capacity of plasma, but, with higher alcohol consumption, plasma demonstrates pro-oxidant activities. Prickett *et al.* [40] reported that one drink of red wine, beer, or stout (5% alcohol) resulted in increased plasma antioxidant activity for up to 4 h after consumption, but consumption of three drinks of wine, beer, or stout made the plasma pro-oxidant and the effect lasted almost 6 h after consumption of alcohol. However, when wine, beer, and stout were tested separately, all showed antioxidant activities *in vitro*, although red wine showed higher antioxidant capacity due to higher levels of antioxidant polyphenolic compounds. Based on these data, the authors hypothesized that plasma antioxidant capacity should be higher after consumption of three drinks than after consumption of one drink because more antioxidant polyphenolic compounds should be absorbed from the gut. The authors concluded that the increased antioxidant capacity of plasma after consumption of one drink was related to polyphenolic compounds and not alcohol, but after consumption of three drinks, the higher blood alcohol level was responsible for making plasma pro-oxidant. This conclusion was supported by the fact that when volunteers consumed alcohol solution corresponding to one drink, no increase in antioxidant capacity of blood was noted. However, when volunteers consumed alcohol solution comparable to consumption of three drinks, plasma showed pro-oxidant properties. Metabolism of alcohol is known to produce reactive oxygen species [40]. Puddey *et al.* [41] also noted that although *in vitro* experiments have indicated that extracts of red wine, white wine, grape juice, and beer can inhibit oxidation of low-density lipoprotein (LDL) and the degree of inhibition is usually directly related to polyphenolic content of the beverages, these *in vitro* antioxidant effects cannot be reproduced *in vivo* after consumption of alcoholic drinks, especially with increasing numbers of drinks. In such cases, enhanced oxidation of LDL has been observed. Such findings are consistent with the possibility that, depending on the alcoholic beverage, the pro-oxidant effect of alcohol may override any antioxidant effect of polyphenolic compounds.

Metabolism of alcohol by liver enzymes causes alcohol-induced oxidative stress. Alcohol is first metabolized by the liver enzyme alcohol dehydrogenase to acetaldehyde, which is an intermediate in the metabolic pathway of alcohol. Then, acetaldehyde is further metabolized by aldehyde dehydrogenase to acetic acid.

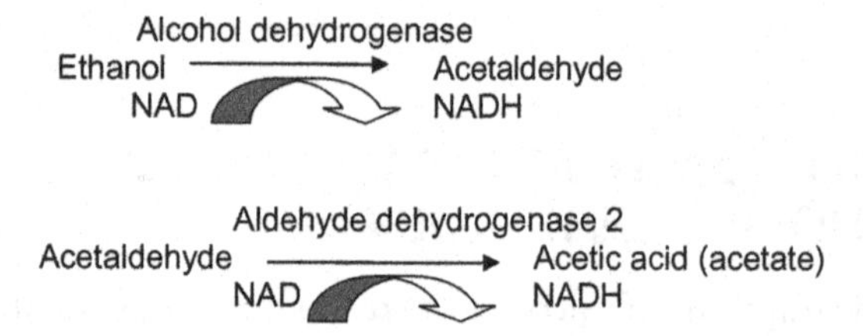

To facilitate both reactions, hydrogen is removed from alcohol and also from acetaldehyde during oxidative metabolism, and the hydrogen atom is transferred to nicotinamide adenine dinucleotide (NAD), a cofactor that is then converted into NADH, a chemically reduced form of NAD. Then NADH can transfer newly acquired hydrogen to other metabolic reactions in order to be converted into NAD, and NADH again becomes available to accept hydrogen. The regeneration of NAD from NADH takes place inside mitochondria, and the ratio of NAD to NADH can determine the redox balance of the cell. Excess production of NADH due to heavy alcohol consumption can lead to an increase in superoxide free radicals, which in turn generate hydroxyl radicals causing lipid peroxidation and damage to mitochondrial DNA. However, the major oxidative stress induced by alcohol consumption is due to an alternative pathway of alcohol metabolism by the CYP2E1 isoenzyme producing acetaldehyde. In normal circumstances, the CYP2E1 pathway represents less than 10% of metabolism of alcohol, but in alcoholics this pathway is activated and excessive alcohol ingestion may cause up to 10-fold upregulation of CYP2E1. Metabolism of alcohol by CYP2E1 generates free radicals, thus inducing oxidative stress [42]. Catalase-induced oxidation of alcohol, a minor pathway, can also produce oxidative stress. Acetaldehyde, although an intermediate in alcohol metabolism, is a toxic chemical that can react with amino, hydroxyl, and sulfhydryl groups to modify the structure of macromolecules such as proteins and enzymes [43]. Lu and Cederbaum [44] noted that the CYP2E1 pathway of alcohol metabolism is mainly responsible for alcohol-induced oxidative stress because CYP2E1 is capable of generating superoxide anion radicals, hydrogen peroxide, and, in the presence of iron, hydroxyl radicals.

14.9 BENEFICIAL EFFECTS OF ALCOHOL ON THE HEART: ARE THESE EFFECTS DUE TO POLYPHENOLIC COMPOUNDS ALONE?

Because alcohol can be a pro-oxidant, it is unclear why there are so many health benefits to drinking in moderation. One logical answer is that polyphenolic compounds present in alcoholic beverages are responsible for the beneficial effects of alcohol when it is consumed in moderation, but with excess alcohol consumption these benefits disappear because the pro-oxidant effects of alcohol cannot be compensated for by the antioxidant effects of polyphenolic compounds. The association between reduced risk of developing cardiovascular diseases and moderate alcohol consumption is very well-established and one of the most cited benefits of drinking in moderation. Mechanisms by which alcoholic beverages exert such effect include lipid regulation and systemic anti-inflammatory effects. The alcohol component increases high-density lipoprotein (HDL) cholesterol, inhibits platelet aggregation, and reduces systemic inflammation. Increases in HDL cholesterol are directly related to the alcohol component of alcoholic beverages and independent of the type of beverage consumed.

Many studies have demonstrated increased HDL cholesterol levels in drinkers compared to nondrinkers. The Honolulu Heart Study showed that men who consumed alcoholic beverages had higher blood levels of HDL cholesterol than nondrinkers. Gordon *et al.* [45] reviewed data from 10 different studies, including the Honolulu Heart Study, and observed that there was a positive correlation between amounts of alcohol consumed and the serum level of HDL cholesterol. In another study, the authors observed that the HDL cholesterol level in blood increased up to 33% in social drinkers compared to nondrinkers [46]. In addition to increasing HDL cholesterol, light to moderate consumption of alcohol reduces the level of apolipoprotein (a), reduces platelet aggregation, and prevents clot formation [47]. Alcohol also diminishes thrombus formation on damaged walls of the coronary artery, and the plausible mechanism of inhibition of platelet aggregation is through inhibiting phospholipase A2 [48]. Therefore, it has been speculated that approximately 50% of the benefit derived from consuming alcohol in moderation is due to an increased level of HDL cholesterol, and the rest is due to the favorable effect on preventing platelet aggregation and other effects. Studies have indicated that wine, especially red wine, provides additional benefits compared to other alcoholic beverages by decreasing blood pressure, inhibiting oxidation of LDL particles, improving endothelial function, inhibiting platelet aggregation, and activating proteins that protect cells from apoptosis. These additional benefits of red wine are due to its higher polyphenolic content compared to that of beer and other alcoholic beverages. When the effects of three alcoholic beverages—vodka, beer, and red wine—were compared, only red wine provided protection against vascular oxidative stress because polyphenolic compounds present in red wine can protect vasculature by reducing the level of reactive oxygen species and by inhibiting endothelin-1 expression. In addition, red wine polyphenolic compounds are capable of improving the functionality and activity of endothelial nitric oxide synthase, thus increasing the vascular level of nitric oxide, which has cardioprotective effects [49]. Wu *et al.* [50] noted that resveratrol, which is more abundant in red wine than in white wine, is capable of inhibiting oxidation of LDL and suppression of platelet aggregation, thus providing a superior cardioprotective effect.

However, there is controversy regarding whether wine, especially red wine, is superior to other alcohol beverages for protecting the heart. Denke [51] noted that several published reports indicate that moderate consumption of any alcoholic beverage (up to two drinks per day for men and up to one drink per day for women) is associated with lower rates of cardiovascular disease. In addition, from a nutritional standpoint, beer contains more vitamin B complex than wine. Moreover, the antioxidant content of beer is comparable to that of wine, but specific antioxidant components are different because barley and hops used in the production of beer contain different flavonoids than those present in the grapes from which wine is prepared. The author concluded that there is no evidence to support consumption of one type of alcoholic beverage over another for cardioprotection. Volcik *et al.* [52] observed that regardless of the type of alcoholic beverage, there were increases in HDL cholesterol, HDL3 cholesterol, and apolipoprotein A-I in both white and African American males and females. Significantly lower levels of LDL cholesterol and apolipoprotein B and triglycerides were only observed in white females. However, with higher alcohol consumption, triglyceride levels were elevated more in African Americans than in whites. Krenz and Korthuis [53] noted that although the link between a reduced risk of cardiovascular disease and moderate consumption of alcohol (one or two drinks per day) was initially attributed to polyphenolic compounds of red wine, subsequent research indicates that the ethanol component of alcoholic beverage also contributes to the beneficial effects associated with moderate drinking, regardless of the type of alcoholic beverage (red vs. white wine, beer, and spirits).

14.10 HEALTH BENEFITS OF DEALCOHOLIZED BEVERAGES

Substantial amounts of beneficial polycyclic phenolic antioxidants are present in both beer and dealcoholized beer. Piazzon *et al.* [4] analyzed dealcoholized beer specimens and found that they contained 366 μg of gallic acid equivalent polyphenolic compounds per milliliter of beer. Bassus *et al.* [54] reported that consumption of dealcoholized beer by young adults inhibited thrombogenic activity, which could also provide cardiovascular benefit. Resveratrol is abundant in both red wine and dealcoholized red wine, and after drinking red wine or dealcoholized red wine for 28 days, resveratrol metabolites were increased in urine regardless of drink type [55]. Noguer *et al.* [56] reported that alcohol-free wine was capable of increasing the activity of antioxidant enzymes, including superoxide dismutase, catalase, glutathione peroxidase, and glutathione reductase. The authors concluded that alcohol-free wine may be an excellent source of antioxidants to protect people with oxidative stress-related diseases such as cancer, diabetes, and Alzheimer's disease. Chiva-Blanch *et al.* [57] reported that dealcoholized red wine decreased systolic and diastolic blood pressure through a nitric oxide-mediated mechanism, and daily consumption of dealcoholized red wine may be useful for prevention of minor to moderate hypertension.

14.11 CONCLUSION

Consuming alcoholic beverages in moderation is beneficial for health. Some benefits derive directly from alcohol content and other from the high content of various polyphenolic compounds in alcoholic beverages that are excellent antioxidants. Low to moderate wine consumption has been linked with reduced risk of heart disease, cancer, and age-related dementia. Resveratrol, which is abundant in wine, is primarily responsible for many beneficial effects of red wine, including protection from various heart diseases. However, dealcoholized drinks also have health benefits due to the presence of antioxidants.

REFERENCES

[1] Dudley R. Ethanol, fruit ripening and the historical origins of human alcoholism in primate frugivory. Integr Comp Biol 2004;44:315–23.

[2] Kerr WC, Greenfield TK, Tujague J, Brown SE. A drink is a drink? Variation in the amount of alcohol contained in beer, wine and spirits drinks in a U.S. methodological sample. Alcohol Clin Exp Res 2005;29:2015–21.

[3] Logan B, Case GA, Distefano S. Alcohol content of beer and malt beverages: forensic consideration. J Forensic Sci 1999;44:1292–5.

[4] Piazzon A, Forte M, Nardini M. Characterization of phenolic content and antioxidant activity of different beer types. J Agric Food Chem 2010;58:327–34.

[5] Granato D, Branco GF, Faria de Assisis Fonseca J, Cruz AG. Characterization of Brazilian lager and brown ale beers based on color, phenolic compounds, and antioxidant activity using chemometrics. J Sci Food Agric 2011;91:563–71.

[6] Stevens JF, Page JE. Xanthohumol and related prenylflavonoids from hops and beer: to your good health. Phytochemistry 2004;65:1317–30.

[7] Magalhaes PJ, Carvalho DO, Cruz JM, Guido LF, et al. Fundamentals and health benefits of xanthohumol, a natural product derived from hops and beer. Nat Prod Commun 2009;4:591–610.

[8] Nardini M, Natella F, Scaccini C, Ghiselli A. Phenolic acids from beer are absorbed and extensively metabolized in humans. J Nutr Biochem 2006;17:14–22.

[9] Ghiselli A, Natella F, Guidi A, Montanari L. Beer increases plasma antioxidant capacity in humans. J Nutr Biochem 2000;11:76–80.

[10] Soleas GJ, Diamandis EP, Goldberg DM. Wine as a biological fluid: history, production and role in disease prevention. J Clin Lab Anal 1997;11:287–313.

[11] Sanchez-Moreno C, Cao G, Ou B, Prior RL. Anthocyanin and proanthocyanidin content in selected white wine and red wines: Oxygen radical absorbance capacity comparison with nontraditional wines obtained from highbush blueberry. J Agric Food Chem 2003;51:4889–96.

[12] Jakubec P, Bancirova M, Halouzka V, Lojek A, et al. Electrochemical sensing of total antioxidant capacity and polyphenol content in wine samples using amperometry online coupled with microdialysis. J Agric Food Chem 2012;60:7836–43.

[13] Fernandez-Panchon MS, Villano D, Troncoso AM, Garcia-Parrilla MC. Antioxidant capacity of plasma after red wine intake in humans. J Agric Food Chem 2005;53:5024–9.

[14] Modun D, Music I, Vukovic J, Brizic I, et al. The increase in human plasma antioxidant capacity after red wine consumption is due to both plasma urate and wine polyphenols. Atherosclerosis 2008;197:250–6.

[15] Brouillard R, George F, Fougerousse A. Polyphenols produced during red wine aging. Biofactors 1997;6:403–10.

[16] Dudley JI, Leki I, Mukherjee S, Das M, et al. Does white wine qualify for French paradox? Comparison of cardioprotective effects of red and white wines and their constituents: resveratrol, tyrosol and hydroxytyrosol. J Agric Food Chem 2008;56:9362–73.

[17] Chanjirakul K, Wang SY, Wang CY, Siriphanich J. Natural volatile treatments increase free-radical scavenging capacity of strawberries and blackberries. J Sci Food Agric 2007;87:1463–72.

[18] U.S. Department of Agriculture, U.S. Department of Health and Human Services. Alcoholic beverages. Dietary guidelines for Americans. Washington, DC: U.S. Government Printing

Office; 2005. pp. 43–6. Available at <http://www.health.gov/DIETARYGUIDELINES/dga2005/document/html/chapter9.htm> [accessed February 2013].

[19] Britton A, Marmot M. Different measures of alcohol consumption and risk of coronary heart disease and all-cause mortality: 11-year follow-up of the Whitehall II Cohort Study. Addiction 2004;99:109–16.

[20] Renaud S, de Lorgeril JM. Wine, alcohol, platelets and the French paradox for coronary heart disease. Lancet 1992;339(8808):1523–6.

[21] Friedman LA, Kimball AW. Coronary heart disease mortality and alcohol consumption in Framingham. Am J Epidemiol 1986;124:481–9.

[22] Tolstrup J, Jensen MK, Tjonneland A, Overvad K, et al. Prospective study of alcohol drinking patterns and coronary heart disease in women and men. Br Med J 2006;332:1244–8.

[23] Snow WM, Murray R, Ekuma O, Tyas SL, et al. Alcohol use and cardiovascular health outcomes: a comparison across age and gender in the Winnipeg Health and Drinking Survey Cohort. Age Aging 2009;38:206–12.

[24] Truelsen T, Gronbaek M, Schnohr P, Boyen G. Intake of beer, wine and spirits and risk of stroke: the Copenhagen City Heart Study. Stroke 1998;29:2467–72.

[25] Ruitenberg A, van Swieten JC, Witterman JC, Mehta KM, et al. Alcohol consumption and risk of dementia: the Rotterdam Study. Lancet 2002;359(9303):281–6.

[26] Pinder RM, Sandler M. Alcohol, wine and mental health: focus on dementia and stroke. J Psychopharmacol 2004;18:449–56.

[27] Gao L, Weck MN, Stegmaier C, Rothenbacher D, et al. Alcohol consumption and chronic atrophic gastritises: population-based study among 9444 older adults from Germany. Int J Cancer 2009;125:2918–22.

[28] Chao C, Slezak JM, Caan BJ, Quinn VP. Alcoholic beverage intake and risk of lung cancer: the California Men's Health Study. Cancer Epidemiol Biomarkers Prev 2008;17:2692–9.

[29] Chao C. Association between beer, wine, and liquor consumption and lung cancer risk: a meta-analysis. Cancer Epidemiol Biomarkers Prev 2007;16:2436–47.

[30] Jiang X, Castelao KE, Cortessis VK, Ross RK, et al. Alcohol consumption and risk of bladder cancer in Los Angeles County. Int J Cancer 2007;121:839–45.

[31] Freedman ND, Schatzkin A, Leitzmann MF, Hollenbeck MF, et al. Alcohol and head and neck cancer risk in a prospective study. Br J Cancer 2007;96:1469–74.

[32] Pelucchi C, Galeone C, Montella M, Polesel J, et al. Alcohol consumption and renal cell cancer risk in two Italian case–control studies. Ann Oncol 2008;19:1003–8.

[33] Kallberg H, Jacobsen S, Bengtsson C, Pedersen M, et al. Alcohol consumption is associated with decreased risk of rheumatoid arthritis: results from two Scandinavian studies. Ann Rheum Dis 2009;68:222–7.

[34] Baliunas DO, Taylor BJ, Irving H, Roereke M, et al. Alcohol as a risk factor for type 2 diabetes: a systematic review and meta-analysis. Diabetes Care 2009;32:2123–32.

[35] Shai I, Wainstein J, Harman-Boehm I, Raz I, et al. Glycemic effects of moderate alcohol intake among patients with type 2 diabetes: a multicenter randomized clinical investigation. Diabetes Care 2007;30:3011–16.

[36] Guo R, Ren J. Alcohol and acetaldehyde in public health: from marvel to menace. Int J Environ Res Public Health 2010;7:1285–301.

[37] Takkouch B, Regueira-Mendez C, Garcia-Closas R, Figueiras A, et al. Intake of wine, beer, and spirits and the risk of clinical common cold. Am J Epidemiol 2002;155:853–8.

[38] Klatsky AL, Friedman GD, Siegekaub AB. Alcohol and mortality: a ten-year Kaiser-Permanente experience. Ann Intern Med 1981;95:139–45.

[39] Camargo CA, Hennekens CH, Gaziano JM, Glynn RJ, et al. Prospective study of moderate alcohol consumption and mortality in U.S. male physicians. Arch Intern Med 1997;157:79–85.

[40] Prickett CD, Lister E, Collins M, Trevithick-Sutton CC, et al. Alcohol: Friend or foe? Alcoholic beverage hormesis for cataract and atherosclerosis is related to plasma antioxidant activity. Nonlinearity Biol Toxicol Med 2004;2:353–70.

[41] Puddey IB, Croft KD, Abdu-Amsha Caccetta R, Beilin LJ. Alcohol free radicals and antioxidants. Novartis Found Symp 1998;216:51–62.

[42] Tanaka E, Terada M, Misawa S. Cytochrome P450 2E1: its clinical and toxicological role. J Clin Pharm Ther 2000;25:167–75.

[43] Ren J, Wold E. Mechanisms of alcoholic heart disease. Ther Adv Cardiovasc Dis 2008;2:497–506.

[44] Lu Y, Cederbaum AI. CYP2E1 and oxidative liver injury by alcohol. Free Radic Biol Med 2008;44:723–38.

[45] Gordon T, Ernst N, Fisher M, Rifkind BM. Alcohol and high-density lipoprotein cholesterol. Circulation 1981;64:III 63–7.

[46] Hulley SB, Gordon S. Alcohol and high-density lipoprotein cholesterol: causal inference from diverse study designs. Circulation 1981;64:III 57–63.

[47] Agarwal DP. Cardioprotective effects of light to moderate consumption of alcohol: a review of putative mechanisms. Alc Alcohol 2002;37:409–15.

[48] Rubin R. Effect of ethanol on platelet function. Alc Clin Exp Res 1999;23:1114–18.

[49] Li H, Forsterman U. Red wine and cardiovascular health. Circ Res 2012;111:959–61.

[50] Wu JM, Wang ZR, Hsieh TC, Bruder JL, et al. Mechanism of cardioprotection by resveratrol, a phenolic antioxidant present in red wine. Int J Mol Med 2001;8:3–17 [Review].

[51] Denke MA. Nutritional and health benefits of beer. Am J Med Sci 2000;320:320–6.

[52] Volcik KA, Ballantyne CM, Fuchs FD, Sharrett AR, et al. Relationship of alcohol consumption and type of alcoholic beverage consumed with plasma lipid levels: differences between whites and African Americans of the ARIC study. Ann Epidemiol 2008;18:101–7.

[53] Krenz M, Korthuis RJ. Moderate ethanol ingestion and cardiovascular protection: from epidemiologic association to cellular mechanism. J Mol Cell Cardiol 2012;52:93–104.

[54] Bassus S, Mahnel R, Scholz T, Wegert W, et al. Effect of dealcoholized beer (Bitburger Drive) consumption on hemostasis in humans. Alcohol Clin Exp Res 2004;28:786–91.

[55] Rotches-Ribalta M, Urpi-Sarda M, Llorach R, Boto-Ordonez M, et al. Gut and microbial resveratrol metabolite profiling after moderate long-term consumption of red wine versus dealcoholized red wine in humans by an optimized ultra-high-pressure liquid chromatography tandem mass spectrometry method. J Chromatogr A 2012;1265:105–113.

[56] Noguer MA, Cerezo AB, Donoso Navarro E, Garcia-Parrilla MC. Intake of alcohol-free wine modulates antioxidant enzyme activities in a human intervention study. Pharmacol Res 2012;65:609–14.

[57] Chiva-Blanch G, Urpi-Sarda M, Ros E, Arranz S, et al. Dealcoholized red wine decreases systolic and diastolic blood pressure and increases plasma nitric oxide: short communication. Circ Res 2012;111:1065–8.

CHAPTER 15

Antioxidant Vitamins and Minerals

CONTENTS

15.1 INTRODUCTION

Vitamins are probably the best-selling dietary supplements in the United States, in which an estimated 35% of the population take multivitamin and mineral supplements [1]. Vitamins and minerals are micronutrients that are required in very small amounts (milligrams to micrograms per day) for proper function of the body. Micronutrients are different from macronutrients such as proteins, fats, and carbohydrates, which are consumed in quantities greater than 100 g per day. Whereas macronutrients provide energy needed to fuel the body, micronutrients enable the use of macronutrients for all physiological purposes and are key regulators of health. Vitamins are organic compounds, whereas minerals are inorganic compounds. Vitamins catalyze numerous biochemical reactions and facilitate energy metabolism, although they are not a direct source of energy. Other than vitamin K and biotin, which are produced by gut flora, and vitamin D, which is synthesized during exposure of the skin to sunlight, the human body cannot synthesize vitamins. Therefore, the body must get these vitamins from diet. Vitamins can be broadly classified into two categories—fat-soluble vitamins (vitamins A, D, E, and K) and water-soluble vitamins (vitamin B complex and vitamin C) (Table 15.1). Whereas fat-soluble vitamins can be stored in the human body, water-soluble vitamins cannot. The recommended daily allowances (RDAs) of vitamins for adults are well-established (Table 15.2), but sick people, children, and pregnant women may require a higher intake of certain vitamins. For detailed discussion of all vitamins as well as the role of vitamins in human physiology, consult any nutrition textbook. This chapter focuses on antioxidant vitamins and minerals only and their role in alleviating oxidative stress.

Trace elements can further be classified as essential trace elements and likely essential trace elements. Iron is a key trace element that is an essential component of iron-containing protein, including hemoglobin, myoglobin,

A. Dasgupta and K. Klein: Antioxidants in Food, Vitamins and Supplements. DOI: http://dx.doi.org/10.1016/B978-0-12-405872-9.00015-X

Table 15.1 Vitamins

Classification	Individual Substance
Fat-soluble vitamins	Vitamin A (antioxidant) Vitamin D Vitamin E (antioxidant) Vitamin K
Water-soluble vitamins	Vitamin B_1 (thiamine) Vitamin B_2 (riboflavin) Vitamin B_3 (niacin) Vitamin B_5 (pantothenic acid) Vitamin B_6 (pyridoxine) Vitamin B_7 (biotin) Vitamin B_9 (folate) Vitamin B_{12} (cyanocobalamin) Vitamin C (antioxidant)

Table 15.2 Recommended Daily Allowance (RDA) of Various Vitamins

Vitamin	Deficiency May Cause	RDA
Vitamin A	Night blindness	900 μg, men 700 μg, women
Vitamin B_1 (thiamine)	Beriberi	1.2 mg, men 1.1 mg, women
Vitamin B_2 (riboflavin)	Canker sore, eczema	1.7 mg
Vitamin B_3 (niacin)	Pellagra, skin rash, dementia	16 mg, men 14 mg, women
Vitamin B_5 (pantothenic acid)	Extremely rare	5 mg, men 6 mg, women
Vitamin B_6 (pyridoxine)	Muscle weakness	1.3 mg, adults < 50 years old 1.7 mg. older men 1.5 mg, older women
Vitamin B_7 (biotin)	Extremely rare	30 μg
Vitamin B_9 (folate)	Weight loss, birth defects	400 μg
Vitamin B_{12} (cyanocobalamin)	Anemia	2.4 μg
Vitamin C (ascorbic acid)	Scurvy, bleeding gums	90 mg, men 75 mg, women
Vitamin D	Rickets, poor bone/muscle strength	5–10 μg
Vitamin E	Muscle fatigue, hair loss	15 mg
Vitamin K	Spontaneous bleeding	70–80 μg, men 60–65 μg, women

Table 15.3 Foods Rich in Antioxidant Vitamins and Selenium

Vitamin	Food Source
Vitamin C	Orange, grapefruit, kiwifruit, strawberry, cantaloupe, red pepper, green pepper, Brussels sprouts, broccoli, tomato, cabbage, cauliflower, spinach, green peas
Vitamin E	Fortified cereals, spinach, asparagus, broccoli, cucumber, avocado, hazelnuts, peanut butter, almonds, vegetable oils (corn, cottonseed, safflower, sunflower wheat germ, etc.), eggs, meat, poultry
Vitamin A	Liver, fish, eggs, milk, ice cream, cheese, fortified cereals, yogurt
β-Carotene (pro-vitamin A)	Spinach, carrot, broccoli, pumpkin, cantaloupe, sweet potato, squash, mango, apricot, nuts, black-eyed peas
Selenium	Beef, turkey, chicken breast, cod, tuna, eggs, bagels, sunflower seeds, oatmeal, wheat bread, brown rice, walnut, cheddar cheese

cytochrome, and specific iron-containing enzymes. Magnesium is required in more than 300 enzyme-mediated reactions responsible for food metabolism. Zinc is required for structure and activities of many enzymes and is also required for synthesis of various proteins and nucleic acids. Chromium is involved in the regulation of glucose, lipid, and protein metabolism by potentiation of action of insulin at the cellular level [2]. Selenium is the only trace metal that is an antioxidant.

Vitamin E, vitamin C, and β-carotene (pro-vitamin A) are the only antioxidant vitamins. Vitamin C is the only water-soluble antioxidant vitamin, whereas other antioxidant vitamins are fat-soluble. These vitamins are found in many different foods (Table 15.3). In this chapter, only antioxidant vitamins and selenium are discussed, with an emphasis on their antioxidant role.

15.2 VITAMIN C: THE ONLY WATER-SOLUBLE ANTIOXIDANT VITAMIN

Vitamin C (ascorbic acid) is an outstanding antioxidant. The beneficial effect of vitamin C was known long before its discovery in 1932 because it had been known for a long time that eating citrus foods prevented scurvy, a disease that killed many sailors between 1500 and 1800. Although many species can synthesize vitamin C, humans, other primates, and a few other species cannot and must get this vitamin from diet. Vitamin C plays many important roles, including protein metabolism, collagen biosynthesis, L-carnitine biosynthesis, and absorption of iron. The intestinal absorption of vitamin C is dose dependent (higher absorption with lower vitamin C intake) and is regulated by active transporters. Genetic variation of vitamin C transporter proteins (SVCT1 and SVCT2) can modify the association between dietary vitamin C and serum vitamin C concentration [3]. Recommended daily vitamin C intake is 90 mg for men and 75 mg for women (see Table 15.2). This requirement can be easily fulfilled by eating a proper diet. Johnston *et al.* [4] demonstrated that drinking an 8-oz (240 mL) glass of orange juice can fulfill most of the daily vitamin C requirement. However, vitamin C in orange juice may degrade under prolonged storage. Therefore, it is recommended that ready-to-drink orange juice be purchased at least 3 or 4 weeks before expiration and be consumed within 1 week to get good intake of vitamin C [5]. Other fruits and fruit juices, such as lemon juice and apple juice, are also rich in vitamin C. Vitamin C deficiency is rare in developed countries. However, smokers may have lower concentrations of serum vitamin C than nonsmokers because smoking induces oxidative stress. Patients with malabsorption and end-stage renal disease may also suffer from vitamin C deficiency.

The antioxidant effect of vitamin C has been well documented [6]. One of the vital roles of vitamin C is to protect cellular components from free radical-induced damage. Vitamin C is a scavenger of free radicals in the aqueous phase of the cell and the circulatory system. In addition, vitamin C is effective in regenerating the antioxidant form of vitamin E by reducing tocopheroxyl radicals, thus indirectly protecting membranes and other hydrophobic compartments from free radical-induced damage. Reduced coenzyme Q10, which is hydrophobic, can interact with vitamin C, and in this process the antioxidant form of coenzyme Q is regenerated [7]. Vitamin C is a reducing substance and an electron donor. During free radical scavenging, vitamin C donates high-energy electrons to neutralize free radicals, and it is oxidized to dehydroascorbic acid. Dehydroascorbic acid may be converted back into ascorbic acid for reuse or may be metabolized, further releasing more electrons [8]. Although vitamin C is absorbed from the gut via a sodium-dependent vitamin C transporter, most cells transport vitamin C in oxidized form (dehydroascorbic acid) via facilitative glucose transporters, including glucose transporter 1. Once inside the cell, dehydroascorbic acid is reduced to generate ascorbic acid, which protects mitochondria from free radical-induced oxidative injury [9]. Oxidized low-density lipoprotein (LDL) has been implicated in the pathogenesis of atherosclerosis [10]. Vitamin C inhibits oxidation of LDL, as evidenced by decreased levels of thiobarbituric acid reactive substances in LDL in the presence of vitamin C. Vitamin C is as potent as the antioxidant probucol in inhibiting copper-catalyzed oxidation of LDL [11]. High-density lipoprotein (HDL) has a cardioprotective effect; an HDL cholesterol level greater than 60 mg/dL is considered to provide a superior protective effect. However, HDL is also susceptible to oxidation. Hillstrom *et al.* [12] reported that vitamin C inhibited copper ion-catalyzed oxidation of human HDL and also preserved the antioxidant activity associated with lipoprotein fraction.

Human diseases such as atherosclerosis, stroke, and cancer are linked to excessive oxidative stress. Therefore, it has been hypothesized that vitamin C supplementation may prevent or reduce the risk of such diseases. Epidemiological studies have clearly shown that diets high in fruits and vegetables are effective at lowering the risk of atherosclerosis, stroke, and cancer. Whether these effects are directly attributable to the antioxidant effect of vitamin C or other antioxidants or a combination thereof has not been fully elucidated. However, intervention studies using vitamin C supplementation failed to show any beneficial effect of vitamin C in preventing diseases or improving biochemical parameters that are representative of oxidative stress. The failure of many clinical trials to demonstrate any effectiveness of vitamin C supplementation, even at high dosage, in preventing cancers, cardiovascular disease, and preeclampsia, led one author to comment that the effectiveness of vitamin C antioxidant supplementation in women's health is a myth that requires urgent burial [13]. Rumbold and Crowther [14] noted that vitamin C supplementation may increase the risk of preterm birth.

Professor Linus Pauling, two-time Nobel Prize winner, advocated in 1970 that intake of large dosages of vitamin C prevents the common cold. His conclusion was based on a single clinical trial of schoolchildren who attended a ski camp in the Swiss Alps. A significant reduction in the incidence of the common cold as well as the duration of symptoms was observed among children who took 1 g of vitamin C supplement each day [15]. In a clinical trial that involved 400 healthy volunteers and lasted for 18 months, Audera *et al.* [16] observed no beneficial effect of 1 g daily intake of vitamin C in reducing the duration or severity of cold symptoms. Douglas *et al.* [17] analyzed data from 29 different clinical trials involving 11,077 subjects and concluded that vitamin C supplementation appears to reduce the severity and duration of colds but cannot prevent the incidence of colds. The authors commented that, due to the failure of vitamin C supplementation to reduce the incidence of the common cold among the normal population, regular megadosage of vitamin C as a prophylaxis is not justified. However, a person may take vitamin C supplement for a short period if subjected to severe physical exercise or a cold environment. Vitamin C may play a role in the respiratory defense mechanism, but a clinical trial in which vitamin C was introduced at the onset of cold as therapy did not show any benefit with dosage up to 4 mg daily except in one trial [17]. If taken in excessive dosage (1 g daily), vitamin C may increase urinary oxalate excretion and the risk of calcium oxalate crystallization in calcium stone-forming patients [18].

Vitamin C uptake is tightly controlled, resulting in wide-ranging bioavailability depending on current health status. Although supplementation of vitamin C may not have any effect on the healthy Western population, which receives sufficient vitamin C from a normal diet, a subpopulation with health problems may obtain some benefit from supplementation; however, such potential benefit has not been investigated [19]. Frei *et al.* [20] noted that, based on the evidence, 200 mg of vitamin C per day is the optimum dietary intake for the majority of the adult population to obtain maximum health benefits from vitamin C and has the lowest risk of adverse health effects. Therefore, taking a vitamin C supplement is not recommended for healthy individuals. Patients with vitamin C deficiency may benefit from vitamin C supplementation, but the supplements should only be taken under medical supervision. However, Das *et al.* [21] noted that dietary supplementation with vitamin C may be a simple therapy for the prevention of pathological cardiovascular events in habitual smokers.

15.3 VITAMIN E: FAT-SOLUBLE ANTIOXIDANT VITAMIN

Vitamin E is a fat-soluble vitamin that was discovered by Evans and Bishop in 1922. However, the role of vitamin E in human nutrition was not fully understood until 1967, when it was demonstrated that vitamin E deficiency caused hemolytic anemia in premature infants [22]. Further evidence that vitamin E deficiency may have severe health consequences was obtained from the study of three different groups of patients showing vitamin E deficiencies: patients with abetalipoproteinemia, those with chronic fat malabsorption syndrome, and those with ataxia with vitamin E deficiency (AVED).

Vitamin E exists in eight different forms—α-, β-, γ-, and δ-tocopherol as well as α-, β-, γ-, and δ-tocotrienol—and all of these molecules are synthesized by plants. Although the human diet contains all eight of these molecules, the human body prefers α-tocopherol, which is the most active form, and the RDA of 15 mg (see Table 15.2) is based on α-tocopherol concentration. Some multivitamin preparations list vitamin E content in international units (IU; 1 IU is approximately 0.67 mg). Many foods contain vitamin E (see Table 15.3). Most dietary supplements that contain vitamin E provide more of it than the RDA.

Only α-tocopherol fulfills the human vitamin E requirement because hepatic α-tocopherol transfer protein poorly recognizes non-α-tocopherols. This transfer protein is responsible for maintaining the plasma α-tocopherol level in human [23]. After intestinal absorption, vitamin E is incorporated into chylomicrons and is secreted into the circulation. By the action of lipoprotein lipase, breakdown of chylomicrons into remnant chylomicrons occurs. Some α-tocopherol is incorporated into extrahepatic tissues, but remnant chylomicrons transport the remaining α-tocopherol and other forms of vitamin E into the liver, in which, by the action of the α-tocopherol transfer protein, only α-tocopherol is incorporated into very low-density lipoprotein (VLDL) and the excess α-tocopherol and other forms of vitamin E are excreted into bile. α-Tocopherol incorporated into VLDL is secreted into the circulation and converted into intermediate-density lipoprotein (IDL) and eventually to LDL by the action of lipoprotein lipase. The excess α-tocopherol is transferred into HDL. Uptake of various lipoproteins by tissues also results in incorporation of α-tocopherol [24].

α-Tocopherol, a fat-soluble chain-breaking antioxidant, can compete for peroxyl radicals much faster than polyunsaturated fatty acid, thus preventing lipid peroxidation. α-Tocopherol scavenges peroxy radicals by donating a hydrogen atom. In this process, α-tocopherol is converted into tocopheroxyl radical, which is converted back into α-tocopherol by vitamin C. Thus, only a small amount of α-tocopherol is needed to prevent lipid peroxidation. However, *in vitro* α-tocopheroxyl radicals may also react with other peroxyl radicals, lipid radicals, or molecular oxygen to form various adducts. These adducts may further reduce to α-tocopherol quinone, which eventually may undergo further reduction to produce α-tocopherol hydroquinone. These quinones and hydroquinone can be detected in human samples and may serve as biomarkers of oxidative stress *in vivo* [25]. α-Tocopherol is not homogeneously distributed in lipid membranes. Instead, it is partitioned into domains that are enriched with polyunsaturated fatty acids and phospholipids where it is most needed to prevent lipid peroxidation [26]. The importance of α-tocopherol for protecting the integrity of lipid structures, especially cell membranes *in vivo*, is well recognized, and it is the only major lipid-soluble chain-breaking antioxidant in plasma (associated with lipoproteins), red cells, and tissues. Although the fat-soluble antioxidant β-carotene has chain-breaking capacity, it is less efficient than α-tocopherol [27]. α-Tocopherol in combination with vitamin C is highly effective because such combination is capable of completely inhibiting the oxidation of lipoprotein in plasma. However, in the absence of vitamin C, α-tocopherol may act as a pro-oxidant. Therefore, in the human body, the combination of α-tocopherol and vitamin C is essential for proper antioxidant defense [28]. Expression of the α-tocopherol transfer protein gene is regulated by oxidative stress. Therefore, both oxidative stress and individual genetic makeup contribute to vitamin E homeostasis in humans [29].

In addition to being an excellent antioxidant, α-tocopherol has other important physiological activities that can be broadly classified as "non-antioxidant activities of vitamin E." At the post-translational level, α-tocopherol inhibits protein kinase C, 5-lipoxygenase, and phospholipase A2, and it activates protein phosphatase 2A and diacylglycerol kinase. Protein kinase C is an important enzyme involved in cell proliferation and differentiation of smooth muscle, platelets, and monocytes. α-Tocopherol can prevent platelet aggregation and is capable of dilating blood vessels by increasing the rate of prostacyclin release. α-Tocopherol can also modulate genes such as scavenger receptors, α-tropomyosin, matrix metalloproteinase, and collagenase [30]. α-Tocopherol is essential for proper neurological function, and both antioxidants and mechanisms other than antioxidants have been proposed to explain the role of α-tocopherol in the proper function of neurons.

Most individuals in the United States obtain sufficient vitamin E from dietary sources, but some individuals with severe vitamin E deficiency due to various disorders may require vitamin E supplementation. The most severe form of vitamin E deficiency is observed in patients with abetalipoproteinemia, a relatively rare genetic disorder involving lipoprotein metabolism. These patients are incapable of synthesizing apolipoprotein B, which is essential for various lipoprotein structures. Due to the absence of or very low levels of chylomicrons, VLDL, and LDL in these patients, absorption and transportation of vitamin E is severely impaired. As a result, the level of vitamin E in serum is usually very low or even undetectable. Severe deficiency of vitamin E results in retarded growth and later development of ataxic neuropathy and retinal pigmentation, which may ultimately cause crippling and blindness in these patients. Treating these patients early in life (age 16 months) with very large doses of vitamin E may prevent the development of these symptoms. Vitamin E deficiency may also occur due to any chronic disorder of fat malabsorption, but it is particularly severe in patients with cholestatic liver disease because bile salts synthesized in the liver are essential for absorption of vitamins. Neurological symptoms may also appear in these patients, and these symptoms can be treated with vitamin E supplementation. Patients with familial isolated vitamin E deficiency, also known as AVED (ataxia with vitamin E deficiency), show neurological symptoms associated with vitamin E deficiency. These patients are capable of absorbing vitamin E from the diet like normal individuals, but they lack hepatic α-tocopherol binding protein, which is necessary for transportation of α-tocopherol to VLDL and its subsequent transfer to the liver. It is important to differentiate the diagnosis of AVED from Friedreich's ataxia, a rare inherited disease that causes nervous system damage and movement problems. Patients with AVED respond to vitamin E supplementation, and a normal serum concentration of α-tocopherol can be achieved with 800 mg/day supplementation of vitamin E. There is no known cure for Friedreich's ataxia. Patients with abetalipoproteinemia may require 100 mg/kg/day of vitamin E [31].

α-Tocopherol is essential for health and can protect against coronary artery disease, cancer, and cognitive decline in older age as well as some ophthalmological disorders. However, healthy people do not obtain any additional benefit from vitamin E supplementation. Research indicates that taking vitamin E supplement may cause more harm than good because at higher dosage vitamin E acts as a pro-oxidant rather than an antioxidant [32]. Clinical trials indicate that daily intake of vitamin E (400 IU/day or more) as supplement is associated with increased all-cause mortality, but even doses greater than 150 IU/day may cause harm [33]. Vitamin E supplement should be taken under medical supervision only by people who have chronic vitamin E deficiency [34].

15.4 VITAMIN A AND β-CAROTENE: FAT-SOLUBLE ANTIOXIDANTS

β-Carotene (pro-vitamin A) belongs to a class of chemicals called carotenoids. It plays an important role in human physiology, and is an excellent antioxidant. There are more than 600 carotenoids in plants and bacteria. β-Carotene is the most abundant form of carotenoid, and it is also a precursor of vitamin A. Vitamin A refers to a group of compounds including retinol, retinal (retinaldehyde), retinoic acid, and retinyl esters. In vertebrates, β-carotene is converted to retinol by β-carotene 15,15′-dioxygenase, which is expressed specifically in intestinal epithelium and liver. This enzyme not only plays an important role in providing vitamin A but also determines whether β-carotene should be converted into vitamin A or should be circulated in the form of β-carotene [35]. Vitamin A is critical for vision because retinal is an essential component of rhodopsin, a protein that absorbs light in retinal receptors and also supports the normal differentiation and function of conjunctival membranes and cornea. Rhodopsin contains a protein moiety called opsin, which is reversibly bound to retinal. When retinal is converted into retinoic acid, it plays an important role in skin health, bone growth, and remineralization. Retinol is produced by reduction of retinal or hydrolysis of retinyl ester. Vitamin A also plays an important role in immune function and cellular communication. Human requirement of vitamin A is expressed as retinol equivalent, where one retinol equivalent equals 1 μg of retinol. RDA for vitamin A is 700 μg for women and 900 μg for men (see Table 15.2). Sometimes, β-carotene or vitamin A concentration is given in international units on package labels. For vitamin A, 1 IU is equivalent to 0.0006 mg of β-carotene. Deficiency of vitamin A is associated with night blindness, which is more common in developing countries due to malnutrition. However, vitamin A deficiency may also be associated with fat malabsorption; for example, vitamin A deficiency causing visual deterioration is a problem after bariatric surgery, but these patients respond to intramuscular vitamin A replacement [36].

In contrast to β-carotene, which is found only in fruits and vegetables, vitamin A is found in egg yolk and liver (see Table 15.3). Gentle cooking generally improves the availability of β-carotene from vegetables. β-Carotene has antioxidant properties, but it is a weaker antioxidant than α-tocopherol. Tsuchihashi *et al.* found that β-Carotene can suppress the free radical-mediated oxidation of methyl linoleate in benzene solution and soybean phosphatidylcholine liposomal membranes in aqueous solution in a dose-dependent manner, but α-tocopherol is significantly more reactive than β-carotene toward peroxyl radicals. When β-carotene and α-tocopherol were present together in homogeneous solution, α-tocopherol was consumed predominantly and β-carotene was spared. However, β-carotene was consumed faster than α-tocopherol when the radicals were generated within the lipophilic compartment of the membranes, indicating that β-carotene may be more favorable than α-tocopherol for scavenging lipophilic radicals within the membrane. In general, β-carotene is less potent than α-tocopherol as an antioxidant [37]. Like ascorbate and α-tocopherol, β-carotene can inhibit oxidative modification of LDL [38].

Retinol is also an antioxidant. The antioxidant effect of retinol *in vivo* has been demonstrated in the nuclear and mitochondrial fractions of liver homogenates and in the nuclear fraction of spleen homogenate. In contrast, the pro-oxidant effect of retinol was observed in the mitochondrial fraction of kidney and spleen homogenate and the microsomal fraction of spleen homogenate [39]. Livrea *et al.* [40] showed that vitamin A-enriched LDL was resistant to oxidation. However, the reactivities of antioxidants with lipoperoxyl radicals were in the following sequence: α-tocopherol (most reactive), retinol, β-carotene, and retinyl ester (least reactive). Schwarz *et al.* [41] observed lower breath ethane values indicating higher antioxidant defense of blood in infants after they received vitamin A supplement, demonstrating antioxidant properties of vitamin A *in vivo*. The authors concluded that premature infants with vitamin A deficiency as evidenced by subnormal serum retinol may obtain some benefits from the antioxidant effect of vitamin A from supplementation.

Although β-carotene and vitamin A are antioxidants, results from clinical trials have shown no benefit of β-carotene supplementation to prevent any disease in healthy individuals. Excess intake of vitamin A may cause toxicity, including increased risk of hip fracture and birth defects. Pregnant women should not consume excess vitamin A: it may cause oxidative stress because at higher dosage vitamin A acts as a pro-oxidant rather than an antioxidant. In a rat model, supplementation with vitamin A during pregnancy and nursing increased oxidative damage of maternal and offspring striatum and hippocampus [42]. Excess intake of vitamin A also interferes with absorption of vitamin D, causing vitamin D deficiency. Interestingly, β-carotene, the precursor of vitamin A, is not toxic. If the body has sufficient vitamin A, then it does not make any vitamin A from β-carotene. However, excess carotene may turn skin yellow [43].

β-Carotene supplementation may cause harm rather than produce benefit in healthy people. Based on a meta-analysis of 13 clinical trials, Druesne-Pecollo *et al.* [44] noted that the results of such trials do not support any protective effect of β-carotene supplementation on the incidences of lung, pancreatic, stomach, colorectal, prostate, breast, and skin cancer. In contrast, β-carotene supplementation could increase the risk of lung and stomach cancer if 20–30 mg supplementation of β-carotene per day was consumed. The authors concluded that β-carotene supplementation is not recommended for preventing cancer. Tanvetyanon and Bepler [45] observed that β-carotene was present in 70% of all formulations (47 common brands of multivitamins sold in the United States), with a median content of 0.3 mg (range, 0–17.2 mg). The β-carotene concentration was significantly higher in multivitamins sold to improve visual health (median daily dose of 3 mg; range, 0–24 mg). The authors recommended that the public should be discouraged from taking β-carotene supplement. However, vitamin A deficiency may occur among children younger than age 5 years in poor and lower socioeconomic groups in developing countries; vitamin A supplementation may prevent mortality, illness, and blindness in these children [46].

15.5 ANTIOXIDANT TRACE ELEMENT SELENIUM

Selenium is an important nutritional element to mammals because it is an essential component of several major metabolic pathways, including antioxidant defense and thyroid hormone metabolism, as well as important for optimal immune function. More than 30 selenoproteins have been identified. Selenium is incorporated as selenocysteine at the active sites of a wide range of selenoproteins. Antioxidant glutathione peroxidase enzymes are important selenoproteins. Other important selenoproteins are thioredoxin reductase and iodothyronine deiodinase, both of which play important roles in maintaining proper thyroid function. Approximately 60% of serum selenium is associated with selenoprotein P, which contains 10 selenium atoms per molecule as selenocysteine and may serve as a transport protein for selenium. Sperm capsule selenoprotein is localized in the midpiece portion of spermatozoa, where selenium is responsible for stability of sperm flagella, thus enhancing male fertility. Decreasing selenium concentration in blood may be a concern because selenium may have a protective effect on certain types of cancer and cardiovascular disease, and it may regulate inflammatory markers in asthma [47]. Selenium can regulate gene expression of these selenoproteins as a cofactor, and there is evidence that selenium supplement can attenuate oxidative stress after cardiac surgery. However, other reports have failed to demonstrate beneficial effects of selenium after cardiac surgery [48].

Selenium is found in whole grains and shellfish, and the daily selenium requirement of 55 μg per day can be easily fulfilled by eating a proper diet. Selenium can be found in meat and seafood. Animals that eat grains that grow in selenium-rich soil have higher selenium content in their muscles. Common sources of selenium are given in Table 15.3. The upper tolerable limit of selenium is 400 μg per day. Therefore, excess selenium is toxic because at higher concentrations selenium induces oxidative stress and other toxic effects in the human body [49]. Excess selenium causes selenosis, which is manifest by hair loss, brittle and discolored nails, nausea, diarrhea, and irritability. Soils of certain areas in the Great Plains and western United States as well as other areas of the world have high amounts of selenium, which is taken up by plants. However, until recently, outbreaks of selenium poisoning have been rare except in certain areas of China. An outbreak of selenium toxicity occurred in 2008 in 10 US states affecting 201 individuals who took a liquid multivitamin supplement that, according to the package insert, contained 200 μg of selenium per ounce but that analysis showed contained 200 times higher amounts of selenium. One person who took this supplement ingested an estimated 41,749 μg of selenium. Although one person was hospitalized, no one died, and the company withdrew the faulty product from the market [50]. Therefore, it is advisable not to take selenium supplement. In fact, excess selenium intake may increase the risk of type 2 diabetes [51].

Selenium deficiency is rare in the United States but may occur in patients with severe gastrointestinal problems such as observed in Crohn's disease. People with iodine deficiency, which is rarely observed in the United States, may benefit from selenium supplement. Keshan disease, an endemic cardiomyopathy with mortality that was first observed in 1953 in Keshan County in the Heliongjiang province of China, is associated with selenium deficiency. The soil in this region is deficient in selenium, and as a result grains grown in this soil are low in selenium content. The levels of blood and hair selenium as well as activity of serum glutathione peroxidase are lower in these individuals compared to normal individuals. As expected, these individuals suffer from excessive oxidative stress. For these people, selenium supplementation can significantly reduce the incidence of Keshan disease [52].

15.6 IS THERE ANY HEALTH BENEFIT TO ANTIOXIDANT VITAMIN/MINERAL SUPPLEMENTATION?

Antioxidant vitamins are important, but incorporating enough fruits and vegetables is the best way to maintain adequate amounts of vitamins in the human body for optimal health. In one study, the authors reported that vitamin/mineral supplement use among toddlers was associated with being a first-born child and a picky eater. The authors concluded that generally healthy infants and toddlers can obtained recommended levels of vitamins and minerals from diet, and caregivers should be encouraged to use food rather than supplements as the primary source of nutrition for children. Vitamins and mineral supplements can help infants and toddlers with special nutrient needs only if advised by physicians [53]. Based on a review of 78 clinical trials involving 296,707 participants, Bjelakovic *et al.* [54] reported that there is no evidence that taking antioxidant supplements prevents any disease. In contrast, β-carotene and vitamin E supplement may increase mortality, and so may vitamin A if taken in higher dosage. Therefore, antioxidant supplements must be considered as medicinal products and should be taken under medical supervision. Lin *et al.* [55] concluded that supplementation with vitamin C (500 mg of ascorbic acid daily), vitamin E (600 IU of α-tocopherol daily), or β-carotene (50 mg daily) offers no overall benefit in primary prevention of total cancer incidence or mortality associated with cancer. Ascherio *et al.* [56] reported that vitamin E and vitamin C supplements and specific carotenoids did not reduce the risk of stroke in a study of 43,738 men aged between 40 and 75 years. Based on a meta-analysis of 50 randomized trials involving 294,478 participants, Myung *et al.* [57] concluded that there is no evidence to support the use of vitamins and antioxidant supplements for prevention of cardiovascular disease.

The Women's Antioxidant Cardiovascular Study also demonstrated no beneficial effect of antioxidant vitamins (vitamins C and E or β-carotene) in preventing cardiovascular events among women with a higher risk of cardiovascular disease [58]. Poston *et al.* [59] commented that although higher oxidative stress is encountered in patients with preeclampsia, antioxidant vitamin supplements (vitamins C and E) cannot prevent preeclampsia.

Vitamin formulations for eyes usually contain antioxidant vitamins (carotenoids and vitamins C and E) and minerals (selenium and zinc) and are available without prescription for the prevention of age-related eye diseases such as cataract and macular degeneration. Based on analysis of data from four randomized controlled trials involving a total of 62,520 people, Evans and Lawrenson [60] concluded that taking vitamin E and β-carotene supplements will not prevent or delay the onset of age-related macular degeneration. Similarly, supplements containing vitamin C, lutein, and zeaxanthin provide no beneficial effect [60]. Also, based on a review of nine randomized controlled trials involving 117,272 individuals, Mathew *et al.* [61] concluded that there is no evidence that antioxidant vitamins (β-carotene and vitamins C and E) can prevent or slow the progression of age-related cataract.

Table 15.4 Tolerable Upper Intake Levels of Antioxidant Vitamins and Selenium

Vitamin/Mineral	Tolerable Upper Intake Level
Vitamin A	2000 μm
Vitamin C	2000 mg
Vitamin E	1000 mg
Selenium	400 μm

15.7 CONCLUSION

Antioxidant vitamins and selenium play important physiological roles, but the daily requirements of such vitamins and selenium can be easily met by a balanced diet. Eating a healthy diet that includes meats, poultry, vegetables, grains, and fruits each day is the best way to fulfill the daily requirements of vitamins and minerals. There are upper tolerable limits for all vitamins (Table 15.4) [62]. Exceeding these limits is potentially dangerous for health. Approximately one-third of the US population regularly takes vitamins without understanding that if taken incorrectly or in excess, such vitamins can be toxic. Vitamins also interact with each other; for example, vitamin C participates in the metabolism of vitamin E and folic acid and also enhances absorption of vitamin A. Unfortunately, vitamin supplements cannot counteract poor diet, such as inadequate intake of carbohydrates, fats, and proteins. When the RDAs for vitamin consumption are exceeded, vitamins act as pharmaceutical agents with toxic effects to the body [63]. High-dose supplementation with carotenoids may cause toxicity due to the formation of carotenoid breakdown products such as very reactive aldehydes and epoxide, which can impair mitochondrial function [64].

In summary, proper diet is the best way to achieve an adequate antioxidant defense for healthy living. Vitamin supplementation may help a select group of people with vitamin deficiency; for example, vitamin A supplementation may help patients with cystic fibrosis [65] or the elderly [66]. Ideally, vitamin and antioxidant supplements should be taken only under medical supervision.

REFERENCES

[1] Rock CL. Multivitamin–multimineral supplements: who uses them? Am J Clin Nutr 2007;85:277S–9S.

[2] Lukaski HC. Vitamin and mineral status: effects on physical performance. Nutrition 2004;20:632–44.

[3] Cahill LE, El-Sohemy A. Vitamin C transporter gene polymorphisms, dietary vitamin C and serum ascorbic acid. J Nutrigenet Nutrigenomics 2009;2:292–301.

[4] Johnston CS, Dancho CL, Strong GM. Orange juice ingestion and supplemental vitamin C are equally effective at reducing plasma lipid peroxidation in healthy adult women. J Am Coll Nutr 2003;22:519–23.

[5] Johnston CS, Bowling DL. Stability of ascorbic acid in commercially available orange juices. J Am Dietetic Assoc 2002;102:525–9.

[6] Frei B, Stocker R, England L, Ames BN. Ascorbate: the most effective antioxidant in human blood plasma. Adv Exp Med Biol 1990;264:155–63.

[7] Cathcart RF. A unique function of ascorbate. Med Hypothesis 1991;35:32–7.

[8] Beyer RE. The role of ascorbate in antioxidant protection of biomembrane: interaction with vitamin E and coenzyme Q. J Bioenerg Biomembr 1994;26:349–58.

[9] Sagun KC, Carcamo JM, Golde DW. Vitamin C enters mitochondria via facilitative glucose transporter 1 (Glut1) and confers mitochondrial protection against oxidative injury. FASEB J 2005;19:1657–67.

[10] Jialal I, Devaraj S. Low-density lipoprotein oxidation, antioxidants and atherosclerosis: a clinical biochemistry perspective. Clin Chem 1996;42:498–506.

[11] Jialal I, Grundy SM. Influence of antioxidant vitamins on LDL oxidation. Ann N Y Acad Sci 1992;669:237–47.

[12] Hillstrom RJ, Yacapin-Ammons AK, Lynch SM. Vitamin C inhibits lipid oxidation in human HDL. J Nutr 2002;133:3047–51.

[13] Talaulikar VS, Manyonda IT. Vitamin C as an antioxidant supplement in women's health: a myth in need of urgent burial. Eur J Obstet Gynecol Reprod Biol 2011;157:10–13.

[14] Rumbold A, Crowther CA. Vitamin C supplementation in pregnancy. Cochrane Database Syst Rev 2005;18(2): CD004073.

[15] Hemila H. Vitamin C supplementation and the common cold: was Linus Pauling right or wrong? Int J Vit Nutr Res 1997;67:329–35.

[16] Audera C, Patulny RV, Sander BH, Douglas RM. Mega dosage vitamin C in treatment of the common cold: a randomized controlled trial. Med J Aust 2001;175:359–62.

[17] Douglas RM, Hemila H, D'Souza R, Chalker EB, et al. Vitamin C for preventing and treating common cold. Cochrane Database Syst Rev 2004;4 CD000980.

[18] Baxmann AC, De OG Mendonca C, Heilberg IP. Effect of vitamin C supplement on urinary oxalate and pH in calcium stone-forming patients. Kidney Int 2003;63:1066–71.

[19] Lykkesfeldt J, Poulsen HE. Is vitamin C supplementation beneficial: lessons learned from randomized controlled trials. Br J Nutr 2010;103:1251–9.

[20] Frei B, Birlouez-Aragon I, Lykkesfeldt J. Authors' perspective: what is the optimum intake of vitamin C in humans? Crit Rev Food Sci Nutr 2012;52:815–29.

[21] Das A, Dey N, Ghosh A, Das S, et al. Molecular and cellular mechanisms of cigarette smoke-induced myocardial injury: prevention by vitamin C. PLoS ONE 2012;7(9):e44151.

[22] Oski FA, Barness LA. Vitamin E deficiency: a previously unrecognized cause of hemolytic anemia in the premature infant. J Pediatr 1967;70:211–20.

[23] Traber MG, Sokol RJ, Burton GW, Ingold KU, et al. Impaired ability of patients with familial isolated vitamin E deficiency to incorporate alpha-tocopherol into lipoprotein secreted by the liver. J Clin Invest 1990;85:397–407.

[24] Herrera E, Barbas C. Vitamin E: action, metabolism, and perspective. J Physiol Biochem 2001;57:43–56.

[25] Shi H, Noguchi N, Nikki E. Comparative study of dynamics of antioxidant action of alpha-tocopherol hydroquinone, ubiquinol, and alpha-tocopherol against lipid peroxidation. Free Radic Biol Med 1999;27:334–46.

[26] Atkinson J, Epand RF, Epand RM. Tocopherols and tocotrienols in membranes: a critical review. Free Radic Biol Med 2008;44:739–64.

[27] Burton GW, Ingold KU. Beta-carotene: an unusual type of lipid antioxidant. Science 1984;224:569–73.

[28] Browry VW, Ingold KU, Stocker R. Vitamin E in human low-density lipoprotein: when and how this antioxidant becomes a pro-oxidant. Biochem J 1992;288:341–4.

[29] Ulatowski L, Dreussi C, Noy N, Barnholtz-Sloan J, et al. Expression of the α-tocopherol transfer protein gene is regulated by oxidative stress and common single-nucleotide polymorphism. Free Radic Biol Med 2012;53:2318–26.

[30] Zingg JM, Azzi A. Non-antioxidant activities of vitamin E. Curr Med Chem 2004;11:1113–33.

[31] Muller D. Vitamin E and neurological function. Mol Nutr Food Res 2010;54:710–18.

[32] Soni MG, Thurmond S, Miller ER, Spriggs S, et al. Safety of vitamins and minerals: controversies and perspective. Toxicol Sci 2010;118:348–55.

[33] Miller 3rd ER, Pastor-Barriuso R, Dalal D, Riemersma RA, et al. Meta-analysis: high-dose vitamin E supplementation may increase all-cause mortality. Ann Intern Med 2005;142:37–46.

[34] Dotan Y, Lichtenberg D, Pinchuk I. No evidence supports vitamin E indiscriminate supplementation. Biofactors 2009;35:469–73.

[35] Nagao A. Oxidative conversion of carotenoids to retinoids and other products. J Nutr 2004;134:237S–40S.

[36] Fok JS, Li JY, Yong TY. Visual deterioration cause by vitamin A deficiency in patients after bariatric surgery. Eat Weight Disord 2012;17:e144–6.

[37] Tsuchihashi H, Kigoshi M, Iwatsuki M, Niki E. Action of beta-carotene as an antioxidant against lipid peroxidation. Arch Biochem Biophys 1995;323:137–47.

[38] Jialal I, Norkus EP, Cristol L, Grundy SM. Beta-carotene inhibits the oxidative modification of low-density lipoprotein. Biochim Biophys Acta 1991;1086:134–8.

[39] Callari D, Billitteri A. Retinol and lipid peroxidation [in Italian]. Acta Vitaminol Enzymol 1976;30:38–43.

[40] Livrea MA, Tesoriere L, Bongiorno A, Pintaudi AM, et al. Contribution of vitamin A to the oxidation resistance of human low-density lipoproteins. Free Radic Biol Med 1995;18:401–9.

[41] Schwarz KB, Cox JM, Sharma S, Clement L. Possible antioxidant effect of vitamin A supplementation in premature infants. J Pediatr Gastroenterol Nutr 1997;25:408–14.

[42] Schnorr CE, da Silve Morrone M, Simose-Pires A, da Rocha RF, et al. Vitamin A supplementation in rats under pregnancy and nursing induces behavioral changes and oxidative stress upon striatum and hippocampus of dams and their offspring. Brain Res 2011;1369:60–73.

[43] Penniston KL, Tanumihardjo SA. The acute and chronic toxic effects of vitamin A. Am J Clin Nutr 2006;83:191–201.

[44] Druesne-Pecollo N, Latino-Martel P, Norat T, Barrandon E, et al. Beta-carotene supplementation and cancer risk: a systematic review and meta-analysis of randomized controlled trials. Int J Cancer 2010;127:172–84.

[45] Tanvetyanon T, Bepler G. Beat-carotene in multivitamins and the possible risk of lung cancer among smokers versus former smokers: a meta-analysis and evaluation of national brands. Cancer 2008;113:150–7.

[46] Mayo-Wilson E, Imdad A, Herzer K, Yakoob MY. Vitamin A supplements for preventing mortality, illness, and blindness in children aged under 5: systematic review and meta-analysis. BMJ 2011;343:d5094.

[47] Brown KM, Arthur JR. Selenium, selenoproteins, and human health: a review. Public Health Nutr 2001;4(2B):593–9.

[48] Guo F, Monsefi N, Moritz A, Beiras-Fernandez A. Selenium and cardiovascular surgery: an overview. Curr Drug Saf 2012;7:321–7.

[49] Uguz AC, Nazioglu M, Espino J, Bejarano I, et al. Selenium modulates oxidative stress-induced cell apoptosis in human myeloid HL-60 cells through regulation of calcium release and caspase 3 and 9 activities. J Membr Biol 2009;232:15–23.

[50] MacFarquhar JK, Broussard DL, Melstrom P, Hutchinson R, et al. Acute selenium toxicity associated with a dietary supplement. Arch Intern Med 2010;170:256–71.

[51] Bleys J, Navas-Acien A, Guallar E. Serum selenium and diabetes in U.S. adults. Diabetes Care 2007;30:829–34.

[52] Li Q, Liu M, Hou J, Jiang C. The prevalence of Keshan disease in China. Int J Cardiol 2013;168(2):1121–6.

[53] Briefel R, Hanson C, Fox MK, Novak T, et al. Feeding Infants and Toddlers Study: do vitamin and mineral supplements contribute to nutrient adequacy or excess among U.S. infants and toddlers. J Am Dietetic Assoc 2006;106(Suppl. 1):S52–6.

[54] Bjelakovic G, Nikolova D, Gluud LL, Simonetti RG, et al. Antioxidant supplements for prevention of mortality in healthy participants and patients with various diseases. Cochrane Database Syst Rev 2012;14(3): CD007176.

[55] Lin C, Cook NR, Albert C, Zaharris E, et al. Vitamin C and E and beta-carotene supplementation and cancer risk: a randomized controlled trail. J Natl Cancer Inst 2009;101:14–23.

[56] Ascherio A, Rimmn EB, Hernan MA, Giovannucci E, et al. Relation of consumption of vitamin E, vitamin C, and carotenoids to risk of stroke among men in the United States. Ann Intern Med 1999;130:963–70.

[57] Myung SK, Ju W, Cho B, Oh SW, et al. Efficacy of vitamin and antioxidant supplements in prevention of cardiovascular disease: systematic review and meta-analysis of randomized clinical trials. BMJ 2013;346:f10.

[58] Cook NR, Albert CM, Gaziano JM, Zaharris E, et al. A randomized factorial trial of vitamin C and E and beta-carotene in the secondary prevention of cardiovascular events in women: results from the Women's Antioxidant Study. Arch Intern Med 2007;167:1610–18.

[59] Poston L, Igosheva N, Mistry HD, Seed PT, et al. Role of oxidative stress and antioxidant supplementation in pregnancy disorder. Am J Clin Nutr 2011;94:1980S–5S.

[60] Evans JR, Lawrenson JG. Antioxidant vitamin and mineral supplementation for preventing age-related macular degeneration. Cochrane Database Syst Rev 2012;13(6): CD000253.

[61] Mathew MC, Ervin AM, Tao J, Davis RM. Antioxidant vitamin supplementation for preventing and slowing the progression of age-related cataract. Cochrane Database Syst Rev 2012;13(6): CD004567.

[62] Murphy S, White KK, Park SY, Sharma S. Multivitamin–multimineral supplements' effect on total nutrient intake. Am J Clin Nutr 2007;85(Suppl):280S–4S.

[63] Blair KA. Vitamin supplementation and megadose. Nurse Pract 1986;11:19–26.

[64] Siems W, Wiswedel I, Salerno C, Crifo C, et al. Beta-carotene breakdown products may impair mitochondrial functions: potential side effects of high-dose beta-carotene supplementation. J Nutr Biochem 2005;15:385–97.

[65] Brei C, Simon A, Krawinkel MB, Naehrlich L. Individualized vitamin A supplementation for patients with cystic fibrosis. Clin Nutr 2013;32:805–10.

[66] Kjeldby IK, Fosnes GS, Liggarden SC, Farup PG. Vitamin B_6 deficiency and disease in elderly people: a study in nursing homes. BMC Geriatr 2013;13:13.

CHAPTER 16

Herbal and Other Dietary Supplements That Are Antioxidants

CONTENTS

16.1 INTRODUCTION

Herbal supplements are considered dietary supplements in the United States and are sold under the Dietary Supplement Health and Education Act of 1994, according to which a manufacturer of a herbal supplement cannot claim any medical benefit of the supplement [1]. Unfortunately, the US Food and Drug Administration (FDA) does not have any authority to stop marketing a supplement even if research indicates that it is unsafe. For example, in 2002, the FDA issued a warning regarding the safety of kava, an antianxiolytic herbal sedative, because prolonged use of kava can cause serious liver damage. In addition, death from using kava has been reported, but even today kava products are freely available in herbal food stores. Canadian authorities also issued warnings against the use of kava, but one report noted that even 2 months after the advisory, kava was being sold in health food stores. The authors concluded that federal Canadian advisories do not affect the sale of unsafe herbal products [2]. However, in Germany, herbal monographs called the German Commission E monographs are prepared by an interdisciplinary committee using historical information; chemical, pharmacological, clinical, and toxicological study findings; case reports; epidemiological data; and unpublished manufacturers' data. If a herbal product has an approved monograph, it can be marketed. In addition, there is a push for standardization of the European market for herbal supplements. European Directive 2004/24/EC, released in 2004 by the European Parliament and the Council of Europe, provides the basis of regulation of herbal supplements in the European market. This directive requires that herbal medicines should receive authorization by the national regulatory authorities of each European country and that these products must be safe. In Canada, the federal government implemented a policy in 2004 to regulate natural health products. However, naturopaths, traditional Chinese medicine practitioners, homeopaths, and herbalists are concerned that this policy may eventually affect their access to the products they need to practice effectively [3].

A. Dasgupta and K. Klein: Antioxidants in Food, Vitamins and Supplements. DOI: http://dx.doi.org/10.1016/B978-0-12-405872-9.00016-1

The popularity of herbal supplements is steadily increasing among the general population in the United States. It is estimate that approximately 20,000 herbal products are available in the US, and, in one survey, approximately one out of five adults reported using a herbal supplement within the past year. The 10 most commonly used herbal supplements are echinacea, ginseng, ginkgo biloba, garlic, St. John's wort, peppermint, ginger, soy, chamomile, and kava [4]. Although some herbal supplements are safe and may have antioxidant properties, other herbal supplements may have adverse health effects. Liver toxicity due to use of kava, comfrey, germander, and chaparral has been well documented. The most common toxic herbal products and their presumed indications of use (or misuse) are listed in Table 16.1. These herbal products must not be used at all. In addition, some herbal supplements, such as St. John's wort, can interact with many drugs, and such interactions result in treatment failure. Therefore, if a person is taking any drug to treat a chronic condition, it is important that he or she discuss and disclose the use of such herbal supplement to his

Table 16.1 Potentially Toxic and Toxic Herbs

Herbal Product	Presumably Used For/As	Toxicity
Aristolochic acid-containing herbs: wild ginger, snakeweed, birthroot, and contaminated Asian herbal weight-loss products	Weight loss	Renal failure
Comfrey	Repairing broken bone	Liver
Coltsfoot	Allergy, cold	Liver
Ephedra	Weight loss	Cardiovascular
Chan Su	Tonic for heart	Cardiovascular
Borage	Rheumatoid arthritis, hypertension	Liver, carcinogenic
Calamus	Depression	Carcinogenic
Chaparral	Cleansing tonic Arthritis	Liver, kidney Carcinogenic
Germander	Asthma, cough	Liver
Kava	Sedative, antidepressant	Liver
Mistletoe	Anxiety, lowering blood pressure	Liver
Lobelia (Indian tobacco, pokeweed)	Purgative, respiratory problems	Respiratory failure
Skullcap	Sedative, treating menstrual cramps	Liver
Wormwood	Digestive tonic Expelling worms from the intestine	Neurological Paralysis
Willow bark	Reducing fever, pain	Severe allergy, toxic to children

or her clinician or pharmacist to avoid any adverse drug–herb interaction or adverse health effect from such supplement.

16.2 ANTIOXIDANT HERBAL SUPPLEMENTS

Because plants are rich in antioxidants, it is assumed that many herbal supplements are also rich in antioxidants. Spices can also be considered as herbs, and most spices are rich sources of antioxidants. Unfortunately, relatively few herbs for which antioxidant properties have been clearly established have been described in the medical literature. In this chapter, only these herbs are discussed. Antioxidant herbs are listed in Table 16.2.

16.2.1 Astragalus

In the United States, astragalus, an immune stimulant, is taken to prevent the common cold and upper respiratory infections. Typically, roots of astragalus are used for preparing extract or brewing tea. Astragalus is also used in combination with other herbs, such as licorice, ginseng, and angelica roots. Research indicates that astragalus in animal models can stimulate immune

Table 16.2 Common Antioxidant Herbal Supplements

Antioxidant Herbal Supplement	Primary Indications for Use
Astragalus	Boosting immune function
Berry extracts	
Blueberry extract	Improving circulation
Bilberry extract	Improving circulation
Chasteberry extract	Treating menstrual problems
Cranberry extract	Preventing urinary tract infection
Elderberry extract	Treating cold and flu
Cat's claw	Preventing inflammation/osteoarthritis
Garlic extract	Lowering blood cholesterol
Ginseng	Tonic for good health/lowering blood sugar
Ginkgo biloba	Sharpening memory
Grape seed extract	Reducing risk of blocked coronary arteries, lowering cholesterol
Lycopene supplement	Cardiovascular disease, preventing cancer
Milk thistle	Treating liver problems
St. John's wort	Herbal antidepressant
Soy supplement	Weight reduction, lowering cholesterol, and relief from symptoms of menopause
Turmeric	Treating digestive problems, arthritis; also has other health benefits

function and also rejuvenate depressed immune function [5]. Astragalus also has anticancer properties. In one study, the authors demonstrated an antitumor property of astragalus against human colon cancer cells and human hepatocellular carcinoma cells [6]. Astragalus can also restore impaired T cell function in cancer patients. In China, astragalus has long been used to treat cardiovascular disease. It is also capable of reducing total cholesterol and low-density lipoprotein (LDL) cholesterol in blood while increasing high-density lipoprotein (HDL) cholesterol. It has been demonstrated that total flavonoids of astragalus are capable of scavenging free radicals, including superoxide and hydroxyl radicals. Therefore, astragalus is rich in antioxidants, and it has been postulated that antioxidant properties explain why astragalus is effective in preventing or treating cardiovascular diseases [7]. Astragalus is also generally considered a safe supplement.

16.2.2 Bilberry and Other Berry Extracts

Historically, bilberry fruits were used to treat diarrhea and scurvy. Today, bilberry fruits are taken to treat diarrhea, menstrual cramps, and circulatory problems. Edible berries such as bilberry, blueberry, cranberry, elderberry, raspberry, and strawberry are good sources of natural anthocyanin antioxidants and other antioxidants. Acai berry, which is full of antioxidants and popular in South America, is now available in the United States. Acai berry extract is also used as a herbal supplement. However, six berry extracts (of wild blueberry, bilberry, cranberry, elderberry, strawberry, and raspberry seeds) are commonly used as herbal supplements. These berry extracts have several benefits that are mostly related to their antioxidant properties, and such extracts may reduce the risk of cancer, cardiovascular disease, and diabetes. In addition, these extracts may help improve night vision and also may have a neuroprotective effects in the elderly [8].

Extracts of the chasteberry of the chaste tree (*Vitex agnus-castus*) are used for treating premenstrual symptoms, irregularities in the menstrual cycle, and to stimulate breast milk production. Studies have shown that chasteberry extracts are beneficial in treating the most common premenstrual symptoms and premenstrual mastodynia. In mastalgia, women release more prolactin, and studies have shown that chasteberry extracts reduce such high prolactin levels in blood, thus providing relief [9]. Essential oil extracted from both berry and leaves was also found to be effective in relieving symptoms of menopause [10]. Chasteberry extract is safe but should be avoided during pregnancy and lactation [11].

Cranberry juice is widely used for symptomatic relief of urinary tract infection (UTI). Cranberry juice is also used to help reduce urine odor. It is also believed that cranberry juice can decrease the rate of formation of kidney stones. Cranberry powder in capsule form is also available as a dietary supplement. Jepson and Craig [12] reviewed data from nine clinical trials and concluded that cranberry products significantly reduced the incidence of UTI during the 12-month study period, particularly in women with recurrent infections. In another study, McMurdo *et al.* [13] concluded that the antibiotic trimethoprim had very little advantage compared to cranberry extract in the prevention of recurrent UTI in older women. Moreover, the antibiotic has more adverse effects than cranberry. Cranberry extract may also have activity against *Helicobacter pylori* (which is responsible for the majority of peptic ulcers), and it is also effective at lowering total cholesterol and LDL cholesterol in diabetic patients [14]. Cranberry juice is also rich in antioxidants. Basu *et al.* [15] reported that cranberry decreased lipid peroxidation products and increased antioxidant capacity of blood in women who ingested cranberry juice [two cups per day (480 mL) for 8 weeks].

16.2.3 Cat's Claw

Cat's claw has been used for many centuries in South American countries for the treatment of various diseases. It has been used especially in traditional Peruvian medicine. It is indicated for boosting the immune system, treating viral infections such as herpes, and preventing pregnancy. It is also promoted for treating cancer and arthritis. Clinical research indicates that cat's claw is a potent antioxidant and may inhibit tumor narcosis factor-α and prostaglandin synthesis, thus producing a beneficial effect in treating osteoarthritis by preventing inflammation [16]. Goncalves *et al.* [17] demonstrated the capability of cat's claw preparation in scavenging superoxide, peroxyl, and hydroxyl radicals as well as other oxidant species, including hydrogen peroxide and hypochlorous acid. Caffeic acid, a very potent antioxidant that is responsible for the antioxidant effect of coffee, is also present in cat's claw [17]. Side effects such as headache, dizziness, and vomiting have been reported from the use of cat's claw at recommended dosage. Women who are pregnant or trying to become pregnant should avoid cat's claw because it has been traditionally used for preventing pregnancy and inducing abortion.

16.2.4 Garlic Extract

Garlic is promoted to lower cholesterol, thus preventing heart attack and stroke. It is also indicated for preventing cancer. Garlic contains various sulfur-containing compounds, which are derived from allicin. Allicin is formed by the action of an enzyme when garlic is chopped. Allicin is responsible for the characteristic odor of garlic and is mostly destroyed during cooking. Clinical trials to determine the lipid-lowering effects of garlic produced mixed results. Although some studies did not find any statistically significant effects of garlic in lowering serum cholesterol and triglycerides, other studies showed beneficial lipid-lowering effects of garlic supplement. Three clinical trials found 6.1–11.5% reductions in cholesterol in garlic-treated patients, which were mainly due to reductions in LDL cholesterol. Triglycerides levels were also decreased after garlic treatment [18].

Antioxidant capacity of garlic extract has been clearly demonstrated by several investigators. Lu *et al.* [19] studied the antioxidant properties of various garlic extracts. Garlic was obtained from four different states (California, Washington, Oregon, and New York), and elephant garlic was also studied. The authors observed similar antioxidant capacities of various garlic extracts, but elephant garlic produced lower antioxidant capacity compared to the other extracts. Park *et al.* [20] observed that aged garlic extract had superior antioxidant properties due to higher total phenolic antioxidant compared to raw garlic extract or heated garlic extract. Garlic can reduce the risk of dementia in elderly people, probably due to its antioxidant properties. Avci *et al.* [21] observed that ingestion of garlic on a regular basis by the elderly significantly lowered plasma and erythrocyte concentrations of malondialdehyde and simultaneously increased the activities of the antioxidant enzymes superoxide dismutase and glutathione peroxidase in erythrocytes. Garlic also lowered LDL cholesterol levels. The authors concluded that garlic extract can decrease oxidative stress. The usual dose is a 300-mg supplement taken two or three times per day.

Chopped garlic and oil mixes left at room temperature can result in fatal botulism food poisoning, according to the FDA. *Clostridium botulinum* bacteria are dispersed throughout the environment but are not dangerous in aerobic conditions. The spores, however, produce a deadly toxin in anaerobic, low-acid conditions, causing botulism. Garlic in oil produces this environment. In February 1989, three cases of botulism were reported in persons who consumed chopped garlic in oil that was used as a spread for garlic bread. Testing of leftover garlic in oil showed a high concentration of the organism and toxin, and the garlic in oil mixture had low acidity (pH 5.7). Refrigeration of chopped garlic in oil is important to avoid such food poisoning [22].

16.2.5 Ginseng

Ginseng is a widely used herbal product in China, other Asian countries, and also the United States, where it is one of the most common herbal supplements taken by the general population. For thousands of years, people in China have used ginseng as a tonic to rescue dying patients. *Ginseng* means "essence of man" in Chinese. Although the term *ginseng* is used loosely to define many different herbs, *Asian ginseng*, which is the most common form of ginseng sold in the United States, is prepared from the root of the plant *Panax ginseng*, which grows naturally in

Manchuria. American ginseng is prepared from *Panax quinquefolius*, whereas Siberian ginseng is from the root of *Eleutherococcus senticosus*. The common preparation of ginseng is ginseng root. In the US market, ginseng is sold as liquid extract, tablets, or capsules. Dried ginseng root is also available in health food stores. Asian ginseng contains saponins known as ginsenosides, whereas Siberian ginseng is devoid of any ginsenosides. In general, the health benefits of various types of ginseng include their antioxidant, anticancer, anti-inflammatory, and hypoglycemic effects. Ginseng may be effective in treating insomnia and may improve memory [23]. It has been shown that American ginseng can prevent complication of diabetic nephropathy through a combination of mechanisms, including its antioxidant and hypoglycemic action [24]. Interestingly, the antioxidant capacity of American ginseng can be increased significantly by heat processing [25]. In addition, Bae *et al.* [26] showed that autoclaving at 130°C significantly increased the antioxidant activity of ginseng, as evidenced by increased DPPH radical (1,1-diphenyl-2-picrylhydrazyl) and ABTS radical [2,2′-aziono-bis(3-ethylbenzthiazolin)-6-sulfonic acid] scavenging activities. Liu [27] reviewed the chemical structures of different types of ginseng and noted that the active components of ginseng include ginsenosides, which are responsible for anticarcinogenic, immunomodulatory, anti-inflammatory, antiallergic, antiatherosclerotic, antidiabetes, and antihypertensive activities. Ginsenosides also have antioxidant properties. Lu *et al.* [28] reported that ginsenoside Rb1 can directly scavenge hydroxyl radicals and hypochlorous acid and protects plasmid DNA from damage induced by hydroxyl radicals.

16.2.6 Ginkgo Biloba

Ginkgo biloba is prepared from dried leaves of the ginkgo tree by organic extraction (acetone/water) and sold as a dietary supplement to improve blood flow in the brain and peripheral circulation, to sharpen memory, and to treat vertigo. Ginkgo biloba is one of the bestselling herbs in the United States, and most people take it to prevent age-related dementia and Alzheimer's disease. Although older studies showed the beneficial effect of ginkgo biloba in improving memory and treating dementia due to Alzheimer's disease, recent clinical trials have shown no beneficial effect of ginkgo. A clinical trial funded by the National Center for Complementary and Alternative Medicine and conducted between 2000 and 2008 at five academic medical centers in the United States involving more than 3000 volunteers age 75 years or older did not show any beneficial effects of ginkgo in lowering the overall incidence of dementia and Alzheimer's disease in these participants. In this study, the volunteers daily took 120 mg of ginkgo twice a day, and they were followed for an average of 6 years [29]. However,

Brondino *et al.* [30] performed a systematic review and meta-analysis of eight clinical trials and concluded that gingko biloba can be used in patients with dementia because it can improve cognitive function. In addition, ginkgo biloba can be used as adjunct therapy for schizophrenic patients because it reduces symptoms in these patients. Mechanisms of action of ginkgo biloba include increasing cerebral blood flow, antioxidant and anti-inflammatory effects, and antiplatelet effects attributed to flavone and terpene lactones. However, ginkgo biloba may interact with monoamine oxidase inhibitors, alprazolam, haloperidol, warfarin, and nifedipine [31].

One commonly reported adverse effect of ginkgo is bleeding. Therefore, concurrent use of ginkgo and aspirin as well as other nonsteroidal anti-inflammatory drugs (NSAIDs) may cause bleeding because ginkgolide B, present in ginkgo extract, is a potent inhibitor of platelet-activating factor. Patients receiving anticoagulant therapy with warfarin should also avoid ginko. Intracranial hemorrhage may occur from chronic use of ginkgo biloba [32].

Flavonoids in ginkgo biloba extract are responsible for the antioxidant capacity of ginkgo biloba [33]. Boveris *et al.* [34] reported that supplementation of ginkgo biloba in rats resulted in a 39.7% reduction in lipid peroxidation as assessed by the generation of lipid radicals measured by electron paramagnetic resonance spectroscopy using spin trap. In addition, a 30% reduction was observed in lipid peroxidation products measured as thiobarbituric acid reactive substances. The authors concluded that ginkgo biloba extracts were able to protect membranes from oxidative damage by scavenging lipid radicals. *In vitro* evidence also indicates that ginkgo biloba extract can scavenge nitric oxide, superoxide, hydroxyl, and oxoferryl radicals. In addition, ginkgo biloba can scavenge peroxyl radicals, which are mainly involved in the propagation step of lipid peroxidation [35].

16.2.7 Grape Seed Extract

Grape seed extract is indicated for reducing the risk of atherosclerosis as well as for reducing complications related to diabetes, lowering cholesterol, preventing cancer, and wound healing. Bagchi *et al.* [36] reported superior antioxidant effect of grape seed extract containing proanthocyanidins compared to other known antioxidants, including vitamin C, vitamin E, and β-carotene. A human clinical trial using hypercholesterolemic subjects showed that grape seed extract supplementation significantly reduced oxidized LDL. Bagchi *et al.* concluded that grape seed extract may serve as a potential therapeutic tool in promoting cardiovascular health [36]. Electron spin resonance spectroscopy data also showed that grape seed extract containing proanthocyanidins strongly scavenged hydroxyl and superoxide radicals [37]. Grape seed extract is a relatively safe and effective supplement.

16.2.8 Lycopene Supplement

Lycopene is a naturally occurring compound that contributes to the red color of fruits and vegetables and is a carotenoid antioxidant. Lycopene is found in high amounts in tomatoes but is also present in watermelons, pink grapefruits, apricots, and pink guavas. One cup (240 mL) of tomato juice provides approximately 23 mg of lycopene. Processing raw tomatoes using heat (e.g., in the making of tomato juice, tomato paste, or ketchup) transforms natural lycopene to a form that is easier to utilize by the human body. People take lycopene to prevent cardiovascular disease and cancer of the prostate, breast, lung, bladder, ovaries, colon, as well as pancreas. Lycopene and lycopene supplementation can lower the risk of coronary artery disease, probably due to its excellent antioxidant properties. When healthy people ingested lycopene in the form of tomato juice, tomato sauce, and as soft-gel capsule supplement for 1 week, a significantly lower concentration of oxidized LDL was observed in these subjects compared to controls. A multicenter 10-country European trial also showed that lycopene can reduce the risk of myocardial infarction. Lycopene can also lower cholesterol levels by inhibiting the levels of enzyme essential for cholesterol synthesis by the human body [38]. However, the protective effect of lycopene against various types of cancer is controversial because various studies have obtained conflicting results.

16.2.9 Milk Thistle

For more than 2000 years, seeds of milk thistle have been used to treat liver disease. In the United States, milk thistle is mainly used to treat viral infection and cirrhosis of the liver. The active component of milk thistle, silymarin (composed of three isomers—silybin, silydianin, and silychristin), protects liver from a variety of liver toxins in animal models. Clinical trials indicate that milk thistle may have anticancer and antidiabetic activities and may also lower serum cholesterol [39]. Abenavoli *et al.* [40] reported that silybin, which makes up 50–70% of silymarin, has the highest magnitude of biological activity and protects liver from various toxins. Silymarin acts as an antioxidant, thus reducing liver injury induced by acetaminophen overdose, alcohol overdose, and other known liver toxins, such as carbon tetrachloride. Silymarin may be useful as adjunct therapy in treating alcoholic liver disease, chronic viral hepatitis, and liver damage induced by various toxins [40]. In a study of 55 patients with hepatitis C infection, Kalantari *et al.* [41] showed that use of silymarin extract derived from milk thistle for 24 weeks reduced the mean concentration of the liver enzyme alanine aminotransferase from 108.7 to 70.3 U/L. In addition, the mean concentration of aspartate aminotransferase in these patients was reduced from 99.4 to 59.7 U/L. After treatment with silymarin, 9 patients were found to be negative for hepatitis C viral RNA. Quality of life was also improved in these patients. Silymarin has metabolic and cell-regulating effects, including scavenging oxygen radicals, inhibition of the 5-lipooxygenease pathway, carrier-mediated regulation of cell permeability, and action on DNA expression via suppression of NF-κB [42]. Interestingly, metabolites of silybin also retained antioxidant properties [43].

16.2.10 St. John's Wort

St. John's wort is prepared from a perennial aromatic shrub with bright yellow flowers that bloom from June to September. The flowers are believed to be most abundant and brightest on approximately June 24, the day traditionally believed to be the birthday of John the Baptist. Therefore, the name St. John's wort became popular for this herbal product. The flowering tops or leaves or stem are used to make liquid extracts or dried powder (capsules). St. John's wort is one of the best-selling herbal supplements in the United States and is used mostly for treating depression, anxiety, or sleep disorder. Many chemicals have been isolated from St. John's wort, but hypericin and hyperforin are mostly responsible for its antidepressant effects. Most commercially available St. John's wort preparations are standardized to contain 0.3% hypericin. Linde *et al.* [44] analyzed data from 23 clinical trials involving St. John's wort (1757 outpatients) and concluded that it is effective in treating mild to moderate depression. In addition, it was suggested that St. John's wort has equivalent efficacy to low-dose treatment with antidepressant medications (amitriptyline or imipramine). Interestingly, St. John's wort has fewer side effects than these antidepressant medications. It is fairly safe, but it may induce photosensitivity, especially in fair-skinned people after prolonged exposure to sunlight or ultraviolet light in tanning salons.

Because of the phenolic antioxidant compounds present in St. John's wort, it is an excellent herbal antioxidant. These compounds are capable of scavenging superoxide radicals and other free radicals, thus preventing lipid peroxidation [45]. In addition, hyperforin, which is partially responsible for the antidepressant property of St. John's wort, has powerful antioxidant activity due to its superior capability to scavenge free radicals. Hyperforin also has anti-inflammatory, antibacterial, and antitumor properties [46].

The major problem with taking St. John's wort supplement is its reported interactions with many Western drugs. Hyperforin, an active component of St. John's wort, also induces many cytochrome P450 liver isoenzymes, including CYP3A4, CYP2E1, and CYP2C9. Ott *et al.* [47] showed that St. John's wort extract containing hyperforin, hypericin, and quercetin decreased P-glycoprotein transport activity in a dose- and time-dependent manner. St John's wort extract and hyperforin directly inhibit P-glycoprotein activity, whereas hypericin and quercetin modulate transporter function through protein kinase C. Therefore, St. John's wort can interact with drugs that are metabolized by liver cytochrome P450 enzymes as well as drugs that are substrate for P-glycoprotein but are not metabolized by liver enzymes, causing treatment failure due to subtherapeutic concentrations of these drugs. Clinically significant interactions between St. John's wort and immunosuppressants, anticoagulants, statins, calcium channel blockers, β-blockers, antianginal drugs, cardiac inotropic drugs, antiretroviral drugs, anticancer drugs, benzodiazepines, antidepressants, antiepileptic drugs, and oral contraceptives have been described. Whereas most drug–St. John's wort interactions involve pharmacokinetic interactions that may cause treatment failure, pharmacodynamic interaction of St. John's wort with selective serotonin reuptake inhibitors such as paroxetine, sertraline, venlafaxine, and nefazodone causing central serotonin syndrome has also been reported [48]. Patients receiving warfarin therapy, transplant recipients, and patients being treated for AIDS with HAART (highly active antiretroviral therapy) should not take St. John's wort due to risk of treatment failure. Major interactions between various drugs and St. John's wort are listed in Table 16.3.

16.2.11 Soy Supplement

Soy is available as a dietary supplement in the form of tablets and capsules, and it may contain soy proteins, isoflavone, or both. Soybeans can be cooked and eaten or can be used to make soy milk, tofu, or other food products. Soy supplement is indicated for lowering cholesterol, weight loss, and relieving symptoms of menopause, such as hot flashes and osteoporosis. Soy protein is considered a complete protein because it contains ample amounts of all essential amino acids and several other nutrients. Soy also contains isoflavone, a group of compounds with many beneficial effects due to their antioxidant properties. In obese people, soy supplement reduces body weight and fat mass in addition to reducing cholesterol and triglycerides [49]. Soy supplement reduces cholesterol by a novel mechanism by activating receptors in the liver that remove LDL cholesterol from blood [50]. Another study indicated that soy consumption reduces the incidence of hot flashes in menopausal women [51]. Studies also indicate that soy diets may reduce the risk of breast cancer. The urinary level of isoprostane, a measure of lipid peroxidation, was reduced in eight women who consumed a diet containing soy milk for 4 months, indicating that soy isoflavones can reduce lipid peroxidation. Interestingly, the antioxidant effect of soy was more significant in older women [52]. Genistein, an isoflavone in soy, can protect both LDL and HDL from lipid peroxidation [53].

16.2.12 Turmeric

Turmeric is used as a spice in many Asian countries, and almost 1 billion people throughout the world use it daily as a spice. Turmeric is an important component of traditional Chinese medicine and Indian Ayurvedic medicine. Turmeric's finger-like underground rhizomes are dried and used as a spice or taken as a powder in capsule form. Liquid extract of turmeric is also commercially available. Turmeric can also be used as paste for application on the skin. The most active component of turmeric is curcumin. Scientific research

Table 16.3 Common Pharmacokinetic and Pharmacodynamic Drug Interactions with St. John's Wort

Medication Class	Drug Names
***Pharmacokinetic interaction*[a]**	
Immunosuppressant agents	Cyclosporine, tacrolimus
Anticoagulant agent	Warfarin
Antiretroviral agents	Indinavir, saquinavir, atazanavir, lamivudine, nevirapine
Antianginal drug	Ivabradine
Calcium channel blockers	Nifedipine, verapamil
β-Blocker	Talinolol
Cardiac ionotropic drug	Digoxin
Anti-inflammatory agent	Ibuprofen
Benzodiazepines	Alprazolam, midazolam, quazepam
Antiepileptics	Mephenytoin, phenobarbital
Hypoglycemic agents	Tolbutamide, gliclazide
Anticancer agents	Imatinib, irinotecan
Antimicrobial agents (antibiotic/antifungal)	Erythromycin, voriconazole
Antiasthmatic agent	Theophylline
Proton pump inhibitor	Omeprazole
Statins	Simvastatin, atorvastatin
Oral contraceptives	Ethinyl estradiol and other formula
Antiparkinson agent	Levodopa
Antidepressant	Amitriptyline
Synthetic opioids	Methadone, fentanyl
***Pharmacodynamic interaction*[b]**	
Selective serotonin reuptake inhibitors	Paroxetine, sertraline, venlafaxine, nefazodone
Anxiolytic	Buspirone
No interaction	
Immunosuppressant	Mycophenolic acid
Anticonvulsant	Carbamazepine[c]

[a]Pharmacokinetic interaction usually reduces serum drug level, causing treatment failure.
[b]Pharmacodynamic interaction usually leads to symptoms of serotonin syndrome.
[c]No interaction between carbamazepine and St. John's wort is surprising because carbamazepine is metabolized by cytochrome P450 liver enzyme.

has shown that turmeric has antibacterial, antiviral, antifungal, antioxidant, and anticancer activities and also has a potential to reduce the risk of various malignant disease, arthritis, Alzheimer's disease, and other chronic illness including rheumatoid arthritis [54]. Studies have shown that curcumin, the major phenolic compound present in turmeric, is particularly effective at reducing the risk of colon, skin, oral, and intestinal cancer and has excellent antioxidant and anti-inflammatory properties. Curcumin is very effective at scavenging peroxyl radicals [55]. Bhullar *et al.* [56] showed that curcumin and many of its carbocyclic analogs have excellent free radical scavenging properties, but curcumin is probably the strongest antioxidant present in turmeric.

16.3 OTHER ANTIOXIDANT SUPPLEMENTS

In addition to antioxidant herbs, many other dietary supplements that are sold in many health food stores have significant antioxidant activities as evidenced by published studies in the medical literature. Here, however, only dietary supplements for which there is substantial scientific evidence of their antioxidant properties are discussed. These supplements are listed in Table 16.4.

16.3.1 Coenzyme Q10

Coenzyme Q10 (2,3-dimethoxy-5-methyl-6-decaprenyl benzoquinone) is a naturally occurring fat-soluble quinone also known as ubiquinone. Coenzyme Q10 is a fat-soluble vitamin-like compound that is usually found in many supplements as a single ingredient or in combination with other ingredients. Coenzyme Q10 is vital for proper transfer of electrons within the mitochondrial oxidative respiratory chain, whose main function is to produce adenosine triphosphate (ATP). In addition, coenzyme Q10 is an antioxidant with excellent free radical scavenging capacity and also an indirect stabilizer of calcium

Table 16.4 Common Nonherbal Antioxidant Supplements

Antioxidant Supplement	Primary Indications for Use
Coenzyme Q10	Heart disease, heart failure, and increasing energy
Glutathione	Antioxidant supplement
Glucosamine	Treating osteoarthritis
α-Lipoic acid	Preventing complications of diabetes
Fish oil supplement (omega-3 fatty acids)	Preventing cardiovascular disease
Melatonin	Correcting sleep–wake cycle

channels to decrease calcium load. In Japan, coenzyme Q10 is approved for treating congestive heart failure. Clinical studies indicate that coenzyme Q10 is beneficial in treating heart failure and high blood pressure but less effective in treating ischemic heart disease. In addition, coenzyme Q10 is promising as an adjunct therapy for neurodegenerative disease such as Parkinson's disease. Coenzyme Q10 may reduce oxidative stress in patients with diabetes [57]. It is an excellent antioxidant, and at a dose of 150 mg per day, it can significantly reduce oxidative stress and increase activities of antioxidant enzymes in patients with coronary artery disease [58]. However, it can interact with many drugs, such as warfarin, cholesterol-lowering drugs, diabeties medications, and antidepressants, and it is important for a person to discuss with his or her physician the use of coenzyme Q10 in order to avoid adverse drug interactions with it. Secondary deficiency of coenzyme Q10 is linked to hydroxymethylglutaryl coenzyme A (HMG-CoA) reductase inhibitor (statin) therapy [59].

16.3.2 Glutathione

Glutathione is a tripeptide found naturally in humans. It is a powerful antioxidant due to its superior free radical scavenging capability. Although manufacturers of oral glutathione supplement claim superior antioxidant property compared to that of endogenous glutathione, Allen and Bradley [60] observed no additional protective effect of oral glutathione supplement against oxidative stress in healthy volunteers. When 40 healthy volunteers took oral glutathione supplement (500 mg twice daily) for 4 weeks, no change in markers of oxidative stress was observed.

16.3.3 Glucosamine

Glucosamine is found in cartilage, and glucosamine for use in a dietary supplement is either prepared synthetically or derived from skeleton of marine animals. Chondroitin is prepared from shark or bovine cartilage. Glucosamine alone or in combination with chondroitin is used in treating osteoarthritis. Osteoarthritis affects joints mostly in the hands, neck, lower back, knees, and hips. Clinical trials have indicated that glucosamine is effective at relieving pain from osteoarthritis, especially knee pain, and can also slow the progression of disease. Glucosamine, which is often sold as glucosamine sulfate, is an excellent antioxidant capable of scavenging free radicals, and it has been postulated that the protective effect of glucosamine against osteoarthritis is probably related to its antioxidant property [61]. Fang *et al.* [62] showed that glucosamine is a scavenger of lipid peroxidation products such as malondialdehyde; thus, it has a novel anticarbonylation function that may provide insight into how it is effective in protecting proteins against carbonyl stress.

16.3.4 α-Lipoic Acid

α-Lipoic acid (1,2-dithiolane-3-pentaenoic acid) is a natural product synthesized enzymatically in animals, humans, and plants in mitochondria from octanoic acid and cysteine. α-Lipoic acid is found in small amounts in spinach, broccoli, peas, Brussels sprouts, and potatoes. It acts as a cofactor for pyruvate dehydrogenase and α-ketoglutarate dehydrogenase activity, and it is also needed for oxidative decarboxylation of pyruvate to acetyl-CoA, an important step bridging glycolysis with the citric acid cycle. α-Lipoic acid exists as R and S optical isomers, where R isomer is bound to protein and acts as an essential cofactor in biochemical pathways. In cells containing mitochondria, α-lipoic acid is reduced by NADH-dependent reaction to dihydrolipoic acid, but unlike antioxidant glutathione, both α-lipoic acid and its reduced form are powerful antioxidants quenching reactive oxygen species. In addition, α-lipoic acid can reduce oxidized forms of other antioxidants, such as vitamin C, vitamin E, and glutathione (these antioxidants are effective in their reduced form and become oxidized when they neutralize free radicals), thus acting as an excellent antioxidant. α-Lipoic acid can also chelate metal ions, repair oxidized proteins, regulate gene transcription, and inhibit the activation of NK-κB. Experimental and clinical studies have shown that α-lipoic acid has beneficial effects in metabolic syndrome, diabetes and its complications, radiation damage, ischemia–reperfusion injury, and neurodegenerative diseases [63].

α-Lipoic acid is indicated for treating diabetes and nerve-related symptoms of diabetes, including burning, pain, and numbness in the legs and arms. High doses of α-lipoic acid are approved in Germany for the treatment of these symptoms. Some people use α-lipoic acid for memory loss, chronic fatigue syndrome, HIV/AIDS, cancer, liver disease, Lyme disease, glaucoma, and cardiovascular disease, although more research is needed to establish the usefulness of α-lipoic acid as an adjunct therapeutic agent in all these diseases. Nebbioso *et al.* [64] reported that α-lipoic acid can be used as adjunct therapy to prevent vision loss in diabetic patients because it prevents micro- and macrovascular damage through normalization of mitochondrial overproduction of reactive oxygen species and preserves pericyte coverage of retinal capillaries. Although 50- to 100-mg supplementation of α-lipoic acid may be adequate, dosage up to 600 mg per day has been reported in various clinical studies involving α-lipoic acid. Li *et al.* [65] studied 63 patients with confirmed acute coronary syndrome who were divided into two groups, with one group receiving intravenous injection of α-lipoic acid (600 mg per day) for 5 days and the other group receiving saline. The authors observed a significant decline in serum levels of 8-isoprostaglandin $F_{2\alpha}$, a marker of lipid peroxidation, in patients who received intravenous α-lipoic acid compared to patients who received saline. In addition, increased levels of aldehyde dehydrogenase 2 enzyme, which is involved in alcohol metabolism but also provides protection against oxidative stress, in sera of patients who received α-lipoic acid, were observed. The authors observed beneficial effects of α-lipoic acid in lowering oxidative stress within 24 h of intervention. The authors concluded that α-lipoic acid may ameliorate oxidative stress through upregulation of acetaldehyde dehydrogenase enzyme.

16.3.5 Fish Oil Supplements

Fish oil is a rich source of omega-3 polyunsaturated fatty acids, especially eicosapentaenoic (EPA; 20-carbon chain with five double bonds) and docosahexaenoic acid (DHA; 22-carbon chain with six double bonds). The American Heart Association recommends consumption of fish at least twice per week, which would provide approximately 430 mg of omega-3 fatty acids per day or 3 g per week. The World Health Organization recommends eating one or two servings of fish per week, and the European Food Safety Agency has proposed a daily intake of 250 mg of EPA and DHA (one or two servings of oily fish per week). Eating fish on a weekly basis or taking fish oil supplements effectively enriches lipids with omega-3 fatty acids. Increased fish or fish oil consumption has been associated with reduced risk of mortality and sudden death from cardiovascular disease by means of membrane stabilization in cardiac myocytes, inhibition of platelet aggregation, decreased systolic and diastolic blood pressure, and reduced inflammatory response of the endothelium. Fish oil reduces thromboxane B_2 production but increases plasma renin activity. Regular consumption of fish or fish oil can reduce the progression of atherosclerosis, thus reducing the risk of myocardial infarction. Fish oil can also be beneficial to patients with heart failure. In addition, fish oil or fish consumption can improve glycemic control and may prevent the development of inflammatory disease, thus reducing NSAID use in patients with rheumatoid arthritis. However, environmental pollution of water with organic mercury compounds is a major concern with regard to consuming large amounts of fish. The concentration of methylmercury is higher in predatory fish such as shark, tuna, swordfish, and orange roughy compared to nonpredator fish such as salmon, sardine, and flounder [66]. Therefore, pregnant women should consult their physicians regarding fish consumption during pregnancy. Kar and Webel [67] noted that omega-3 fatty acids decrease the risk of cardiovascular disease and also reduce fatal and nonfatal myocardial infarction, sudden death, stroke, and coronary artery disease as well as all-cause mortality. In addition, omega-3 fatty acid supplementation may be useful in patients with depressive mood disorder [68].

Fish oil supplementation may attenuate oxidative stress by increasing antioxidant defense. Using a rat model, Songur *et al.* [69] showed that malondialdehyde and nitric oxide levels in tissue of hypothalamus were significantly reduced after rats were fed fish oil and concluded that supplementation with omega-3 fatty acid protects membranes from lipid peroxidation and enhances antioxidant defense. Kesavulu *et al.* [70] reported that serum lipid peroxide levels (measured by malondialdehyde levels) were significantly decreased after 2-month supplementation with fish oil. In addition, fish oil supplementation had beneficial effects on serum triglycerides, HDL cholesterol, and antioxidant enzymes, which may be related to the decreased rate of vascular complications in patients with diabetes. Using rotenone (a neurotoxin that can induce oxidative stress) and a rat model, Denny Joseph and Muralidhara [71] observed that fish oil supplementation significantly attenuated rotenone-induced mitochondrial dysfunction. The authors concluded that protection by fish oil was related to its capacity to reduce glutathione levels, thus increasing the antioxidant mechanism, which could offset protein oxidation. In addition, fish oil may have specific modulatory effects on brain mitochondria of rats.

16.3.6 Melatonin

Melatonin is a hormone secreted by the pineal gland of the brain, which regulates the body's 24-h circadian rhythm. Melatonin aids in the sleep process. Darkness stimulates production of melatonin, whereas light suppresses its production. Melatonin also controls the timing of the release of female reproductive hormones, thus playing a role in regulating the menstrual cycle. Although results are mixed, some studies have shown that, if taken for a short period of time, melatonin supplement helps people to fall sleep. Melatonin is remarkably effective at preventing or reducing jet-lag, especially for adult travelers flying across five or more time zones, particularly flying in the eastern direction. It may also be helpful for those traveling across two to four time zones [72]. Melatonin is an excellent antioxidant that is capable of scavenging both oxygen and nitrogen free radicals. Several metabolites of melatonin are also excellent antioxidants. In addition to treating abnormal sleep—wake rhythm, melatonin may be effective for improving immune function as well as in cancer therapy with other drugs [73]. Bai *et al.* [74] reported that, in addition to working as a direct free radical scavenger, melatonin can elicit cellular signaling pathways that are protective against oxidative stress-induced cataracts.

16.4 CONCLUSION

Several herbal supplements are excellent antioxidants. In addition, nonherbal dietary supplements such as coenzyme Q10, α-lipoic acid, and melatonin are also excellent antioxidants. However, healthy individuals may not need these supplements to boost their antioxidant defense provided they eat a balanced diet with generous servings of fruits, vegetables, lean meat, and fish (once or twice per week). Individuals with chronic diseases such as cardiovascular disease, diabetes, and other illness may benefit from these supplements, but such supplements must be taken only under medical supervision.

REFERENCES

[1] Bent S, Ko R. Commonly used herbal medicines in the United States: a review. Am J Med 2004;116:478–85.

[2] Mills E, Singh R, Ross C, Ernst E, et al. Impact of federal safety advisories on health food store advice. J Gen Intern Med 2004;19:269–72.

[3] Moss K, Boon H, Ballantyne P, Kachan N. New Canadian natural health product regulations: a qualitative study on how CAM practitioners perceive they will be impacted. BMC Comp Altern Med 2006;6:18.

[4] Bent S. Herbal medicine in the United States: review of efficacy, safety and regulation. J Gen Intern Med 2008;23:854–9.

[5] Cho WC, Leung KN. In vitro and in vivo immunomodulating and immunorestorative effects of Astragalus membranaceus. J Enthopharmacol 2007;113:132–41.

[6] Auyeung KK, Law PC, Ko JK. Astragalus saponins induce apoptosis via an ERK-independent NF-κB signaling pathway in human hepatocellular HepG2 cell line. Int J Mol Med 2009;23:189–96.

[7] Wang D, Zhuang Y, Tian Y, Thomas GN, et al. Study of the effects of total flavonoids of Astragalus on atherosclerosis formation and potential mechanism. Oxid Med Cell Longev 2012;2012:282383.

[8] Zafra-Stone S, Yasmin T, Bagchi M, Chatterjee A, et al. Berry anthocyanins as novel antioxidants in human health and disease prevention. Mol Nutr Food Res 2007;51:675–83.

[9] Wuttke W, Jarry H, Christoffel V, Spengler B, et al. Chase tree (Vitex agnus-castus): pharmacology and clinical indications. Phytomedicine 2003;10:348–57.

[10] Lucks BC, Sorensen J, Veal L. Vitex agnus-castus essential oil and menopausal balance: a self-care survey. Complement Ther Nurs Midwifery 2002;8:148–54.

[11] Daniele C, Thompson Coon J, Pittler MH, Ernst E. Vitex angus-cestus: a systematic review of adverse events. Drug Saf 2005;28:319–32.

[12] Jepson RG, Craig JC. A systematic review of the evidence for cranberries and blueberries in UTI prevention. Mol Med Food Res 2007;51:738–45.

[13] McMurdo ME, Argo I, Phillips G, Daly F, et al. Cranberry or trimethoprim for the prevention of recurrent urinary tract infection? A randomized controlled clinical trial in older women. J Antimicrob Chemother 2009;63:389–95.

[14] Lee IT, Chan YC, Lin CW, Lee WJ, et al. Effect of cranberry extract on lipid profiles in subjects with type 2 diabetes. Diabetes Med 2008;25:1473–7.

[15] Basu A, Betts NM, Ortiz J, Simmons B, et al. Low-energy cranberry juice decreases lipid peroxidation and increases plasma antioxidant capacity in women with metabolic syndrome. Nutr Res 2011;31:190–6.

[16] Hardin SR. Cat's claw: an Amazonian vine decreases inflammation in osteoarthritis. Complement Ther Clin Pract 2007;13:25–8.

[17] Goncalves C, Dinis T, Batista MT. Antioxidant properties of proanthocyanidins of Uncaria tomentosa bark decoction: a mechanism for anti-inflammatory activity. Phytochemistry 2005;66:89–98.

[18] O'Hara M, Kiefer D, Farrell K, Kemper K. A review of 12 commonly used medicinal herbs. Arch Fam Med 1998;7:523–36.

[19] Lu X, Rodd CF, Powers JR, Aston E, et al. Determination of total phenolic content and antioxidant capacity of garlic (Allium sativum) and elephant garlic (Allium ampeloprasum) by attenuated total reflectance Fourier transform infrared spectroscopy. J Agric Food Chem 2011;59:5215–21.

[20] Park JH, Park YH, Park E. Antioxidant and antigenotoxic effects of garlic (Allium sativum L.) prepared by different processing methods. Plant Foods Hum Nutr 2009;64:244–9.

[21] Avci A, Atli T, Erguder IB, Varli M, et al. Effects of garlic consumption on plasma and erythrocyte antioxidant parameters in elderly subjects. Gerontology 2008;54:173–6.

[22] Morse DL, Pickard LK, Guzewich JJ, Devine BD, et al. Garlic-in-oil associated botulism: episode leads to product modification. Am J Public Health 1990;80:1372–3.

[23] Kiefer D, Pantuso T. Panax ginseng. Am Fam Physician 2002;68:1539–42.

[24] Sen S, Chen S, Feng B, Wu Y, et al. Preventive effects of North American ginseng (Panax quinquefolius) on diabetic nephropathy. Phytomedicine 2012;19:494–505.

[25] Kim KT, Yoo KM, Lee JW, Hwang IK, et al. Protective effects of steamed American ginseng (Panax quinquefolius L.) on V7904 cells induced by oxidative stress. J Ethnopharmacol 2007;111:443–50.

[26] Bae HM, Kim SS, Cho CW, Yang DC, et al. Antioxidant activities of ginseng seeds treated by autoclaving. J Ginseng Res 2012;36:411–17.

[27] Liu ZQ. Chemical insights into ginseng as a resource for natural antioxidants. Chem Rev 2012;112:3329–55.

[28] Lu JM, Weakley SM, Yang Z, Hu M, et al. Ginsenoside Rb 1 directly scavenges hydroxyl radical and hypochlorous acid. Curr Pharm Des 2012;18:6339–47.

[29] DeKosky ST, Williamson JD, Fitzpatrick AL, Kronmal RA, et al. Ginkgo biloba for prevention of dementia: a randomized controlled trial. JAMA 2008;300:2253–62.

[30] Brondino N, De Silvestri A, Re S, Lanati N, et al. A systematic review and meta-analysis of ginkgo biloba in neuropsychiatric disorders: from ancient tradition to modern-day medicine. Evid Based Complement Altern Med 2013;2013:915691.

[31] Diamond BJ, Bailey MR. Ginkgo biloba: indications, mechanism and safety. Psychiatr Clin North Am 2013;36:73–83.

[32] Pedroso JL, Henriques Aquino CC, Escorcio Benzerra M, Baiense RF. Ginkgo biloba and cerebral bleeding: a case report. Neurology 2011;17:89–90.

[33] Kaur P, Chaudhury A, Singh B, Gopichand. An efficient microwave-assisted extraction of phenolic compounds and antioxidant potential of ginkgo biloba. Nat Prod Commun 2012;7:203–6.

[34] Boveris AD, Galleano M, Puntarulo S. In vivo supplementation with ginkgo biloba protects membranes against lipid peroxidation. Phytother Res 2007;21:735–40.

[35] Maitra I, Marcocci L, Droy-Lefaix MT, Packer L. Peroxy radical scavenging activity of ginkgo biloba extract EGb 761. Biochem Pharmacol 1995;49:1649–55.

[36] Bagchi D, Sen CK, Ray SD, Das DK, et al. Molecular mechanisms of cardioprotection by a novel grape seed proanthocyanidin extract. Mutat Res 2003;523–524:87–97.

[37] Li J, Liu H, Ramchandran S, Waypa GB, et al. Grape seed proanthocyanidins ameliorate doxorubicin-induced cardiotoxicity. Am J Chin Med 2010;38:569–84.

[38] Rao AV. Lycopene, tomatoes and the prevention of coronary heart disease. Exp Biol Med 2002;227:908–13.

[39] Tamayo C, Diamond S. Review of clinical trials evaluating safety and efficacy of milk thistle (Silybum marianum [L.] Gaertn.). Integr Cancer Ther 2007;6:146–57.

[40] Abenavoli L, Capasso R, Milic N, Capasso F. Milk thistle in liver disease: past, present, and future. Phytother Res 2010;24:1423–32.

[41] Kalantari H, Shahshahan Z, Hejazi SM, Ghafghazi T, et al. Effects of Silybum marianum on patients with chronic hepatitis C. J Res Med Sci 2011;16:287–90.

[42] Saller R, Meier R, Brignoli R. The use of silymarin in the treatment of liver diseases. Drugs 2001;61:2035–63.

[43] Abourashed FA, Mikell JR, Khan IA. Bioconversion of silybin to phase I and II microbial metabolites with retained antioxidant activity. Bioorg Med Chem 2012;20:2784–8.

[44] Linde K, Ramirez G, Mulrow CD, Pauls A, et al. St. John's wort for depression: an overview and meta-analysis of randomized clinical trials. Br Med J 1996;313:253–8.

[45] Orcic DZ, Mimica-Dukic NM, Franciskovic MM, Petrovic SS, et al. Antioxidant activity relationship of phenolic compounds in Hypericum perforatum L. Chem Cent J 2011;5:34.

[46] Meinke MC, Schanzer S, Haag SF, Casetti F, et al. In vivo photo protective and anti-inflammatory effect of hyperforin is associated with high antioxidant activity in vitro and ex vivo. Eur J Pharm Biopharm 2012;81:346–50.

[47] Ott M, Huls M, Cornelius MG, Fricker G. St. John's wort constituents modulate P-glycoprotein transport activity at the blood–brain barrier. Pharm Res 2010;27:811–22.

[48] Borrelli F, Izzo AA. Herb–drug interactions with St. John's wort (Hypericum perforatum): an update on clinical observations. AAPS J 2009;11:710–27.

[49] Velasquez MT, Bhathena SJ. Role of dietary soy in obesity. Int J Med Sci 2007;26:72–82.

[50] Sirtori CR, Lovati MR. Soy proteins and cardiovascular disease. Curr Atheroscler Rep 2001;3:47–53.

[51] Kurzer MS. Soy consumption for reduction of menopausal symptoms. Inflammopharmacology 2008;16:227–9.

[52] Nhan S, Anderson KE, Nagamani M, Grady JJ, et al. Effect of a soymilk supplement containing isoflavones on urinary F2 isoprostane levels in premenopausal women. Nutr Cancer 2005;53:73–81.

[53] Ferretti G, Bacchetti T, Menanno F, Curatola G. Effect of genistein against copper-induced lipid peroxidation of human high-density lipoprotein. Atherosclerosis 2004;172:55–61.

[54] Aggarwal BB, Sundrram C, Malani N, Ichikawa H. Curcumin: the Indian solid gold. Adv Exp Med Biol 2007;595:1–75.

[55] Marchiani A, Rozzo C, Fadda A, Delogu G, et al. Curcumin and curcumin-like molecules: from spices to drugs. Curr Med Chem 2013;21(2):204–22.

[56] Bhullar KS, Jha A, Youssef D, Rupasinghe HP. Curcumin and its carbocyclic analogs: structure–activity in relation to antioxidant and selected biological properties. Molecules 2013;18:5389–404.

[57] Bonakdar RA, Guarneri E. Coenzyme Q10. Am Fam Physician 2005;72:1065–70.

[58] Lee BJ, Huang YC, Chen SJ, Lin PT. Coenzyme Q10 supplementation reduces oxidative stress and increases antioxidant enzyme activity in patients with coronary artery disease. Nutrition 2012;28:250–5.

[59] Potgieter M, Pretorius E, Pepper MS. Primary and secondary coenzyme Q10 deficiency: the role of therapeutic supplementation. Nutr Rev 2013;71:180–8.

[60] Allen J, Bradley RD. Effects of oral glutathione supplementation on systematic oxidative stress biomarkers in human volunteers. J Altern Comp Med 2011;17:827–33.

[61] Mendis E, Kim MM, Rajapakse N, Kim SK. Sulfated glucosamine inhibits oxidation of biomolecules in cells via a mechanism involving intracellular free radical scavenging. Eur J Pharmacol 2008;579:74–85.

[62] Fang C, Peng M, Li G, Tian J, et al. New functions of glucosamine as a scavenger of the lipid peroxidation product malondialdehyde. Chem Res Toxicol 2007;20:947–53.

[63] Golbidi S, Badran M, Laher I. Diabetes and α-lipoic acid. Front Pharmacol 2011;2:69.

[64] Nebbioso M, Pranno F, Pescosolido N. Lipoic acid in animal models and clinical use in diabetic retinopathy. Expert Opin Pharmacother 2013;14:1829–38.

[65] Li RJ, Ji WQ, Pang JJ, Wang JL, et al. Alpha-lipoic acid ameliorates oxidative stress by increasing dehydrogenase-2 activity in patients with acute coronary syndrome. Tohoku J Exp Med 2013;229:45–51.

[66] Tur JA, Bibiloni MM, Sureda A, Pons A. Dietary sources of omega-3 fatty acids: public health risks and benefits. Br J Nutr 2012;107(Suppl. 2):S23–52.

[67] Kar S, Webel R. Fish oil supplementation and coronary artery diseases: does it help? Mol Med 2012;109:142–5.

[68] Hagarty B, Parker G. Fish oil as a management component for mood disorders: an evolving signal. Curr Opin Psychiatry 2013;26:33–40.

[69] Songur A, Sarsilmaz M, Sogut S, Ozyurt H, et al. Hypothalamic superoxide dismutase, xanthine oxidase, nitric oxide, and malondialdehyde in rats fed with omega-3 fatty acids. Prog Neuropsychopharmacol Biol Psychiatry 2004;28:693–8.

[70] Kesavulu MM, Kameswararao B, Apparaco CH, Kimat EG, et al. Effect of omega-3 fatty acids on lipid peroxidation and antioxidant enzyme status in type 2 diabetic patients. Diabetes Metab 2002;28:20–6.

[71] Denny Joseph KM, Muralidhara M. Fish oil prophylaxis attenuates rotenone-induced oxidative impairments and mitochondrial dysfunction in rat brain. Food Chem Toxicol 2012;50:1529–37.

[72] Herxheimer A, Petrie KJ. Melatonin for the prevention and treatment of jet lag. Cochrane Database Syst Rev 2002; CD001520.

[73] Carpentieri A, Diaz de Barbozz G, Areco V, Peralta-Lopez M, et al. New perspectives in melatonin uses. Pharmacol Res 2012;65:437–44.

[74] Bai J, Dong L, Song Z, Ge H, et al. The role of melatonin as an antioxidant in human lens epithelial cells. Free Radic Res 2013;47:635–42.

CHAPTER 17

Combating Oxidative Stress with a Healthy Lifestyle

17.1 INTRODUCTION

Stress in classical terms indicates a situation in which an individual has to decide whether to fight or flee. For both humans and animals, stress is an adaptive mechanism in response to an external threat. Psychological stress, which is also commonly known as "stress," usually arises from an unpleasant situation, but stress can also occur from a good situation such as a job interview. Common sense dictates that if the source of the stress (stressor) can be avoided, stress can be eliminated from human life. In reality, certain stressors, such as the death of a family member or caring for a critically ill family member, cannot be avoided. Therefore, a healthy lifestyle is the best way to combat stress. Various healthy habits to fight stress are listed in Box 17.1, including 10 approaches recommended by the American Heart Association.

17.2 INTERACTION WITH FRIENDS/FAMILY MEMBERS AND LAUGHTER TO REDUCE STRESS

Interaction with family members and friends is a good way to reduce stress. McDaniel *et al.* [1] observed that blogging may improve new mothers' well-being because they feel more connected to friends and extended family members as well as the outside world through the Internet. According to the buffer hypothesis, social support or social network may improve mental health by buffering the negative effect of stress. Olstad *et al.* [2] studied this hypothesis and concluded that total social support/social network buffered the deteriorating effect of total stressor score on the mental health of study participants. The effect was weak but stronger in women than in men. Interestingly, social support/social network also has significant buffer effect against work-related stress. Coker *et al.* [3] conducted a cross-sectional survey of 1152 women aged 18–65 years who were recruited from family practice clinics. These women were screened for intimate partner violence (IPV)

CONTENTS

A. Dasgupta and K. Klein: Antioxidants in Food, Vitamins and Supplements. DOI: http://dx.doi.org/10.1016/B978-0-12-405872-9.00017-3

Box 17.1 APPROACHES TO REDUCE STRESS BY DEVELOPING HEALTHY HABITS

Approaches Recommended by the American Heart Association

- Talking to family members and friends
- Daily exercise/physical activity
- Learning to accept things that cannot be changed
- Laughter
- Avoiding smoking and consuming too much alcohol and caffeine
- Time management (plan ahead to get important tasks done on time)
- Six to eight hours of sleep each night
- Getting organized
- Volunteering
- Fewer worries

Other Approaches

- Social networking
- Pet ownership
- Meditation and yoga
- Aromatherapy and massage
- Listen to music
- Sex with spouse/partner
- Vacation

during brief clinic interviews. IPV is associated with poor perceived mental and physical health, substance abuse, current depression, anxiety, and intentions for committing suicide. The authors concluded that among women experiencing IPV, higher social support scores were associated with a significantly reduced perception of poor mental and physical health as well as lower levels of anxiety and depression.

A good sense of humor and laughter are associated with good health and stress reduction. Laughter provides a physical release for accumulated tension. Laugher can effectively reduce stress and also has a positive effect on the immune system [4]. Based on a study of 10 male healthy subjects who viewed a 60-min humor video, Berk *et al.* [5] observed that mirthful laughter can reduce stress by reducing the blood levels of cortisol and other stress hormones. Hassed [6] noted that laughter has a role in reducing blood levels of stress hormones such as cortisol, thus improving mood, enhancing creativity, and improving immunity while simultaneously reducing blood pressure and pain perception. It has been demonstrated that laughter can improve mood directly by stimulation of particular cerebral regions involved in depression pathogenesis and also by normalization of the hypothalamic–pituitary–adrenocortical system, thus counteracting depressive symptoms. In addition, laughter has a favorable effect in stimulating social interaction, thus reducing stress levels [7]. In a study of 70 children undergoing surgery, Fernandes and Arriaga [8] observed that children accompanied by their parents and a clown showed reduced preoperative worries and emotional stress compared to children who were accompanied by their parents but no clown. In addition, the presence of a clown also reduced the emotional stress level of parents. Laughter therapy may also improve the quality of life of cancer patients [9].

17.3 INSOMNIA AND OXIDATIVE STRESS: THE NEED FOR A GOOD NIGHT'S SLEEP

A good night's sleep is essential for a healthy life. Many studies have documented the link between various diseases and lack of a good night's rest, including cardiovascular disease, high blood pressure, and diabetes. However, stress level may also directly affect sleep quality. Petersen *et al.* [10] concluded that even moderate daily stress has a negative effect on sleep quality. Several studies have also demonstrated a link between insomnia and increased oxidative stress. Hachul de Campos *et al.* [11] studied 38 postmenopausal women with insomnia and observed higher concentrations of lipid peroxides in the blood of these women measured as thiobarbituric acid reactive substances. Gulec *et al.* [12] demonstrated that 30 patients with insomnia had a higher level of lipid peroxides (measured as malondialdehyde) and lower level of antioxidant enzyme glutathione peroxidase in blood compared to 30 healthy volunteers with no insomnia. The authors concluded that sleep plays an important role in attenuating oxidative stress. In another study involving 18 poor sleepers and 18 good sleepers matched for sex, age, height, and weight, Adam *et al.* [13] observed that poor sleepers tended to have higher urinary cortisol, a stress hormone, and adrenaline excretion.

Poor sleep quality also adversely affects preschool children. Hatzinger *et al.* [14] studied 67 kindergarten children (35 boys and 32 girls; mean age, 5.34 years) and observed that, compared to girls, boys showed significantly more rapid eye movement sleep time. In addition, children who were poor sleepers showed significantly increased morning cortisol levels in saliva compared to good sleepers. Moreover, children who were good sleepers showed a lower degree of impulsivity and a lower degree of social inhibition. The authors concluded that poor sleep quality in preschool children was associated with increased hypothalamic–pituitary–adrenocortical activity, which may lead to more behavior and psychological problems later in life.

17.4 HEALTH BENEFITS OF PET OWNERSHIP AND HUMAN–ANIMAL INTERACTION

Pet ownership has many health benefits, including a significant reduction in stress levels as evidenced by many published articles in the literature. A pet acts as a "social catalyst" and facilitates interpersonal interactions. Wells [15] studied the behavior of 1800 strangers toward a female experimenter and observed that when the experimenter was alone, strangers ignored her, but when she was in the company of a dog, she received attention from these strangers, indicating that a pet may act as a catalyst for social interaction. One of the health benefits of dog ownership is dog walking, which correlates with benefits of daily exercise. Lentino *et al.* [16] observed that dog owners who walked their dogs had significantly lower incidences of hypertension, diabetes, and depression compared to non-dog owners and dog owners who did not walk their dogs. In addition, dog owners in general had lower blood cholesterol than non-dog owners. The authors concluded that owning and walking a dog contributes to a healthy lifestyle.

Social support and pet ownership have been associated with better coronary artery disease survival. Based on a study of 369 patients who suffered from myocardial infarction, dog owners were significantly less likely to die within 1 year of heart attack (only 1 of 87 dog owners died) compared to non-dog owners (19 of 282 patients died). The authors concluded that both pet ownership and social support are significant predictors of survival after heart attack regardless of physiological status. Arhant-Sudhir *et al.* [18] noted that ownership of a pet, particularly a dog, is associated with positive health benefits. Wood *et al.* [19] reported that there are therapeutic, psychological, physical, and psychosocial benefits to human–pet interactions, and pet ownership can significantly reduce stress and the risk of developing depression.

Cortisol is a stress hormone, and the blood concentration of cortisol usually increases in response to stress. Human–animal interaction can reduce cortisol levels in the circulation. Odendaal and Meintjes [20] observed a reduction in plasma cortisol in subjects after they petted their own dog or even an unfamiliar dog, but when they read quietly, there was no reduction of plasma cortisol. Cole *et al.* [21] compared a visit with a dog to a visit without a dog among adults hospitalized with heart failure and observed that significantly lower epinephrine and norepinephrine levels were measured during and after dog visits, indicating a stress-reducing effect of dogs. Cat ownership also has a beneficial effect on stress reduction and reduction of blood pressure. Based on a study of 4435 patients, of whom 2435 patients (55%) were previous or current cat owners, Qureshi *et al.* [22] observed a decreased risk of death due to myocardial infarction and cardiovascular disease among cat owners. The authors commented that acquisition of a cat as a domestic pet may represent a novel strategy for reducing the risk of cardiovascular diseases in high-risk individuals. Interestingly, owning a cat before age 18 years protects against the development of asthma and atopy as an adult [23].

In a review article, Beetz *et al.* [24] noted that the major benefits of pet ownership include social attention; improved social behavior; and reduced stress level and stress-related parameters, such as cortisol, heart rate, and blood pressure. In addition, human–animal interaction also reduces anxiety and fear, and it has a beneficial effect on mental health. It has been postulated

Box 17.2 MAJOR BENEFITS OF PET OWNERSHIP

Physiological/Therapeutic Benefits

- Reduction in stress-related parameters, including heart rate, blood pressure, and serum and salivary cortisol
- Reduction in other stress-related parameters, such as epinephrine and norepinephrine, in some studies
- Lower levels of cholesterol and triglycerides
- Improved immune function
- Improved general health and fewer visits to doctor's office
- Reduced chance of getting a headache, stomach upset, or difficultly getting to sleep
- Higher level of physical activity and benefits of daily exercise for dog owners who walk their dogs every day

Psychological/Psychosocial Benefits

- Improvement in social interaction, mood, and interpersonal interaction
- Reduction of fear and anxiety
- Feeling of calmness
- Increased trustworthiness and trust toward other pet owners
- Improved pain management
- Reduced aggression
- Enhanced empathy
- Improved learning among children
- Children who have house pets have better self-esteem.

Therapeutic Benefits

- Pets have a therapeutic effect on reducing cardiovascular risk factors and outcome after heart attack.
- The presence of a pet can provide a better quality of life for elderly patients.
- Pet dogs (and, less often, other animals) are used for pet therapy in children's hospitals.
- Pet therapy can improve quality of life in patients with AIDS, age-related dementia, Alzheimer's disease, and other chronic illnesses.

that activation of the oxytocin system plays a key role in the beneficial effects of human–animal interactions because oxytocin has a calming effect on the body and can counteract secretion of the stress hormone cortisol [24]. Major benefits of pet ownership are listed in Box 17.2.

17.5 MEDITATION AND YOGA TO COMBAT STRESS AND IMPROVE ANTIOXIDANT DEFENSE

During meditation, a person achieves a very relaxed but alert state of mind. The goal of meditation from the ancient times of Vedic, Buddhist, and Taoist practice is to achieve *samadhi*, a wakeful but tranquil state of total bliss also known as "pure awareness" or "enlightenment." This state of mind produces altered brain waves such as alpha waves and theta waves, increases cerebral blood flow, and evokes other physiological changes [25]; achieving this state requires many years of dedicated practice. However, meditating for even 15–30 min has health benefits, including lowering blood cholesterol, improving cardiac and immune functions, and reducing stress. Dr. Jon Kabat-Zinn developed the mindfulness-based stress reduction (MBSR) program at the University of Massachusetts Medical Center. MBSR is an 8-week training program in which participants meet for approximately 90 min once a week and then can practice this meditation at home. The goal of MBSR practice is to learn to focus attention on moment-by-moment experiences with an attitude of curiosity, openness, and acceptance. Many studies have shown the beneficial effects of such practice. Since its inception in 1979, this form of complementary and alternative medicine has gained popularity, and MBSR is currently offered in more than 200 hospitals in the United States. Mindful meditation, also known as "vipassana meditation," has its roots in ancient Zen meditation practice, in which a person develops present-moment awareness without being judgmental [26]. However, MBSR is secular in approach and is a clinically based method that employs manual and standardization technique. Marchand [27] noted that evidence suggests MBSR is beneficial for general psychological health and stress management not only in people with medical or psychiatric illness but also in healthy individuals. MBSR may also be effective as a complementary therapy in pain management. In a study of 229 women who underwent surgery, chemotherapy, and radiation treatment for breast cancer, Hoffman *et al.* [28] observed that MBSR improved mood, breast and endocrine-related quality of life, and well-being, more effectively than standard care in women with stage 0 to III breast cancer. Another study observed that MBSR is effective at reducing salivary cortisol levels not only in advanced-stage cancer patients but also in their caregivers. The authors concluded that MBSR is a beneficial intervention for reducing stress, anxiety, and cortisol levels in both advanced-stage cancer patients and their caregivers [29]. Edenfield and Saeed [30] have reviewed the benefits of MBSR.

In loving-kindness meditation practice, during meditation the person cherishes all living beings on the earth. Hutcherson *et al.* [31] investigated the effect of brief loving-kindness meditation exercise and concluded that even a few minutes of loving-kindness mediation increased feelings of social connection and positive attitude. It has been well-established that feeling socially connected confers mental and physical health benefits [31]. A brief lifestyle education program based on "asana" (yoga) and "pranayama" reduced fasting glucose, serum total cholesterol, and low-density lipoprotein cholesterol and increased the concentration of high-density lipoprotein cholesterol in patients with chronic diseases, including diabetes, hypertension, and coronary heart diseases [32]. Transcendental meditation is also effective for improving antioxidant defense of the body and lowering stress levels. Schneider *et al.* [33] demonstrated that subjects who practiced transcendental meditation had lower lipid peroxide levels in blood and also experienced less stress than control subjects who did not practice meditation.

Repetition of "mantram" (e.g., "Om" or a holy word) is a useful way to reduce stress. In a study in which outpatient veterans participated in a 5-week (90 min per week) program of mantram repetition, the authors found significantly reduced anxiety and stress and improved quality of life and spiritual well-being among the patients. The authors noted that silent mantram repetition can be easily taught and may be used by nurses and patients for managing stress and increasing feelings of well-being [34]. In a report based on a study of 42 hospital workers, Bormann *et al.* [35] observed improvement in emotional and spiritual well-being mediated by frequent mantram repetition. Studies have indicated that spiritual well-being is inversely related to psychological distress and rates of disease progression in HIV-infected patients. Using a randomized design, Bormann *et al.* [36] divided HIV-infected individuals into a mantram intervention group ($n = 36$) and a control group ($n = 35$), and observed that faith levels increased following 5 weeks of mantram intervention. In addition, a significant reduction in cortisol was observed in HIV-infected individuals in the mantram intervention group compared to the control group. The authors suggested that a decreased cortisol level could potentially benefit immune functioning among HIV-infected individuals.

Yoga, an ancient discipline that began in India as early as 3000 BC, is designed to create harmony among physical, mental, and spiritual dimensions in a person. Yoga as practiced in India entered the Western world during approximately the 19th century following translations of Sanskrit text. Many people in Western countries have adopted yoga for its health benefits, and adoption of Hindu religion or philosophy is not necessary for obtaining the benefits of yoga because it can be practiced in a secular environment. Studies have indicated that yoga can improve health through downregulation of the hypothalamus–pituitary–adrenal axis and the sympathetic nervous system. Yoga enhances antioxidant defense of the body by increasing the blood level of glutathione as well as activities of antioxidant enzymes, including superoxide dismutase, catalase, glutathione peroxidase, and glutathione reductase. Tai chi practices also lower lipid peroxidation product in blood and improve the concentration of the antioxidant enzyme superoxide dismutase [37]. Sinha *et al.* [38] reported that healthy male volunteers ($n = 30$) who practiced yoga had higher total antioxidant capacity of their blood as well as increased blood concentration of glutathione and increased activity of antioxidant enzyme glutathione reductase compared to male volunteers ($n = 21$) who did not practice yoga. Hegde *et al.* [39] reported that practice of yoga for 3 months reduced markers of oxidative stress in the blood (lower concentration of malondialdehyde) and improved concentrations of antioxidants (glutathione and vitamin C). It also improved glycemic control in these patients.

Box 17.3 BENEFITS OF MEDITATION AND YOGA

Benefits of Meditation

- Better antioxidant defense of the body as evidenced by a lower amount of lipid hydroperoxide in blood
- Improved immune function
- Lower concentration of blood cholesterol
- Higher levels of high-density lipoprotein cholesterol
- Reduction of anxiety and stress
- May help to improve mood in depressed patients
- Reduced levels of salivary cortisol, a stress hormone
- Improved mood

Benefits of Yoga

- Better antioxidant defense of the body as evidenced by increased levels of antioxidant enzymes, including superoxide dismutase and catalase
- Improved glycemic control
- Reduced stress
- Reduced plasma levels of cholesterol
- Reduced inflammation as evidenced by lower concentrations of interleukin-6 and tumor necrosis factor-α

Practicing yoga for even a brief amount of time was effective in reducing stress, as evidenced by a lower level of plasma cortisol and increased level of β-endorphin [40]. In addition, concentrations of markers for inflammation (interleukin-6 and tumor necrosis factor-α) were also reduced following 10 days of yoga and pranayama (breathing exercise) practice. Mindful yoga practice was also effective in reducing stress in school-age girls [41]. In a study of 81 subjects, Smith *et al.* [42] compared exercised-based practice of yoga with more comprehensive yoga practice (one with an ethical/spiritual component) for reducing stress and anxiety. They observed that both yoga practices reduced stress and anxiety. However, only those who practiced yoga with ethical and spiritual components showed decreased salivary level of cortisol and a significant decrease in anxiety-related symptoms. The benefits of meditation and yoga are summarized in Box 17.3.

17.6 BENEFITS OF PHYSICAL ACTIVITY AND EXERCISE

Many studies have clearly demonstrated that being physically active reduces the risk of all-cause mortality. Woodcock *et al.* [43] reported that daily 30-min moderate-intensity activity for 5 days a week (2.5 h per week) was associated with a 19% reduction in all-cause mortality compared to no physical activity. In addition, 7 h/week moderate activity reduced all-cause mortality by 24%. In 1995, the American College of Sports Medicine and the Centers for Disease Control and Prevention published national guidelines for physical activity; later, the American Heart Association endorsed these guidelines. The primary guideline is that all healthy adults aged 18–65 years need moderate-intensity physical activity for a minimum of 30 min per day, 5 days per week or vigorous-intensity physical exercise at least 20 min per day, 3 days per week. A combination of moderate- and vigorous-intensity activity is also acceptable. For example, a person can walk briskly for 30 min twice per day and then jog 20 min for two other days during the week. Physical activity as well as vigorous exercise increases heart rate and is beneficial for health [44]. Based on a study of 31 healthy sedentary subjects 50–65 years old, Tully *et al.* [45] reported that brisk walking for 30 min, 5 days per week significantly reduced systolic and diastolic blood pressure and reduced the risk of stroke. A short-term moderate-intensity aerobic exercise such as brisk daily walking for 12 weeks can restore vascular endothelial function in previously sedentary middle-aged and older adults, probably by reducing oxidative stress. Seals *et al.* [46] concluded that habitual exercise can reduce the risk of cardiovascular disease. Older men also receive benefit from habitual exercise, as evidenced by the observation that older men who exercise regularly do not demonstrate vascular endothelial oxidative stress, which also reduces the risk of cardiovascular diseases [47]. Interestingly, exercise is effective in reducing craving for smoking, and 15 min of running also reduced the salivary cortisol level [48]. Similarly to brisk walking, tai chi, a moving meditation, effectively reduces stress because salivary cortisol levels decline following tai chi practice, as they also do for brisk walking. Therefore, both tai chi and brisk walking are effective tools for reducing stress. In addition, heart rate, blood pressure, and urinary catecholamine changes for tai chi were found to be similar to those for subjects walking at a speed of 6 km/h (3.8 miles/h) [49].

Whereas moderate exercise strengthens the immune system, long and intensive exercise weakens the immune system [50]. Strenuous physical activity including exercise also increases oxidative stress because, as oxidative phosphorylation increases in response to exercise, there is also an increase in the production of free radicals. In addition to production of free radicals due to increased oxygen consumption by mitochondria during exercise, other sources of increased free radicals during exercise include prostanoid metabolism, xanthine oxidase, as well as release of free radicals by macrophages in order to repair damaged tissues due to strenuous exercise. However, if people eat a balanced diet that includes generous servings of fruits and vegetables each day, additional free radicals generated during strenuous physical activity may be neutralized by the antioxidant defense of the body. There is no evidence that taking antioxidant supplements has any benefit for athletes. It is recommended that athletes eat a balanced diet rich in antioxidants [51].

17.7 AROMATHERAPY AND MASSAGE FOR STRESS REDUCTION

Aromatherapy can significantly reduce stress. In one study, 22 healthy volunteers who sniffed lavender or rosemary for 5 min showed higher free radical scavenging capacity in saliva and lower cortisol level in saliva, indicating that aromatherapy can reduce stress and improve antioxidant defense of the human body. Interestingly, lavender can increase the antioxidant capacity of saliva at much lower concentrations than rosemary [52]. Cooke *et al.* [53] reported that emergency room nurses were more anxious during winter than during summer, but aromatherapy massage with music significantly reduced their level of anxiety during both seasons. The higher level of anxiety during the winter months among these nurses may be related to the fact that, in general, emergency rooms are busier during the winter months due to admission of patients with cardiac and respiratory dysfunctions [53]. Massage also has a direct relationship with health outcome because patients' perceptions of stress and anxiety were significantly reduced after massage. The simple act of touch-focused care and even 5-min hand and foot massage can be useful in lowering a patient's perceived level of stress [54]. Corbin [55] reported that massage therapy can be easily incorporated into the care of cancer patients because there is strong evidence that massage therapy can reduce stress and anxiety levels in these patients.

17.8 MUSIC FOR STRESS REDUCTION

In recent years, significant research has been performed on the beneficial effects of music on hospital patients, and music therapy has been accepted as a "complementary and alternative therapy" with virtually no side effects. Research has clearly demonstrated a role of music in the regulation of the hypothalamic–pituitary axis, sympathetic nervous system, and immune system, and music therapy can reduce stress [56]. Cervellin and Lippi [57] reported that increasing experimental evidence suggests that some kind of music might modulate different cardiac and neurological functions as well as reduce stress. This effect, also known as the "Mozart effect," may be useful as a therapeutic tool for reducing stress in both healthy and ill individuals.

Bauer *et al.* [58] studied the effect of music therapy on pain and anxiety relief in postoperative cardiac surgical patients (music group, 49 patients; control group, 51 patients) and observed that music therapy twice daily was associated with a significant reduction in pain score and anxiety on the second day in the music group compared to the control group. Based on a review of 30 trials involving 1891 cancer patients, Bradt *et al.* [59] concluded that music therapy may have beneficial effects on reducing anxiety and pain as well as improving mood and quality of life in cancer patients. Based on a review of 29 studies, Evans [60] concluded that music played via headphones reduced anxiety in patients during normal care delivery and also appeared to improve mood and tolerance of patients. Kushnir *et al.* [61] studied 60 healthy women waiting to undergo cesarean section and observed that those who listened to preselected favorite music for 40 min had a significant increase in positive emotion and decreased perceived threat of the procedure compared to the control group. In addition, women who listened to the music also exhibited significant reductions in systolic blood pressure and respiratory rate compared to the control group. In contrast, women in the control group showed significant increases in diastolic blood pressure and respiratory rate. In another study, Moris and Linos [62] observed that listening to music was associated with reductions in stress, anxiety, and demand for analgesic and anesthetic drugs. In addition, music was found to reduce the heart rate, blood pressure, and muscle effort of surgeons while simultaneously increasing the accuracy of the surgical task [62]. Music therapy for only 15 min also reduced anxiety levels and systolic blood pressure in oncology patients prior to receiving radiotherapy [63]. Music therapy also benefits patients in intensive care units by reducing stress levels, as evidenced by significant reductions in cortisol levels of the music group compared to the control group. However, classical and meditative music produce the most beneficial effects, in contrast with heavy metal music, which is not only ineffective but also may produce adverse effects in these patients [64].

Music therapy was also effective in reducing aggressive behavior in children with highly aggressive behavior. Cho *et al.* [65] used 50 min of music intervention therapy twice per week for children in a music intervention group and observed that after 15 weeks, this group showed a significant reduction in aggression and improved self-esteem compared to the control group. The authors concluded that music therapy may reduce aggressive behavior and improve self-esteem in children with highly aggressive behavior.

17.9 SEX FOR STRESS REDUCTION

Having sex with a spouse or partner is a beneficial way to reduce stress. A study based on 58 middle-aged women (mean age, 47.6 years), who recorded physical affection, different sexual behavior, stressful events, and mood ratings every morning for 36 weeks, demonstrated that physical affection or sex with a partner on one day correlated with higher positive mood the following morning, indicating that physical intimacy for these women was associated with experiencing less stress. However, this relation did not hold when a woman achieved orgasm without a partner [66]. Ditzen *et al.* [67] exposed 67 women (ages 20–37 years) who had been married or cohabiting with a male partner for at least 12 years to a standardized psychosocial laboratory stressor and measured salivary cortisol and blood oxytocin levels, as well as heart rate. Women who had positive physical contact with their husband/partner before the stress test exhibited significantly lower levels of cortisol and heart rate than women who did not have positive physical contact with their husband/partner. Emotional support alone did not have the same effect.

Men also benefit from sex. Studies indicate that the frequency of penile–vaginal intercourse (PVI), but no other forms of sex, is associated with lower stress level, better blood pressure control, better cardiac health, and decreased mortality in both men and women. Based on a study of 24 women and 22 men who recorded their daily PVI activity, Brody [68] demonstrated that both men and women who reported PVI activities (but not other forms of sex) reacted significantly better (better blood pressure control under stress) to controlled stress situations (public speaking and verbal mathematical challenge) than men and women who did not have PVI. Based on a study of 29,342 men, Leitzman *et al.* [69] observed that men who experienced 21 or more ejaculations per month had significantly reduced risk of prostate cancer compared to men who had 4–7 ejaculations per month. However, note that although sexual activity with a spouse or stable partner is a stress buster and also has health benefits, sexual misadventure or changing partners on a regular basis is detrimental to the human psyche because research indicates that chemicals secreted from the brain (mostly oxytocin for females and vasopressin for males) during sex also bring partners emotionally close to each other. Vasopressin and oxytocin are the neuroendocrinological basis for why humans prefer stability in relationships [70]. Sexual misadventure increases the risk of sexually transmitted diseases, including infection with the HIV virus.

Oxytocin can alleviate stress. Although this hormone is secreted during childbirth and breast-feeding and plays an important role in mother–child bonding, it is secreted in both males and females during intimacy and sex. Oxytocin in conjunction with estrogen plays an important role in trust building between partners and eventually in long-term relationships. Although oxytocin has more pronounced psychological effects in women, men also secrete oxytocin. However, in men, vasopressin (also known as arginine vasopressin), which is secreted in response to sex, plays an important role in bond formation with women and eventually helps in the development of long-term relationships [71].

17.10 LEISURELY ACTIVITIES AND VACATION FOR STRESS REDUCTION

Leisurely activities and vacation can reduce stress, at least on a short-term basis. Based on a study of 53 participants, Strauss-Blasche *et al.* [72] observed that vacation may improve the feeling of well-being on a short-term basis. Three days after vacation, quality of sleep and mood improved in participants compared to those on pre-vacation days. Five weeks after vacation, participants still reported fewer physical complaints compared to those reported on pre-vacation days. Based on a study of 87 blue-collar workers, Westman and Etzion [73] observed that vacation alleviated perceived job stress and burnout. The decline in burnout was observed immediately after vacation, and a return to pre-vacation levels took approximately 4 weeks. Gump and Matthews [74] performed a large 9-year study of 12,338 participants on the effect of vacation on all-cause mortality in middle-aged men with a high risk of cardiovascular disease. They concluded that frequency of vacation by these men was associated with a reduced risk of all-cause mortality and, specifically, mortality associated with cardiovascular diseases. The authors concluded that vacationing is beneficial to health.

17.11 CONCLUSION

Free radicals are produced not only by the body's normal metabolism causing oxidative stress but also from other sources, such as psychological stress, exposure to noise, sunlight, and air pollution. Various diseases can also increase oxidative stress, including some normal physiological processes, such as menopause and aging. Sanchez-Rodriguez *et al.* [75] observed elevated lipid peroxidation products (as measured by thiobarbituric acid reactive substances) in postmenopausal women ($n = 93$) compared to premenopausal women ($n = 94$). The authors speculated that depletion of estrogen during menopause may contribute to this added oxidative stress. The best way to combat oxidative stress is to eat a healthy diet that includes fruits and vegetables each day. Various relaxation methods are very helpful in reducing psychological stress-induced oxidative stress. A healthy lifestyle is the best way to counteract oxidative stress induced by various external factors.

REFERENCES

[1] McDaniel BT, Coyne SM, Holmes EK. New mothers and media use: associations between blogging, social networking and maternal well-being. Matern Child Health J 2012;16:1509–17.

[2] Olstad R, Sexton H, Sogaard AJ. The Finnmark study: a prospective population study of the social support buffer hypothesis, specific stressors and mental distress. Soc Psychiatr Epidemiol 2001;36:582–9.

[3] Coker AL, Smith PH, Thompson MO, McKeown RE, et al. Social support protects against the negative effects of partner violence on mental health. J Women's Health Gend Based Med 2002;11:465–76.

[4] Wooten P. Humor: an antidote for stress. Holist Nurs Pract 1996;10:49–56.

[5] Berk LS, Tan SA, Fry WF, Napier BJ, et al. Neuroendocrine and stress hormone changes during mirthful laughter. Am J Med 1989;298:390–6.

[6] Hassed C. How humor keeps you well. Aust Fam Physician 2001;30:25–8.

[7] Fonzi L, Matteucci G, Bersani G. Laughter and depression: hypothesis of pathogenic and therapeutic correlation [in Italian]. Riv Psichiatr 2010;45:1–6.

[8] Fernandes SC, Arriaga P. The effects of clown intervention on worries and emotional responses in children undergoing surgery. J Health Psychol 2010;15:405–15.

[9] Noji S, Takayanagi K. A case of laughter therapy that helped improve advanced gastric cancer. Jpn Hosp 2010;29:59–64.

[10] Petersen H, Kecklund G, D'Onofrio P, Nilsson J, et al. Stress vulnerability and the effects of moderate daily stress on sleep polysomnography and subjective sleepiness. J Sleep Res 2013;22:50–7.

[11] Hachul de Campos H, Brandao LC, D'Almeida V, Grego BH, et al. Sleep disturbances, oxidative stress and cardiovascular risk parameters in postmenopausal women complaining of insomnia. Climacteric 2006;9:312–19.

[12] Gulec M, Ozkol H, Selvi Y, Tuluce Y, et al. Oxidative stress in patients with primary insomnia. Prog Neuropsychopharmacol Biol Psychiatry 2012;37:247–51.

[13] Adam K, Tomeny M, Oswald I. Physiological and psychological differences between good and poor sleepers. J Psychiatr Res 1986;20:301–4.

[14] Hatzinger M, Brand S, Perren S, Stadelmann S, et al. Electroencephalographic sleep profiles and hypothalamic–pituitary–adrenocortical (HPA) activity in kindergarten children: early indication of poor sleep quality associated with increased cortisol secretion. J Psychiatr Res 2008;42:532–43.

[15] Wells DL. The facilitation of social interactions by domestic dogs. Anthrozoos 2004;17: 340–52.

[16] Lentino C, Visek AJ, McDonnell K, Dipietro L. Dog walking is associated with a favorable risk profile independent of a moderate to high volume of physical activity. J Phys Act Health 2012;9:414–20.

[17] Friedman E, Thomas SA. Pet ownership, social support, and one year survival after acute myocardial infarction in the Cardiac Arrhythmia Support Trial (CSAT). Am J Cardiol 1995;76:1213–17.

[18] Arhant-Sudhir K, Arhant-Sudhir R, Sudhir K. Pet ownership and cardiovascular risk reduction: supportive evidence, conflicting data and underlying mechanism. Clin Exp Pharmacol Physiol 2011;38:734–8.

[19] Wood L, Giles-Corti B, Bulsara M. The pet ownership as a conduit for social capital. Soc Sci Med 2005;61:1159–73.

[20] Odendaal JS, Meintjes RA. Neurophysiological correlation of affiliative behavior between humans and dogs. Vet J 2003;165:296–301.

[21] Cole KM, Gawlinski A, Steers N, Kotleman J. Animal assisted therapy in patients hospitalized with heart failure. Am J Crit Care 2007;16:575–85.

[22] Qureshi AL, Memon MZ, Vazquez G, Suri MF. Cat ownership and the risk of fatal cardiovascular diseases: results from the second National Health and Nutrition Examination Study mortality follow-up study. J Vasc Interv Neurol 2009;2:132–5.

[23] De Meer G, Toelle BG, Ng K, Tovey E, et al. Presence and timing of cat ownership by age 18 and the effect on the atopy and asthma at age 28. J Allergy Clin Immunol 2004;113: 433–8.

[24] Beetz A, Uvnas-Moberg K, Julius H, Kotrschal K. Psychosocial and psychophysiological effects of human–animal interactions: the possible role of oxytocin. Front Psychol 2012;3: 234.

[25] Jevning B, Wallace RK, Beidebach M. The physiology of mediation: a review. Neurosci Biobehav Rev 1992;16:415–23.

[26] Edwards L. Meditation as medicine: benefits go beyond relaxation. Adv Nurse Pract 2003;11:49–52.

[27] Marchand WR. Mindfulness-based stress reduction, mindfulness-based cognitive therapy and Zen meditation for depression, anxiety and psychological distress. J Psychiatr Pract 2012;18:233–52.

[28] Hoffman CJ, Ersser SJ, Hopkinson JB, Nicholls PG, et al. Effectiveness of mindfulness-based stress reduction in mood, breast and endocrine related quality of life and well-being in stage 0 to III breast cancer: a randomized controlled trial. J Clin Oncol 2012;30:1335–42.

[29] Lengacher CA, Kip KE, Barta MK, Post-White J, et al. A pilot study evaluating the effect of mindfulness-based stress reduction on psychological status, physical status, salivary cortisol and interleukin-6 among advanced stage cancer patients and their caregivers. J Holist Nurs 2012;30:170–85.

[30] Edenfield TM, Saeed SA. An update on mindfulness meditation as a self-help treatment for anxiety and depression. Psychol Res Behav Manag 2012;5:131–41.

[31] Hutcherson CA, Seppala EM, Gross JJ. Loving-kindness meditation increases social connectedness. Emotion 2008;8:720–5.

[32] Abbott AV, Snyder C, Vogel ME. Type A behavior and coronary heart disease. Am Fam Physician 1984;30:209–17.

[33] Schneider RH, Nidich SI, Salerno JW, Sharma HM, et al. Lower lipid peroxide levels in practitioners of the transcendental meditation program. Psychosom Res 1998;60:38–41.

[34] Bormann JE, Smith TL, Becker S, Gershwin M, et al. Efficacy of frequent mantram repetition on stress, quality of life and spiritual well-being in veterans: a pilot study. J Holist Nurs 2005;23:395–414.

[35] Bormann JE, Becker S, Gershwin M, Kelly A, et al. Relationship of frequent mantram repetition on emotional and spiritual well-being in healthcare workers. J Contin Educ Nurs 2006;37:218–24.

[36] Bormann JE, Aschbacher K, Wetherell JL, Roesch S, et al. Effects of faith/assurance on cortisol levels are enhanced by a spiritual mantram intervention in adults with HIV: a randomized trial. J Psychosom Res 2009;66:161–72.

[37] Mahagita C. Roles of meditation on alleviation of oxidative stress and improvement of antioxidant system. J Med Assoc Thai 2010;93(Suppl. 6):S242–53.

[38] Sinha S, Singh SN, Monga YP, Ray US. Improvement of glutathione and total antioxidant status with yoga. J Altern Complement Med 2007;13:1085–90.

[39] Hegde SV, Adhikari P, Kotian S, Pinto VJ, et al. Effect of 3-month yoga on oxidative stress in type 2 diabetes with or without complications. Diabetes Care 2011;34:2208–10.

[40] Yadev RK, Magan D, Mehta N, Sharma R, et al. Efficacy of a short-term yoga-based lifestyle intervention in reducing stress and inflammation: preliminary results. J Altern Complement Med 2012;18:662–7.

[41] White LS. Reducing stress in school-age girls through mindful yoga. J Pediatr Health Care 2012;26:45–56.

[42] Smith JA, Greer T, Sheets T, Watson S. Is there more to yoga than exercise? Altern Ther Health Med 2011;17:22–9.

[43] Woodcock J, Franco OH, Orsini N, Roberts I. Non-vigorous physical activity and all-cause mortality: systematic review and meta-analysis of cohort studies. Int J Epidemiol 2011;40: 121–38.

[44] Haskell WL, Lee IM, Pate RR, Powell KE, et al. Physical activity and public health: updated recommendation for adults from the American College of Sports Medicine and the American Heart Association. Med Sci Sports Exerc 2007;39:1423–34.

[45] Tully MA, Cupples ME, Chan WS, McGlade K, et al. Brisk walking and cardiovascular risk: a randomized controlled trial in primary care. Prev Med 2005;41:622–8.

[46] Seals DR, Desouza CA, Donato AJ, Tanaka H. Habitual exercise and arterial aging. J Appl Physiol 2008;105:1323–32.

[47] Pierce GL, Donato AJ, LaRocca TJ, Eskurza I, et al. Habitually exercising older men do not demonstrate endothelial oxidative stress. Aging Cell 2011;10:1032–7.

[48] Scerbo F, Faulkner G, Taylor A, Thomas S. Effects of exercise on cravings to smoke: the role of exercise intensity and cortisol. J Sports Sci 2010;28:11–19.

[49] Jin P. Efficacy of Tai Chi, brisk walking, meditation, and reading in reducing mental and emotional stress. J Psychosom Res 1992;36:361–70.

[50] Hejazi K, Hosseini SR. Influence of selected exercise on serum immunoglobulin, testosterone and cortisol in semi-endurance elite runners. Asian J Sports Med 2012;3:185–92.

[51] Urso ML, Clarkson PM. Oxidative stress, exercise and antioxidant supplementation. Toxicology 2003;189:41–54.

[52] Atsumi T, Tonosaki K. Smelling lavender and rosemary increases free radical scavenging activity and decreases cortisol level in saliva. Psychiatry Res 2007;150:89–96.

[53] Cooke M, Jolzhauser K, Jones M, Davis C, et al. The effect of aromatherapy massage with music on the stress and anxiety levels of emergency nurses: comparison between summer and winter. J Clin Nurs 2007;16:1695–703.

[54] Labrique-Walusis F, Keister KJ, Russell AC. Massage therapy for stress management: implications for nursing practice. Orthop Nurs 2010;29:254–7.

[55] Corbin L. Safety and efficacy of massage therapy for patients with cancer. Cancer Control 2005;12:158–64.

[56] Yamasaki A, Booker A, Kapur V, Tilt A, et al. The impact of music on metabolism. Nutrition 2012;28:1075–80.

[57] Cervellin G, Lippi G. From music-beat to heart-beat: a journey in the complex interactions between music, brain and heart. Eur J Intern Med 2011;22:371–4.

[58] Bauer BA, Cutshall SA, Anderson PG, Prinsen SK, et al. Effect of the combination of music and nature sounds on pain and anxiety in cardiac surgical patients: a randomized study. Altern Ther Health Med 2011;17:16–23.

[59] Bradt J, Dileo C, Grocke D, Magil L. Music interventions for improving psychological and physical outcomes in cancer patients. Cochrane Database Syst Rev 2011; August 10(8): CD006911.

[60] Evans D. The effectiveness of music as an intervention for hospital patients: a systematic review. J Adv Nurs 2002;37:8–18.

[61] Kushnir J, Friedman A, Ehrenfeld M, Kushnir T. Coping with preoperative anxiety in cesarean section: physiological, cognitive, and emotional effects of listening to favorite music. Birth 2012;39:121–7.

[62] Moris DN, Linos D. Music meets surgery: two sides to the art of "healing." Surg Endosc 2013;27:719–23.

[63] Chen LC, Wang RF, Shih YN, Wu LJ. Fifteen minute music intervention reduces pre radiotherapy anxiety in oncology patients. Eur J Oncol Nurs 2013;17:436–44.

[64] Trappe HJ. Role of music in intensive care medicine. Int J Crit Illn Inj Sci 2012;2:27–31.

[65] Cho AN, Lee MS, Lee JS. Group music intervention reduces aggression and improves self-esteem in children with highly aggressive behavior: a pilot controlled study. Evid Based Complement Alternat Med 2010;7:213–17.

[66] Burleson MH, Trevathan WR, Todd M. In the mood for love or vice versa? Exploring the relations among sexual activity, physical affection, affect and stress in the daily lives of mid-aged women. Arch Sexual Behav 2007;36:357–68.

[67] Ditzen B, Neumann ID, Bodenmann G, von Dawans B, et al. Effects of different kinds of couple interactions on cortisol and heart rate response to stress in women. Psychoneuroendocrinology 2007;32:565–74.

[68] Brody S. Blood pressure reactivity to stress is better for people who recently had penile–vaginal intercourse than for people who had other or no sexual activity. Biol Psychol 2006;71:214–22.

[69] Leitzman MF, Platz EA, Stammpfer MJ, Willett WC. Ejaculation frequency and subsequent risk of prostate cancer. JAMA 2004;291:1578–86.

[70] Young LJ, Wang Z, Insel TR. Neuroendocrine base of monogamy. Trends Neurosci 1998;21:71–5.

[71] Meyer-Lindenberg A, Domes G, Kirsch P, Heinrichs M. Oxytocin and vasopressin in the human brain: social neuropeptide for translational medicine. Nat Rev Neurosci 2011;12:524–38.

[72] Strauss-Blasche G, Ekmekcioglu C, Marktl W. Does vacation enable recuperation? Changes in well-being associated with time away from work. Occup Med (Lond) 2000;50:167–72.

[73] Westman M, Etzion D. The impact of vacation and job stress on burnout and absenteeism. Psychol Health 2001;16:595–606.

[74] Gump BB, Matthews KA. Are vacations good for your health? The 9-year mortality experience after the Multiple Risk Factor Intervention Trial. Psychosom Med 2000;62:608–12.

[75] Sanchez-Rodriguez MA, Zacarias-Flores M, Arronte-Rosales A, Correa-Munoz E, et al. Menopause as a risk factor for oxidative stress. Menopause 2012;19:361–7.

Index

Note: Page numbers followed by "*f*" "*t*" and "*b*" refer to figures, tables and boxes respectively.

A

B

C

D

L

M

N

O

P

Q

R

S

Printed in the United States
By Bookmasters